ECONOMICS AND ECOLOGY IN AGRICULTURE AND MARINE PRODUCTION

Economics and Ecology in Agriculture and Marine Production

Bioeconomics and Resource Use

Clem Tisdell

Professor of Economics, School of Economics, University of Queensland, Australia

Edward Elgar
Cheltenham, UK • Northampton, MA, USA

© Clem Tisdell 2003

All rights reserved. No part of this publication may be reproduced, stored in a retrieval system or transmitted in any form or by any means, electronic, mechanical or photocopying, recording, or otherwise without the prior permission of the publisher.

Published by
Edward Elgar Publishing Limited
Glensanda House
Montpellier Parade
Cheltenham
Glos GL50 1UA
UK

Edward Elgar Publishing, Inc.
136 West Street
Suite 202
Northampton
Massachusetts 01060
USA

A catalogue record for this book
is available from the British Library

Library of Congress Cataloguing in Publication Data

Tisdell, C. A. (Clement Allan)
  Economics and ecology in agriculture and marine production : bioeconomics and resource use / Clem Tisdell.
    p. cm.
  Includes bibliographical references.
  1. Environmental economics. 2. Sustainable agriculture. 3. Mariculture. I. Title.

HC79.E5T573 2003
338.1—dc21

2003040780

ISBN 1 84376 020 7

Printed and bound in Great Britain by MPG Books Ltd, Bodmin, Cornwall

Contents

Preface viii
Acknowledgements ix

PART I   AN OVERVIEW

1. An Overview of Economics and Ecology in Agriculture and Marine Production 3

PART II   INFLUENCES OF ECOLOGICAL CONDITIONS ON AGRICULTURAL PRODUCTION, SUSTAINABILITY, AND ECONOMIC DECISIONS

2. 'The Optimal Choice of a Variety of a Species for Variable Environmental Conditions', *Journal of Agricultural Economics*, **34** (2), May 1983, pp. 175–85. 23
3. 'The Biological Law of Tolerance, Average Biomass and Production in a Variable Uncontrolled Environment', *International Journal of Ecology and Environmental Sciences*, **9** (2), 1983, pp. 99–109. 34
4. Clem Tisdell and Mohammad Alauddin, 'New Crop Varieties: Impact on Diversification and Stability of Yields', *Australian Economic Papers*, **28** (52), June 1989, pp. 123–40. 45
5. C.A. Tisdell and N.T.M.H. De Silva, 'The Economic Spacing of Trees and Other Crops', *European Review of Agricultural Economics*, **10** (3), 1983, pp. 281–93. 63
6. C.A. Tisdell and N.T.M.H. De Silva, 'Supply-maximising and Variation-minimising Replacement Cycles of Perennial Crops and Similar Assets: Theory Illustrated by Coconut Cultivation', *Journal of Agricultural Economics*, **37** (2), May 1986, pp. 243–51. 76
7. 'Conserving and Planting Trees on Farms: Lessons from Australian Cases', *Review of Marketing and Agricultural Economics*, **53** (3), December 1985, pp. 185–94. 85
8. 'Integrating Economics within the Framework for the Evaluation of Sustainable Land Management (FESLM)', *International Board for Soil Research and Management*, IBSRAM Special Paper No. 1, IBSRAM, Bangkok, 1995, pp. 2–23. 95
9. 'Agricultural Sustainability and Conservation of Biodiversity: Competing Policies and Paradigms', in Andrew K. Dragun and Kristin M. Jakobsson (eds), 1997, *Sustainability and Global Environmental Policy: New Perspectives*, Edward Elgar, Cheltenham, UK, pp. 97–118. 118

10. C.A. Tisdell and C. Wilson, 'Genetic Selection of Livestock and
    Economic Development', 2001.   140

PART III   PEST AND DISEASE CONTROL AND AGRICULTURAL
           PRODUCTION – BIOECONOMIC ASPECTS

11. C.A. Tisdell, B.A. Auld and K.M. Menz, 'Crop Loss Elasticity in
    Relation to Weed Density and Control', *Agricultural Systems*, **13** (3),
    1984, pp. 161–6.   153
12. B.A. Auld and C.A. Tisdell, 'Economic Threshold/Critical Density
    Models in Weed Control', *Economic Weed Control*, Proc. EWRS
    Symposium, 1986, pp. 261–8.   159
13. B.A. Auld and C.A. Tisdell, 'Influence of Spatial Distribution of Weeds
    on Crop Yield Loss', *Plant Protection Quarterly*, **3** (2), 1988, p. 81.   167
14. 'Interdependent Pests: The Economics of Their Control', *Environmental
    Systems*, **12** (2), 1982, pp. 153–61.   170
15. B.A. Auld and C.A. Tisdell, 'Biological Weed Control – Equilibria
    Models', *Agriculture, Ecosystems and Environment*, **13**, 1985, pp. 1–8. 179
16. C.A. Tisdell, B.A. Auld and K.M. Menz, 'On Assessing the Value of
    Biological Control of Weeds', *Protection Ecology*, **6**, 1984, pp. 169–79. 187
17. 'Economic Evaluation of Biological Weed Control', *Plant Protection
    Quarterly*, **2** (1), 1987, pp. 10–12.   198
18. C.A. Tisdell and B.A. Auld, 'Evaluation of Biological Control Projects',
    in E.S. Delfosse (ed.), *Proceedings of the VII International Symposium
    on Biological Control of Weeds*, Federconsorzi Conference Hall, Rome,
    Italy, 6–11 March 1988, pp. 93–100.   203
19. 'Economic Impact of Biological Control of Weeds and Insects', in
    M. Mackauer, J. Ehler and J. Roland (eds), *Critical Issues in Biological
    Control*, Intercept Ltd, Andover, Hants, UK, 1990, pp. 301–16.   212
20. C.A. Tisdell, S.R. Harrison and G.C. Ramsay, 'The Economic Impacts
    of Endemic Diseases and Disease Control Programmes', *Revue
    Scientifique et Technique: The Economics Of Animal Disease Control*,
    18 (2), August 1999, pp. 380–94, 396–8.   228
21. C.A. Tisdell and C. Wilson, 'Genetically Modified (Transgenic) Crops
    and Pest Control Economics'.   246

PART IV   MARINE PRODUCTION – BIOECONOMIC ASPECTS

22. 'Economic Problems in Managing Australia's Marine Resources',
    Economic Analysis and Policy, **13** (2), September 1983, pp. 113–41.   259
23. R. Sathiendrakumar and C.A. Tisdell, 'Optimal Economic Fishery Effort
    in the Maldivian Tuna Fishery: an Appropriate Model', *Marine Resource
    Economics*, **4**, 1987, pp. 15–44.   277
24. 'Development of Aquaculture and the Environment: Coastal Conflicts,
    and Giant Clam Farming as a Case', *International Journal of
    Environmental Studies*, **39**, 1991, pp. 35–44.   307

25. Clement A. Tisdell, William R. Thomas, Luca Tacconi and John S. Lucas, 'The Cost of Production of Giant Clam Seed *Tridacna gigas*', *Journal of the World Aquaculture Society*, **24** (3), September 1993, pp. 352–60. 317
26. C.A. Tisdell, L. Tacconi, J.R. Barker and J.S. Lucas, 'Economics of Ocean Culture of Giant Clams, *Tridacna gigas*: Internal Rate of Return Analysis', *Aquaculture*, **110**, 1993, pp. 13–26. 326
27. Carunia Firdausy and Clem Tisdell, 'Economic Returns from Seaweed (*Eucheuma cottonii*) Farming in Bali, Indonesia', *Asian Fisheries Science*, **4**, 1991, pp. 61–73. 340
28. Aquaculture, Capture Fisheries and Available Marine Resources: Ecological and Economic Interdependence 353

Name index 361

Preface

The articles in this book have been published in scattered outlets in the last decade or so. Consequently, most readers are unlikely to have been aware of them as a whole. This volume will increase their accessibility. Hopefully, it will also add further value by enabling readers to more easily appreciate the way in which the themes in the different articles are interconnected. Chapter 1 provides background information on the origins of many articles in this book and an extensive overview of the book's contents. The latter should facilitate the appreciation of the interdependence of these articles. The whole should, therefore, be more valuable than its fragmented parts.

Ecological and environmental economics is an exciting and challenging subject. It deals with matters at the forefront of knowledge. It addresses extremely important issues for the welfare of humankind and for the future of the Biosphere.

Ecological economics is a particularly fascinating subject because it is not bound in its inquiries by traditional paradigms nor by settled theory. As pointed out by Robert Costanza, it is eclectic in nature and requires pragmatism in the formulation of policies. It is also a pleasing subject because it is multidisciplinary or interdisciplinary in character. This can be a basis of continuing vitality for the subject and a source of renewable enthusiasm for those who research in this area or must apply ecological economics. I hope this book adequately conveys the importance of ecological and environmental economics and stimulates enthusiasm for its study.

Some of the articles in this book are co-authored and I would like to thank the co-authors for their contributions and for enthusiastically sharing research efforts with me. Their efforts have added to the depth of coverage of the articles concerned.

Nearly all of the articles in this book were written after I came to the University of Queensland. I thank the University of Queensland and in particular the School of Economics for the support it has provided for my research and writing efforts. Dr Clevo Wilson was kind enough to read and comment on new materials included in this book. Furthermore, assistance from Alison Mohr and Genevieve Larsen and other office staff in processing of materials for the manuscript for this book is much appreciated. Thanks too to Isoa Korovulavulu, a postgraduate student at the University of Queensland from Fiji, for pointing out a printer's error in the original version of a diagram used in Chapter 16.

I am also grateful to Edward Elgar and staff for supporting this endeavour. Furthermore, I thank my dear wife Mariel for her encouragement; and my whole family for their patience as I continued with my efforts to produce the manuscripts for this book.

Clem Tisdell
Brisbane

Acknowledgements

Most of the articles/contributions in this book have been previously published. Unless otherwise stated, I am the sole author. I am grateful to the appropriate editors and/or publishers for permission to reproduce articles or material as specified below.

Clem Tisdell

2. 'The Optimal Choice of a Variety of a Species for Variable Environmental Conditions', *Journal of Agricultural Economics*, **34** (2), May 1983, pp. 175–85. Reprinted with permission of the editor.
3. 'The Biological Law of Tolerance, Average Biomass and Production in a Variable Uncontrolled Environment', *International Journal of Ecology and Environmental Sciences*, **9** (2), 1983, pp. 99–109. Reprinted with permission of the editor.
4. Clem Tisdell and Mohammad Alauddin, 'New Crop Varieties: Impact on Diversification and Stability of Yields', *Australian Economic Papers*, **28** (52), June 1989, pp. 123–40. Reprinted with permission of Blackwell Publishers.
5. C.A. Tisdell and N.T.M.H. De Silva, 'The Economic Spacing of Trees and Other Crops', *European Review of Agricultural Economics*, 1983, **10** (3), pp. 281–93, by permission of Oxford University Press.
6. C.A. Tisdell and N.T.M.H. De Silva, 'Supply-maximising and Variation-minimising Replacement Cycles of Perennial Crops and Similar Assets: Theory Illustrated by Coconut Cultivation', *Journal of Agricultural Economics*, **37** (2), May 1986, pp. 243–51. Reprinted with permission of the editor.
7. 'Conserving and Planting Trees on Farms: Lessons from Australian Cases', *Review of Marketing and Agricultural Economics*, **53** (3), December 1985, pp. 185–94. Reprinted with permission of the editor.
8. 'Integrating Economics within the Framework for the Evaluation of Sustainable Land Management (FESLM)', *International Board for Soil Research and Management*, IBSRAM Special Paper No. 1, IBSRAM, Bangkok, 1995, pp. 2–23. Reprinted with permission of the International Water Management Institute.
11. Reprinted from *Agricultural Systems*, **13** (3), C.A. Tisdell, B.A. Auld and K.M. Menz, 'Crop Loss Elasticity in Relation to Weed Density and Control', pp. 161–6. Copyright (1984), with permission from Elsevier Science.
12. B.A. Auld and C.A. Tisdell, 'Economic Threshold/Critical Density Models in Weed Control', *Economic Weed Control*, Proc. EWRS Symposium, 1986, pp. 261–8. Reprinted with permission of the European Weed Research Society.
13. B.A. Auld and C.A. Tisdell, 'Influence of Spatial Distribution of Weeds on Crop Yield Loss', *Plant Protection Quarterly*, **3** (2), 1988, p. 81. Reprinted with permission of the editor.
14. C.A. Tisdell, 'Interdependent Pests: The Economics of their Control', *Environmental Systems*, **12** (2), 1982, pp. 153–61. Reprinted with permission of Baywood Publishing.

15. Reprinted from *Agriculture, Ecosystems and Environment*, **13**, B.A. Auld and C.A. Tisdell, 'Biological Weed Control – Equilibria Models', pp. 1–8. Copyright (1985), with permission from Elsevier Science.
16. Reprinted from *Protection Ecology*, **6**, C.A. Tisdell, B.A. Auld and K.M. Menz, 'On Assessing the Value of Biological Control of Weeds', pp. 169–79. Copyright (1984), with permission from Elsevier Science.
17. 'Economic Evaluation of Biological Weed Control', *Plant Protection Quarterly*, **2** (1), 1987, pp. 10–12. Reprinted with permission of the editor.
18. C.A. Tisdell and B.A. Auld, 'Evaluation of Biological Control Projects', in E.S. Delfosse (ed.), *Proceedings of the VII International Symposium on Biological Control of Weeds*, Federconsorzi Conference Hall, Rome, Italy, 6–11 March 1988, pp. 93–100. Reproduced from Proceedings of the VII International Symposium on Biological Control of Weeds (ed. E.S. Delfosse, 1990) with permission of CSIRO Publishing.
19. 'Economic Impact of Biological Control of Weeds and Insects', in M. Mackauer, J. Ehler and J. Roland (eds), *Critical Issues in Biological Control*, Intercept Ltd, Andover, Hants, UK, 1990, pp. 301–16. Reprinted with permission of Intercept Ltd.
20. C.A. Tisdell, S.R. Harrison and G.C. Ramsay, 'The Economic Impacts of Endemic Diseases and Disease Control Programmes', *Revue Scientifique et Technique: The Economics Of Animal Disease Control*, **18** (2), August 1999, pp. 380–98. Reprinted with permission of the Office International des Epizooties.
22. 'Economic Problems in Managing Australia's Marine Resources', *Economic Analysis and Policy*, **13** (2), September 1983, pp. 113–41. Reprinted with permission of the editor.
23. R. Sathiendrakumar and C.A. Tisdell, 'Optimal Economic Fishery Effort in the Maldivian Tuna Fishery: An Appropriate Model', *Marine Resource Economics*, **4**, 1987, pp. 15–44. Reprinted with permission of the editor.
24. 'Development of Aquaculture and the Environment: Coastal Conflicts, and Giant Clam Farming as a Case', *International Journal of Environmental Studies*, **39**, 1991, pp. 35–44. Reprinted with permission from Taylor and Francis www.tandf.co.uk
25. Clement A. Tisdell, William R. Thomas, Luca Tacconi and John S. Lucas, 'The Cost of Production of Giant Clam Seed *Tridacna gigas*', *Journal of the World Aquaculture Society*, **24** (3), September 1993, pp. 352–60. Reprinted with permission of the managing editor.
26. Reprinted from *Aquaculture*, **110**, C.A. Tisdell, L. Tacconi, J.R. Barker and J.S. Lucas, 'Economics of Ocean Culture of Giant Clams, *Tridacna gigas*: Internal Rate of Return Analysis', pp. 13–26. Copyright (1993), with permission from Elsevier Science.
27. Carunia Firdausy and Clem Tisdell, 'Economic Returns from Seaweed (*Eucheuma cottonii*) Farming in Bali, Indonesia', *Asian Fisheries Science*, **4**, 1991, pp. 61–73. Reprinted with permission of the editor.

PART I

AN OVERVIEW

[1]
An overview of economics and ecology in agriculture and marine production

Introduction
Agricultural and marine production based on living resources is principally determined by the combined influences of economic and ecological factors. Their combined influences on the level of this production and its sustainability have interested me for several decades. The articles comprising chapters in this collection outline several of these influences. The first of these articles was originally published in 1982. While most articles in this book have been previously published, three have been written especially for this book. Also this is the first time these articles (previously published ones appeared in very diverse outlets) have been included in one volume and this should increase their accessibility. Nevertheless, it is also appropriate to point out that this collection is only a partial set of my articles concerned with economics and ecology in relation to agriculture and marine production. It does not, for example, include those in Dragun and Tisdell (1999), relevant ones that appeared in the series, *Research Papers and Reports in Animal Health Economics*, published by the Department of Economics, The University of Queensland, plus others, some of which seemed to be too specialized for this collection.

The articles in this book have been arranged by chapters into three groups. Those in Part II consider general influences of economic decisions and ecological conditions on the level of agricultural production and its sustainability. Part III concentrates on pest and disease control as a means to further agriculture production. Part IV explores several issues involving capture fisheries and economic production from marine aquaculture. Let us consider the particular coverage of each of these parts.

An overview of Part II – influences of economic decisions and ecological conditions on the level of agricultural production and its sustainability
Chapter 2 explores the optimal choice of a variety of a species used in agriculture when environmental conditions are variable and when production from different species exhibits differing degrees of tolerance to altered environmental conditions. It explores the rules for determining which variety of a crop or species will provide the highest level of yield on average. Concave production functions (those involving diminishing marginal productivity as a function of the relevant environmental influence on production) are explored initially and subsequently convex production possibilities are also considered. This is because the biological law of tolerance (Owen, 1975) indicates that varieties of species occurring towards the limits of their range of tolerance of an environmental condition experience increasing marginal productivity. Some of the findings reported are not intuitively obvious; for example, that varieties with less

yield tolerance to variable environmental conditions can on average give higher yields than those with greater tolerance in marginal zones. This article was the subject of a comment by Mellor (1985) and a rejoinder by myself (Tisdell, 1985).

Chapter 3 expounds the biological law of tolerance that provides the basis of the analysis in Chapter 2. It helps connect this law more directly to the results reported in Chapter 2, provides a broadened ecological basis for the productivity results, and considers the implications of the results for the average level of profits of economic agents relying on the biomass of a species for their profit. However, unlike in the case of Chapter 2, choice of species or of different varieties of a species is ruled out. The relevant species or its variety is taken for granted.

It is widely believed that modern varieties of crops and other cultivars, livestock and other domesticated species show less environmental tolerance than traditional ones (Conway, 1985a, b). However, depending on the extent to which environmental conditions can be controlled, average levels of yield or production and profit may be higher for modern varieties than traditional ones. In fact, as new varieties of crops and other species used in agriculture are introduced to agriculture, it often happens that simultaneously human environmental control of agricultural conditions increases. The presence of such increased environmental control is often a critical factor favouring adoption of the modern variety of a crop, a particular livestock breed and so on by farmers. For example, the presence of irrigation water, chemical fertilizers and pesticides able to create modified (or artificial) environments for the growing of rice, played an important role in the introduction of high yielding varieties (HYVs) of rice. If HYVs are to sustain their high yields on average then the ability of farmers to sustain the favourable environmental conditions under which they are grown must be maintained.

Typically the yield function from a modern variety of a species used in agriculture as a function of its environment might be like that shown by curve ABC in Figure 1, and for a traditional variety in the same locality, like that shown by curve DEF. When environmental fluctuation is considerable, yields on average will be highest when the traditional variety is used. But if the relevant environmental condition can be controlled so that it does not diverge too far from values corresponding to B, the modern variety will maximize yields on average and, at the same time, yields can also display a low degree of variability. However, if this position is to be sustained, it must be possible to maintain the required degree of control of relevant environmental conditions. In the long term, this may prove to be impossible for a variety of reasons; for example, an unexpected disease may reduce the vigour of the HYV or pest controls may fail.

The impact of new varieties on crop diversification and the variability of yields are discussed in Chapter 4 using empirical data from Bangladesh. The introduction of HYV foodgrains is considered. Both short-term effects and long-term ecological risks are taken into account, such as those arising from loss of genetic diversity of crops. The portfolio of crop varieties selected by farmers for cultivation may not be the optimal one from a social long-term point of view because it is likely to entail greater irreversible genetic loss than is socially desirable.

An important factor that influences the yields from many crops is the spacing between the plants. This affects the availability to each of scarce natural resources such as sunlight, nutrient uptake and so on. Economic conditions for the optimal spacing of crops are specified in Chapter 5. If the spacing of all trees or plants can be completely

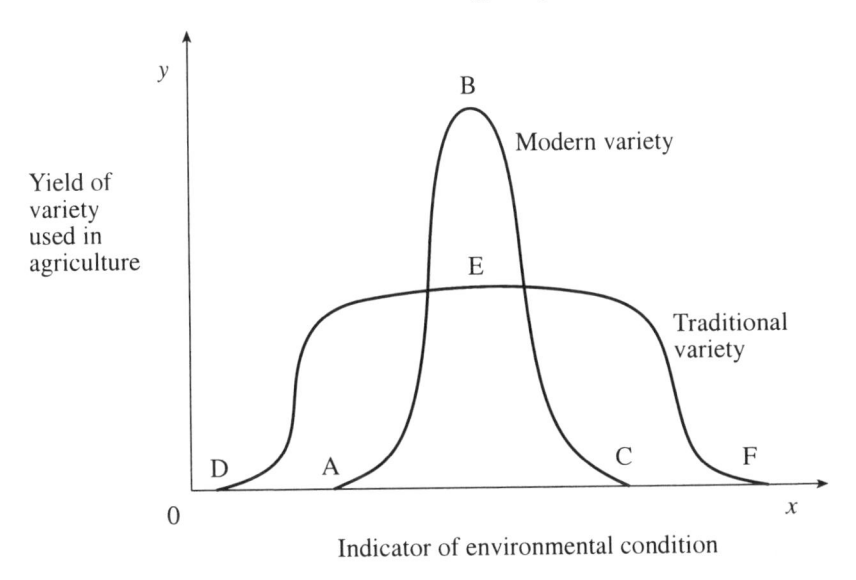

Figure 1 Compared to traditional varieties of crops and other species used in agriculture, modern varieties usually show little tolerance to variations in environmental conditions

controlled, the problem is easily solved. But if some trees or plants are already established, there are impediments to varying the spacing of all these plants. In such cases, when actual densities fall in a particular band or range, leaving the existing density unchanged maximizes profit. Pre-existing conditions create an 'inert' area or band of densities for which intervention is unprofitable.

Chapter 6 considers supply-maximizing and variation-minimizing replacement cycles for production from perennial crops and illustrates the theory by an application to coconut production in Sri Lanka. It is also pointed out that the theory can be applied to other biological assets such as dairy cattle. In this analysis, only private yields from the crops themselves are considered.

While Chapters 5 and 6 involve mostly trees used for cropping, Chapter 7 concentrates on environmental spillovers from trees on farms. Both environmental benefits to farmers, such as those arising from the use of trees to provide shelter belts, and favourable environmental spillovers beyond the farm are considered. Topics touched on include private and social net benefits from trees on farms, private farm benefits from shelter belts, trees on farms as an economic mixed good, agroforestry and trees in relation to dryland salting in Australia.

Chapter 8 is primarily intended to integrate economic considerations into a framework for sustainable land (soil) management in agriculture. It gives particular attention to economic viability as a criterion for the sustainability of a land management technique. Economic viability of a land management system is shown to be influenced by natural or biophysical factors and by market-related factors. The economic literature relating to the sustainability of agricultural techniques is summarized. In addition, various economic approaches for deciding between different techniques of agricultural production are outlined, the relevance of economic break-even points for the sustained use of land management techniques is highlighted, and further contributions that economics could make to the development of an informed framework for the evaluation of sustainable land management are identified. Particular farm systems for land management in Zimbabwe, the Philippines and Australia are assessed in this context.

Broad issues involving agricultural sustainability and the consequences of agriculture for biodiversity are discussed in Chapter 9. The coverage includes different approaches to environmental policy formation and implementation, alternative concepts of sustainability and biodiversity and their policy implications, and a review of concepts of agricultural sustainability. Furthermore, the sustainability and environmental implications of alternative types of agricultural techniques are examined. These types include modern versus traditional agricultural techniques, conservation versus conventional farming, high external input agriculture (HEIA) versus low external input agriculture (LEIA), organic agriculture versus non-organic agriculture, and extensive versus intensive use of land for agriculture. This chapter then considers the impact of agriculture on biodiversity conservation, the benefits to agriculture from conservation of biodiversity, and constraints placed on agriculture by decisions and policy measures designed to conserve biodiversity. It ends with an assessment of some socio-economic methods for sustaining biodiversity and a discussion of their agricultural applications.

Changes that occur in biodiversity of livestock as economic development occurs are considered in Chapter 10. Particular attention is given to economic and ecological factors that have reduced livestock diversity in modern times. Considerable loss of livestock varieties has already occurred in most developed countries, and the process of loss in developing countries may accelerate with economic globalization and their own economic development. This can create an economic or environmental dilemma for less developed countries; a dilemma made no easier in some instances by the importation of foreign breeds tied to foreign aid. A case study involving genetic improvements in Vietnam's stocks of pigs illustrates potential economic benefits of genetic improvement in its pig stock. It also highlights some potential long-term drawbacks and negative development consequences from using imported breeds to improve the genetics of its pig herd.

Genetic 'improvements' in livestock can bring substantial economic benefits. However, a farmed animal contains many genetic attributes, and improvements in one attribute can be at the expense of another. Furthermore, the demand for different genetic attributes changes with economic development and as humans alter their demands for different types of products produced by livestock, as well as with humankind's ability to manage the environments faced by livestock. Therefore, economic development and change have substantial influences on the evolution of livestock breeds and their survival. In turn, improvements in livestock breeds can add significantly to human welfare. But it should be noted that deciding what is an improvement in the genetic composition of livestock involves a value judgement.

Chapter 10 specifies factors, particularly economic ones, which influence the evolution and survival of livestock breeds. In addition, factors making for market failure in the conservation of livestock breeds are outlined. Particular attention is given to the challenges that developing countries face in conserving local herds. Vietnam has been actively trying to improve the genetic composition of its pig herd and, as noted above, its efforts in this regard are given particular attention as a case study.

Pest and disease control and agricultural production – an overview of bioeconomic aspects

Part III is concerned with living organisms (pests) that have negative effects on agricultural production, and the economics of control of such pests. While most chapters concentrate on weeds as pests, some deal with pests generally, and one includes consideration of the control of insects and another discusses animal diseases. It should be noted that often models applied to one class of pest have application to others; for example, those applied to control of weeds may have application to control of other pests, such as insects.

My publications on pest control are much more numerous than those included in Part IIII. Publications not included comprise for instance, several on vertebrate pest control (e.g. Tisdell, 1982; Bandara and Tisdell, 2002), further work on weed control economics (e.g. Auld, Menz and Tisdell, 1987) and control of insect pests of livestock (e.g. Davis and Tisdell, 2002), as well as papers on animal diseases and their control in the series *Research Papers and Reports in Animal Health Economics*, Department of Economics, The University of Queensland.

The chapters in Part III are arranged so that the earlier ones concentrate principally on the economics of private gains from pest control whereas the latter ones involve both private and social economic dimensions. Consider the main points covered in each of the articles comprising the chapters in this part.

In Chapter 11, relationships between yield loss and density of a weed are considered and the concept of elasticity of yield loss in relation to increased weed density is introduced. It is suggested that typically the elasticity of yield for increased weed density is positive but less than unity. This indicates that overall loss in crop yield will tend to be relatively greater, other things equal, if weeds are spread more evenly in a crop rather than being clumped or occurring in patches in the crop. The simplest economic threshold or break-even model is outlined for determining whether a herbicide or mechanical control is justified.

This model is extended further in Chapter 12. In this case, allowance is made for reinvasion and carryovers of weeds in crops, for uncertainty about weed densities, and uncertainty of the value of yield losses arising from weeds, and quality penalties when weeds contaminate the harvest. The presence of wild oats in wheat provides an example.

Note that the basic economic threshold model for weed control introduced in Chapter 11 and extended in Chapter 12 assumes that the value of the loss averted (say by the treatment of a crop with herbicides to control weeds) is equal to the value of the yield loss that would arise from the presence of weeds if weeds are not controlled. It therefore assumes that the treatment is fully effective in removing weeds and that the control used, for example the herbicide application, has no deleterious effects on the yield and quality of the crop. Nevertheless, if the weed control is only partially effective, this can be readily allowed for in the threshold model.

8 *Economics and Ecology in Agriculture and Marine Production*

Controlling a weed by means of a standard application of a herbicide may, for example, be only partially effective because it kills only a fraction of the weeds expected to grow in the absence of weed control. At the same time, the vigour of remaining weeds may be so reduced that they compete less effectively with the crop. The larger the fraction of weeds destroyed by weed control and the greater the reduced vigour of remaining weeds, the greater are the economic gains to be had from weed control, other things equal. For a given weed density, the likelihood of weed control rises, other things constant, the higher is the proportion of weeds killed and the greater is the reduction in the competitiveness of remaining weeds.

Adjustment of the standard threshold weed control model to allow for varying degrees of effectiveness of weed control can be illustrated by Figure 2. In this figure, line CE represents the total cost of the weed control technique. Curve OBD corresponds to the standard case in which the weed control is fully effective in eliminating the weed and yields reach their full potential and quality. Curve OFG indicates economic gains when control of the weed is only partial and OHJ represents a case in which even less control of the weed occurs. It can be seen that as the effectiveness of the weed control technique declines, weed density in the absence of control must be greater before control becomes economic. Thus when the gross benefit function is OHJ, weed density in the absence of control must exceed W_3, but it must exceed W_2 when OFG applies, and must be greater than W_1 when OBD applies, if control is to be profitable.

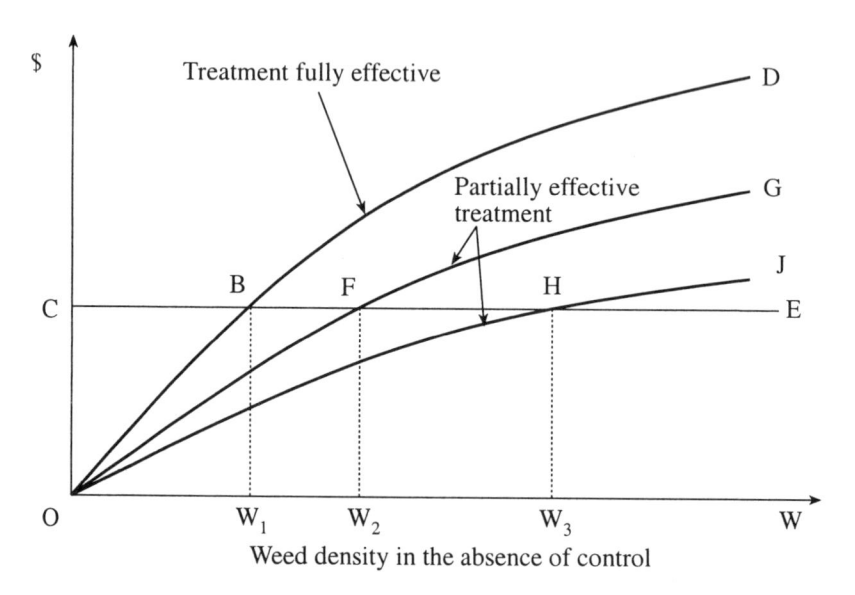

Figure 2 Adjustment of economic threshold model to allow for partial control of weeds

A further economic complication that should be allowed for is that higher yields will often involve extra harvesting costs and so the appropriate curves for economic benefit (excluding weed control costs) should allow for these costs. For example, the curve OBD should be extra revenue obtained as a result of greater yield, less the cost of bringing this extra production to market. Allowing for this does not, however, alter the basic qualitative conclusions.

Chapter 13 provides further proof (by using mathematical theory involving convex functions) that yield losses in crops from the presence of weeds tend to be greater when the weeds are spread relatively evenly than when they occur in patches. It is also shown that crop yield losses based on mean densities of weeds over a whole field will usually be an overestimate if the weeds have a clumped distribution in the field.

Multiple pest species are not uncommon and often their population levels are interdependent. Their ecological interdependence affects the economics of their control. Some pest species exhibit predator–prey relationships (such as the wild dog and the wild pig in Australia), other pest species exhibit competitive relationships (such as some species of kangaroo in Australia) and still other pests exhibit symbiotic or complementary relationships. Using vertebrate pests as examples, these relationships are outlined in Chapter 14 in terms of population response curves and equilibria. The consequences of such interdependencies for the optimal economic control of target species are outlined.

Chapter 15 also makes use of equilibria models to discuss biological weed control. Classical biological weed control involves the use of a predator to control the population of its prey or host. In this case, the weed to be controlled is the prey or host. The possibility of discontinuously stable systems is demonstrated, and the shape of the predator consumption function is shown to be important for the nature of the equilibria in the model; for example, the possible existence of multiple equilibria and their stability. The existence of unstable equilibria implies that supplementation of the prey or control population can, in certain situations, have a dramatic effect on the population dynamics of the system. At particular points, such supplementation can result in the biological agent becoming much more effective in reducing the population of the pest. In addition, the importance of a reserve population of the weed or pest of sufficient magnitude to prevent the population of the predator being wiped out is illustrated.

Following consideration of equilibria models, attention is focused in several chapters on the economic impact and evaluation of biological control of pests. Both augmentative and classical methods of biological control are considered but most attention is given to the latter. The subject is introduced in Chapter 16 and extended in Chapter 17 by considering Australian policies for biological weed control and methods of economic evaluation used by Australian public authorities. A more extensive discussion of methods for economic evaluation of biological control of projects is provided in Chapter 18. The application of social cost–benefit analysis, using supply and demand curves, is explored as a means of providing an economic evaluation and its application is illustrated by the biological control of *Echium spp.* in Australia. Also the scope for using a multiple characteristics approach to such assessment is considered, taking into account the attributes that Conway (1985a, b, 1987) argues are important in assessing alternative agricultural techniques. The types of attributes emphasized by Conway are levels of yields, their instability, equitability of income distribution, and sustainability of yields. Evaluations at both the pre- and post-release stages of a biological control agent are discussed. Chapter 19 provides some historical background on the use of biological controls, a further discussion of conventional indicators of economic benefits from biological control of pests, two case studies and consideration of the scope for additional economic gains from further applications of biological pest controls in the future. Genetically modified organisms (GMOs) open up expanded avenues for biological control of pests but also bring new risks. Some of the economic issues raised by the use of GMOs in pest control are discussed in Chapter 21.

Diseases of livestock involve pests in most cases. Hence, most of the economic analysis of pest control is also applicable to disease control in animals. The pest causing a livestock disease may, for example, be a parasite, bacteria, a fungus or a virus. Chapter 20 outlines methods for evaluating the economic impact of endemic livestock diseases and provides an overview of techniques for assessing the economics of control of such diseases. Case studies include the economic impact and control of helminthiasis (caused by helminths, parasitic worms) and of tick-borne diseases of livestock. Chapter 20 also contains a substantial discussion of the economics of livestock disease control.

The introduction of GMOs to agriculture is a controversial topic. In recent years, genetic engineering of crops for the purpose of pest control has assumed growing importance, particularly in America. A wide range of crops has been genetically modified for this purpose. These include potatoes, corn, cotton, canola (rape), squash and soya beans. Some are genetically engineered to produce endotoxins to kill insects, some are modified to resist plant viruses and other infections by microbes, and still others are engineered to be more tolerant to applications of pesticides or herbicides.

While human-controlled genetic selection has a long history, modern genetic engineering involves direct relocation by humans of selected genetic material from one organism to another, often from entirely unrelated species. For example, genetic material from *Bacillus thuringiensis* (Bt), a soil bacteria, has been engineered into a number of major crops. Such transgenic engineering creates new organisms that are unlikely to evolve naturally or by gradual co-evolutionary change. The fact that the organisms created both arise by 'unnatural' means and involve quantum jumps rather than gradual co-evolutionary change is, understandably, one reason for some public concern about these techniques.

Chapter 21 provides background information on the genetic modification of crops for pest control and analyses evidence about private returns from such crops. Furthermore, it takes into account ecological considerations, and considers possible negative and positive environmental spillovers from their use, as well as other market failures.

Marine production – an overview of bioeconomic aspects

In Part IV, attention turns to economic and ecological influences on marine production. Just as in the case of agricultural output, combined economic and ecological influences must be taken into account when modelling and predicting marine production.

Chapter 22 deals with issues involving marine production in Australia considered to be important at the time the original article was written. It discusses Australian government policies for managing the commercial fisheries, government attitudes and policies towards foreign fishing in Australian waters, aspects of recreational fishing and other uses of the sea. Its analysis of the management of a transboundary migratory species, in this case the southern bluefin tuna, *Thunnus maccoyi* (Castlenau) is of general interest. Game theory is applied to analyse economic conflict between nations (in this case Australia and Japan) about their joint exploitation of a common fisheries stock of southern bluefin tuna. A Nash game theory model is used that allows for possible threat behaviour by the participating fishing nations. The type of issues discussed arise from many fish species where their populations are jointly exploited by different nations. Both socio-economic and ecological factors must be combined in considering the production consequences of joint exploitation of migratory stocks of fish, and in assessing the prospects for co-operation to maximize sustainable economic yield, or where considered appropriate, to maximize sustainable physical yield.

These yield problems become especially important when different nations have marine property rights that enable them to capture migratory fish at different stages of their life-cycle. If, for example, relatively immature migrating fish pass initially through the exclusive fishing zone of one nation, these may be so heavily exploited by this nation in their immature stage that global sustainable physical and economic catch is severely reduced. Furthermore, given the highly migratory nature of some fish species, they are in effect common property or fugitive resources. In the absence of effective co-operation between all nations exploiting a common fugitive fishing resource, there is a high probability that it will be fished to extinction. As with open-access resources, this is more likely, other things equal, the higher the price of the fish involved. As it so happens, southern bluefin tuna is very highly priced in the Japanese market. In fact, scope for selfish economic gains has resulted in its over-exploitation, and has enticed an increasing number of distant water nations to exploit its stock in international waters. As a result, parental biomass of southern bluefin tuna is now reduced to very low levels and some scientists believe that the species is in danger of extinction.

The problem can be illustrated by Figures 3 and 4. Figure 3 involves a simple linear case of two nations exploiting a common migratory fishery resource. Nation one has access to the stock at a more critical stage in its life-cycle than nation two so that nation one's catch has a greater impact on the level of the parental biomass than does nation two's catch. Therefore, if it is decided to keep the parental biomass of the migratory fish species at a particular level, a trade-off curve in terms of catches by each of the nations like that shown by DBF applies. The trade-off function has a downward slope in excess of minus one. Thus the total biomass catch consistent with sustaining the parental biomass at its desired level would be maximized at point D. This is because the isoquants in this case are lines with a downward slope of minus unity. At point D, country two accounts for the whole catch. However, it is unlikely that country one will agree to such an arrangement unless adequate compensation is paid to it by country two. Agreement on limiting joint catch might be much easier if the trade-off functions for maintaining desired levels of parental biomass have a slope of minus unity as illustrated by line ABC in Figure 3. In such a case, if the two nations were equally efficient in economic terms in fishing, they would be able to gain equally from a joint decision to equally limit their joint catch in order to sustain production. They might, therefore, agree to equal quotas for their catch.

In practice, the type of trade-off curve shown in Figure 3 will probably be non-linear and, of course, different trade-off curves will apply for different levels of desired parental biomass. In general, the lower is the combined catch of the two

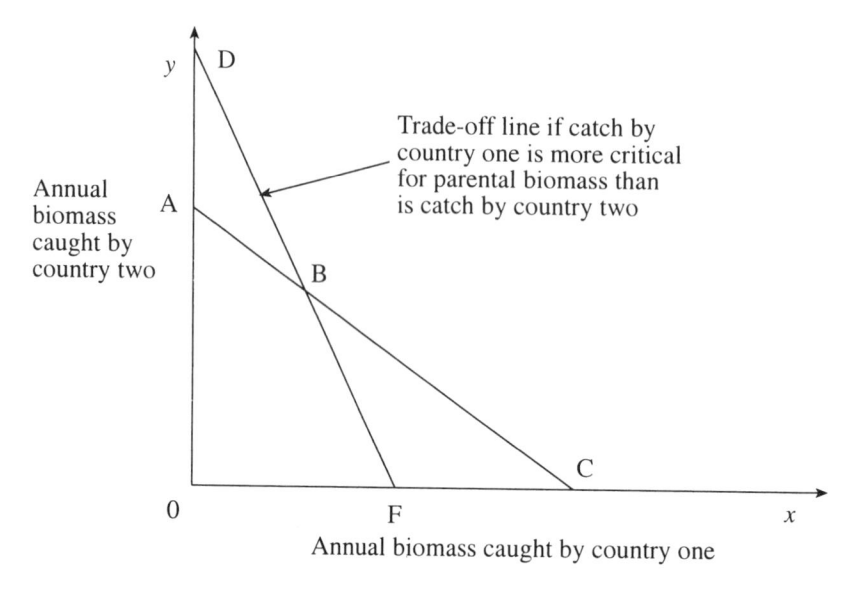

12   Economics and Ecology in Agriculture and Marine Production

Figure 3   Illustration of a sustainability problem involving fugitive or transboundary fishing stocks

nations (with the shares of the two nations held constant) the higher will be the level of parental biomass. Nevertheless, total yield from the fishery, if a particular level of parental biomass is to be sustained, can require very unequal quotas for the nations sharing the fugitive resource. This is illustrated in Figure 4. Here, curve ABC is the non-linear trade-off curve for sustaining a particular level of parental biomass. Line HBJ represents the highest isoquant attainable and is tangential to the trade-off curve at point B. At this point, the rate of technical substitution of the catch by the nations is equal to the rate of their contribution to the total catch. This involves a higher catch quota for country two compared to one. Depending on the circumstances, the quota may be much higher for country two than one. Thus this can be a serious source of international conflict. The conflict may be extremely difficult to overcome, especially if nation one has pre-emptive property rights due to the pattern of migration of the fish species involved.

Thus Chapter 22 touches on complex issues involving socio-economics and ecology. It does not, however, completely resolve issues involved in management of transboundary migratory fish to maximize global economic gain. In particular, note that in Figures 3 and 4, the allocation of the catch between countries to maximize joint sustainable catch by weight so as to achieve the desired biomass of the parental stock, may not be the one that maximizes their aggregate economic gain. If the market price of each unit of catch is the same irrespective of which country catches it, the optimal economic allocation of the catch between nations will be influenced by any difference in the per unit costs of capturing biomass experienced by the nations. For example, if the per unit cost of capture of biomass by country two is lower than for country one, the economically optimal allocation of the catch in Figure 4 will be on the trade-off curve to the left of B, if the Kaldor-Hicks criterion is applied.

Economics and Ecology in Agriculture and Marine Production 13

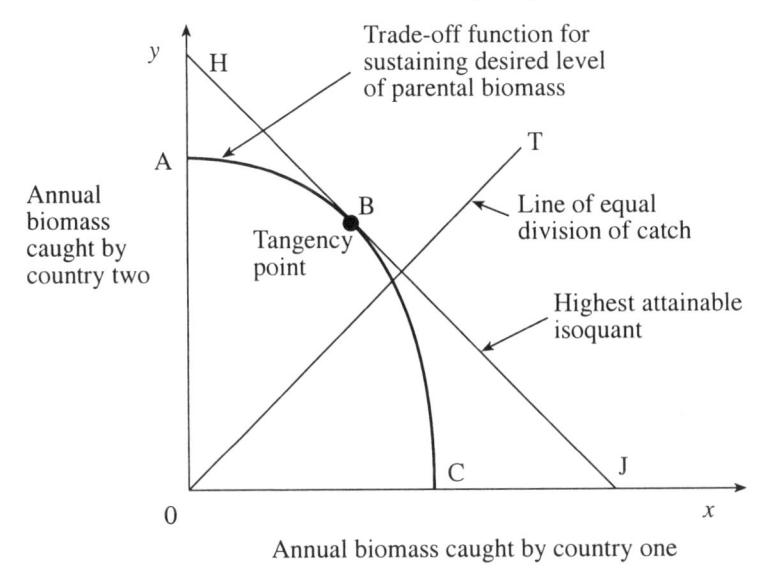

Figure 4 Another example in which unequal catch quotas maximize the total catch consistent with sustaining a given level of parental biomass

However, cases can occur where exploitation of transboundary fugitive (fish) resources by one nation makes virtually no difference to the size and composition of the parental biomass of the fish involved. This can happen, for instance, if the fish migrating through the fishing area of a small nation form only a very small fraction of their total migratory population. In the 1980s, and possibly still today, this seems to have been the situation in the Maldives for the exploitation of tuna by domestic fishers. This fishing in the Maldives is confined to inshore areas. It is also the case for some other small island nations, for example in the Pacific Ocean.

The situation envisaged is illustrated in Figure 5. The arrows in this figure represent the migratory path of a large population of fish migrating (on a broad front) past an island shown as a dark disk. However, this island's community only catches fish in a small area of the migratory path shown in Figure 5. For example, its catch may be restricted to an area in the broken circle. Thus, even if the island community catches all the biomass of the migratory fish passing through the island's fishing area, this catch can have virtually no impact on the size of parental biomass.

Given the above situation, catch–effort production models such as the Schaefer model (Schaefer, 1957), may have little applicability. That is, little applicability to this particular type of bioeconomic situation, even though they are useful in other contexts. The Schaefer model, and similar bioeconomic fishing models, assume eventually diminishing total catch from increased fishing effort because this effort decreases the parental biomass and, eventually, the available biomass of fish to be caught. However, such models clearly do not apply to the type of situation envisaged in Figure 5. Nevertheless, fishers in the island community do compete for the scarce fish migrating through their fishing area. In doing so, their catch increases but at a decreasing rate. Furthermore, their limits are determined by the amount of biomass

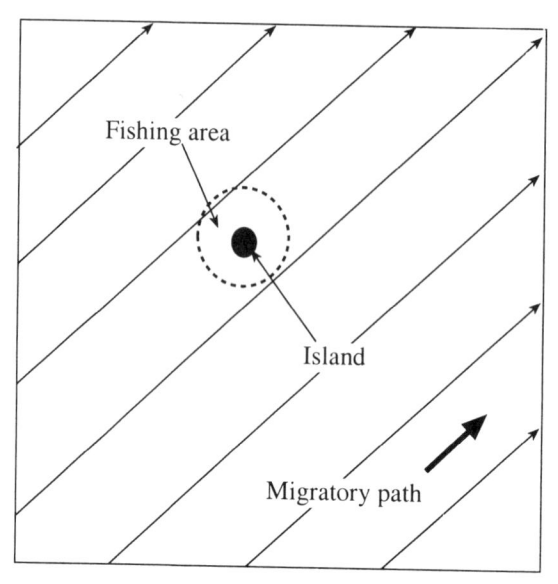

Figure 5 A case in which an island community fishes only in a very small area of the migratory path of a fish species and captures little of the biomass of the species

of the targeted fish migrating through their fishing area in a given unit of time. In such cases, open-access or common-property failure is liable to occur but it does not arise from a reduction in the parental biomass, as is the case of models such as those of Schaefer (1954) and Fox (1970). An alternative model is suggested in Chapter 23 for modelling the type of situation illustrated by Figure 5. The model is empirically estimated for tuna in the Maldives and policy implications for the tuna industry in the Maldives are considered. This model was the subject of a note by Campbell and Lindner (1989) and a response by Sathiendrakumar and Tisdell (1989).

Marine aquaculture (mariculture) is of growing importance as a source of marine production. The development of mariculture has the potential to add to food availability for humans of fish and other organic economic supplies. It can also help conserve marine species and reduce harvesting pressures on wild-caught stocks of fish. Nevertheless, these results are by no means assured. Indeed, the development of mariculture can have effects opposite to those just mentioned. This possibility is discussed in Chapter 24 and analysed in Chapter 28.

Chapter 24 discusses generally the compatibility, or otherwise, of global aquacultural development with conservation and environmental goals, examines coastal land-use conflicts, outlines the development of mariculture in Australia, and considers the environmental consequences of giant clam farming as an example. Although the Brundtland Report (World Commission on Environment and Development, 1987) favours the expansion of aquaculture, it is claimed that it gives insufficient attention to the environmental consequences of such development. While aquaculture development, including seafarming, can help to save farmed species from extinction, it is not without serious consequences for the natural environment and may well reduce overall biodiversity.

Chapters 25–27 examine the economics of mariculture of some specific marine species, mainly giant clams of the species *Tridacna gigas* and seaweed of the species *Eucheuma cottonii*.

Some forms of aquaculture involve two separate stages or processes of production; namely, the production of seed or fingerlings, and an ocean growout phase. This is usually the case for the husbandry of giant clams. Methods for farming giant clams developed rapidly in the 1980s. The mariculture of giant clams was seen by organizations funding the basic scientific research for its development (such as the Australian Centre for International Agricultural Research) as a means to prevent these species becoming extinct, as an avenue for supplementing subsistence food supplies in the Pacific islands, and as a species having commercial possibilities that could add to cash incomes in some developing countries, especially small island Pacific nations (Tisdell, 1992; Tisdell et al., 1994). Market possibilities for giant clams exist for the aquarium trade, for sale of their meat for food and for sale of their shells. Chapter 25 provides estimates of the cost of producing giant clam seed and Chapter 26 provides estimates of economic returns from the ocean growout of giant clams.

Seaweed is a major non-fishing export of the Philippines and Indonesia. Such farming has been able to reduce income inequality in some coastal villages (Firdausy and Tisdell, 1993) and reduce the incidence of poverty (Firdausy and Tisdell, 1992). Chapter 27 provides an estimate of economic returns on *E. cottonii* production based on a farm in Bali. In addition to the potentially high private return from *E. cottonii* farming, it is attractive to developing countries because it is relatively labour-intensive and does not require significant amounts of secondary inputs or import of inputs such as fertilizers, chemicals, fuel and food. Furthermore, harvested seaweed is not a highly perishable commodity. However, the cultivation of seaweed does involve significant risks in areas subject to cyclones.

An issue not discussed in this book is production externalities or spillovers from sessile aquacultured organisms. This can be of importance for production from cultivated marine organisms that are sessile or immobile at the stage of their life-cycle when they are farmed and that extract nutrients or food particles from the passing water column. Seaweeds extract nutrients from the water column that flows past them and many molluscs (such as oysters and mussels) extract food particles from this source too. Depending on the direction and nature of currents and food availability in the shared water body, increased density of farmed organisms in one location can reduce the yields of the same (or related organisms) cultivated in another location. This is because the organisms at the first location reduce the availability of food to those in the second location. However, the productivity interconnections can be quite complex. They depend on the direction of currents, changes in their direction, and other factors, such as the level of the nutrient load in the water body and variations in nutrients with changes in the currents. Nevertheless, a simple case can be used to illustrate the production problem arising from such spillovers.

In Figure 6, a steady current is assumed to flow from west to east with a fixed food content when entering the rectangular marine area shown. There is no addition or natural loss of the nutrient-content of the seawater in the rectangular area but

16 *Economics and Ecology in Agriculture and Marine Production*

mariculture is practised in the areas indicated by AB and CD using sessile organisms, and these extract nutrients from the water column. It is possible that the organisms farmed in area AB may be of a density that extracts sufficient nutrient from the water column to reduce the growth of those organisms found in area CD. If we suppose that many farmers privately own the aquaculture farms in each of the areas, there is a risk that farmers in area AB will overstock from the point of view of maximizing the net value of production from the farmed organisms in the rectangular region enclosed.

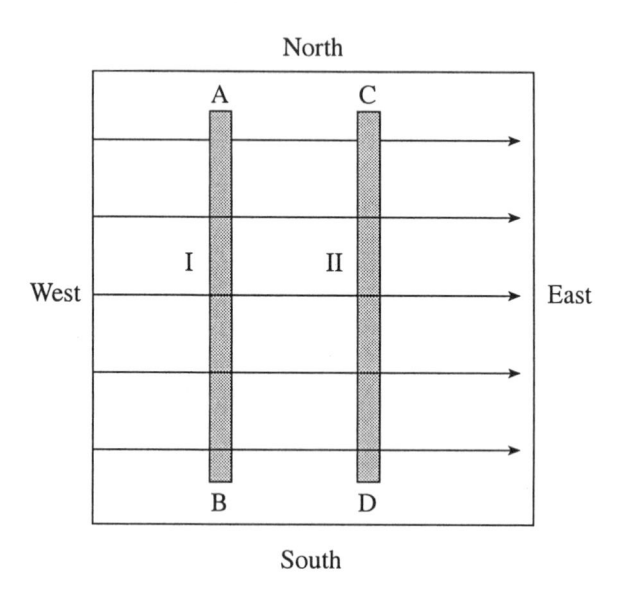

Figure 6 Aquaculturalists farming organisms that extract nutrients from shared seawater can create unfavourable external effects on aggregate production of the farmed organisms, as illustrated by the above diagram. Close proximity of farms increases the likelihood of adverse production spillovers

Letting the farmed area AB be denoted by I and that of CD by II, aquaculturalists in area I will not (in the absence of government intervention) take into account the impact of their stocking density on the yields of aquaculturalists in area II. Therefore, they are liable to overstock their farms from the point of view of maximizing aggregate production from the farmed organisms in the rectangular area. This is illustrated by Figure 7. In this figure, curve ABC shows for a representative farmer in area I the net value of his or her marginal product from the aquacultured species as a function of his or her stocking density of the species. Line DEC represents the marginal cost of altering the stocking rate. Therefore, this farmer in area I will stock the species at the rate x_2. But beyond a stocking rate of x_0, the yield of the species on farms in area II is reduced. Consequently, the social net marginal value of the production from stocking by the farmer in area I may be as indicated by curve ABE. The difference between the portions BC and BE of the relevant curves in Figure 7 represents the marginal negative spillover or externality effect from increased stocking by the aquaculturalist in area I on the production of aquaculture farms in area II. When this negative externality is taken into account, the socially optimal density of stocking on the farm in region I is x_1. This is lower than the density that would be privately chosen. Therefore, a case can exist for public intervention to regulate stocking rates.

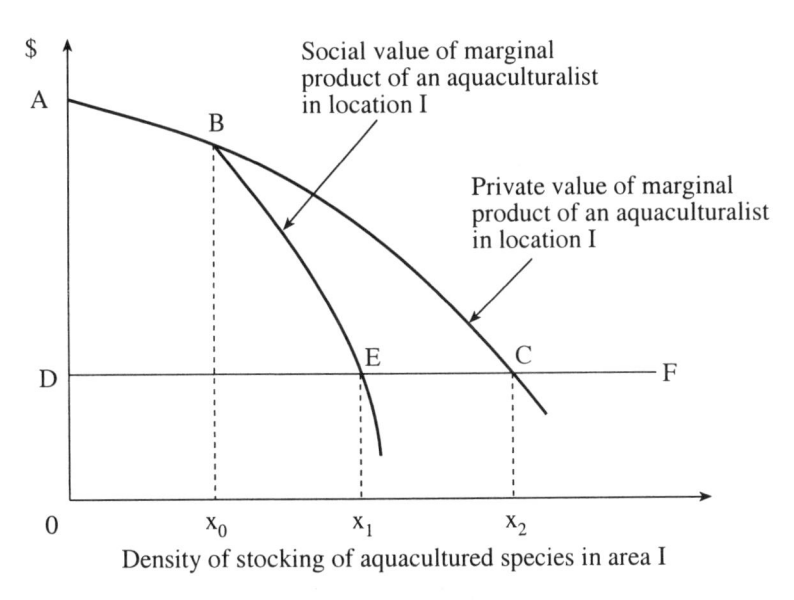

Figure 7 An illustration of market failure in aquaculture due to negative production externalities arising from density of stocking by aquaculturalists in an advantaged location, I

Note that in the type of situation envisaged above, it is inputs that are mobile (fugitive) rather than the animals or the organisms being harvested. The mobility of either can cause market failure. Observe also that, as in the case of open-access fisheries, there is a tendency for economic producers (aquaculturalists) to overcrowd the more productive farming area in cases like that illustrated by Figure 7, that is to overcrowd this area from the point of view of maximizing the aggregate production and economic benefits from use of the limited marine resources available and employed.

Often conditions affecting the capture fisheries and aquaculture are considered independently. This may be a reasonable approximation in some cases, but not others. However, it is becoming less reasonable as an assumption as aquaculture expands and as the supplies of wild caught fish reach their maximum sustainable limits or, in many cases, exceed these. Usually supplies of fish from the capture fisheries influence the development of mariculture and, in turn, development of mariculture may impact in varied ways on supplies from the capture fisheries and the availability of other marine resources. Chapter 28 analyses some of these interconnections. In particular, it qualifies the view, expounded for example by Anderson (1985), that the aquaculture of a species normally favours the conservation of its wild stocks, if these are commercially harvested under conditions of open access. Both economic and ecological factors must be taken into account in considering such interconnections.

Concluding comment

The importance of taking both economics and ecology into account when studying agricultural and marine production is not surprising because both agriculture and to a large extent, marine production are based on the utilization of living resources. A concerted effort is made in this book to combine both economics and ecology in such a study. This combination provides better understanding of many issues involved in the management of living resources. The importance of using ecological economics increases as environmental problems involving agriculture and marine areas grow. Loss of biodiversity and new developments, such as those in genetic engineering, add weight to this interdisciplinary approach. While we cannot be experts in every discipline, there is no advantage in wrapping ourselves in the cocoon of a single field of study when it comes to trying to understand issues that obviously involve multiple academic disciplines. Nonetheless, as we increasingly adopt a more holistic perspective in economics, we should continually remind ourselves of the limits to our knowledge generally and individually; that is, remind ourselves of our bounded rationality (Tisdell, 1996).

References

Anderson, J.L. (1985), 'Market interaction between aquaculture and the common-property commercial fishery', *Marine Resource Economics*, **2**, 1–24.

Auld, B.A., Menz, K.M. and Tisdell, C.A. (1987), *Weed Control Economics*, London: Academic Press.

Bandara, R. and Tisdell, C. (2002), 'Asian elephants as agricultural pests: damages, economics of control and compensation in Sri Lanka', *Natural Resources Journal* (in press).

Campbell, M.F. and Lindner, R.K. (1989), 'A note on optimal effort in the Maldivian tuna fishery', *Marine Resource Economics*, **6**, 173–6.

Conway, G.R. (1985a), 'Agroecosystems analysis', *Agricultural Administration*, **20**, 31–55.

Conway, G.R. (1985b), 'Agricultural ecology and farming systems research', in S. Remenyi (ed.), *Agricultural Systems Research for Developing Countries*, Canberra: Australian Centre for International Agricultural Research, pp. 43–59.

Conway, G.R. (1987), 'Agroecosystems analysis', *Agricultural Systems*, **24**, 95–117.

Davis, R. and Tisdell, C.A. (2002), 'Alternative specifications and extensions of the control of livestock pests', in D.C. Hall and L.J. Moffitt (eds), *Economics of Pesticides, Sustainable Food Production and Organic Food Markets*, Amsterdam: JAI, Elsevier Science, pp. 35–79.

Dragun, A.K. and Tisdell, C. (1999), *Sustainable Agriculture and Environment*, Cheltenham, UK and Lyme, US: Edward Elgar.

Firdausy, C. and Tisdell, C.A. (1992), 'Rural poverty and its measurement: a comparative study of villages in Nusa Peninda, Bali', *Bulletin of Indonesian Studies*, **28**(2), 75–93.

Firdausy, C. and Tisdell, C.A. (1993), 'The effects of innovation on inequality of income distribution: the case of seaweed cultivation in Bali, Indonesia', *The Asian Profile*, **21**(5), 393–408.

Fox, W.W. Jr (1970), 'An exponential surplus yield model for optimizing exploited fish populations', *American Fisheries Society Transactions*, **100**, 80–8.

Mellor, C.J. (1985), 'The optimal choice of a variety of a species: a comment', *Journal of Agricultural Economics*, **36**(2), 263–4.

Owen, O.S. (1975), *Natural Resource Conservation: An Ecological Approach*, New York: Macmillan Publishing.

Sathiendrakumar, R. and Tisdell, C.A. (1989), 'Comments on a note on optimal effort on the Maldivian tuna industry', *Marine Resource Economics*, **6**, 177–8.

Schaefer, M.B. (1954), 'Some aspects of the dynamics of populations important to the management of the commercial marine fisheries', *Inter-American Tropical Tuna Commission Bulletin*, **1**(2), 25–56.

Schaefer, M.B. (1957), 'A study of the dynamics of the fishery for yellowfin tuna in the eastern tropical Pacific Ocean', *Inter-American Tropical Tuna Commission Bulletin*, **2**(6), 245–85.

Tisdell, C.A. (1982), *Wild Pigs: Environmental Pest or Economic Resource?*, Sydney: Pergamon Press.

Tisdell, C.A. (1985), 'The optimal choice of a variety of a species: a rejoinder', *Journal of Agricultural Economics*, **36**(2), 265.

Tisdell, C.A. (1992), *Giant Clams in the Sustainable Development of the South Pacific*, Canberra: Australian Centre for International Agricultural Research.
Tisdell, C.A. (1996), *Bounded Rationality and Economic Evolution*, Cheltenham, UK and Brookfield, US: Edward Elgar.
Tisdell, C.A., Shang, Y.C. and Leung, P.S. (1994), *Economics of Commercial Giant Clam Mariculture*, Canberra: Australian Centre for International Agricultural Research.
World Commission on Environment and Development (1987), *Our Common Future*, Oxford: Oxford University Press.

PART II

INFLUENCES OF ECOLOGICAL CONDITIONS ON AGRICULTURAL PRODUCTION, SUSTAINABILITY, AND ECONOMIC DECISIONS

[2]

THE OPTIMAL CHOICE OF A VARIETY OF A SPECIES FOR VARIABLE ENVIRONMENTAL CONDITIONS

Clem Tisdell*
University of Newcastle[†]

> *When yields of a variety are a non-linear function of a relevant environmental condition, variability of this condition affects the average level of yields, and may favour one variety in comparison to another. When the yield function is strictly concave (exhibits diminishing marginal yield as a function of the environmental condition), heightened environmental variability increases the likelihood of the variety with the smallest change in marginal yield, the more tolerant one, maximising yield on average. However, if the yield function is strictly convex (increasing marginal yields are the case) increased environmental variability raises the likelihood of the variety with the greatest change in marginal yield, the less tolerant one, maximising yields on average. Examples are given, and simple rules are stated for choosing the variety that maximises average yields when yield functions are quadratic. The analysis can be adapted to decision-making in which average profit or expected utility, rather than average yield, is to be maximised.*

Introduction

It is well known that many varieties of crops that give high yields under ideal environmental conditions can be very intolerant to environmental conditions outside the ideal range. Yields of many varieties of species, some of which spearheaded the 'green revolution', decline markedly for any deviation away from their ideal set of environmental conditions. Some varieties are more susceptible than others to moisture stress, to temperature changes and to the availability of sunlight for example. Where a species or a crop is likely to be growing under variable environmental conditions, its tolerance to this variation can be an important consideration in whether it or an alternative variety gives the greatest yield on average and/or highest level of profit on average.

While some attention has been given to this matter in the literature, it is less than one might have expected. The main contributions on tolerance, adaptability and stability of genotypes are briefly reviewed by Evanson and co-authors (1978). These contributions, such as the seminal one by Finlay and Wilkinson (1963), concentrate on measuring, by means of selected coefficients or indicators, the stability and adaptability of genotypes in the face of environmental variation. Evanson and co-authors (1978) point out that a clear distinction has not been made in this literature between (a) the *stability* of a genotype, that is, its changing performance with respect to changing

---

* I would like to thank Professor John Dillon, Dr Bruce Auld, Mr N. M. H. de Silva, the editor, and an anonymous referee for suggestions that were useful in revising the draft of this paper.

[†] Department of Economics, University of Newcastle, N.S.W., 2308, Australia.

environmental factors over time and (b) its *adaptability*, that is, its performance with respect to environmental factors that change across locations. They believe this has happened because most writers have considered these two aspects of variability of the performance of a variety to be highly correlated. They produce evidence that in the case of different varieties of rice this close correlation may not exist.

In this context, the paper by Evanson and co-authors (1978, p. 15) is important because it gives concrete evidence of differences in the stability of yields for different types of rice. Their selected coefficient for measuring stability indicates, for example, that yields of local varieties of rice in India tend to remain more stable as environmental conditions in a locality alter between years than do yields for newer varieties such as IR-8. It is, however, difficult to see the *direct* decision-making or operational significance of their particular coefficient measuring stability, which is not, for example, a variance of yields. They do, however, point out that the variability issue is important in deciding on the direction of crop improvement research. It is important also in deciding on what crop varieties to release and promote for use by farmers.

This paper considers the problem of selecting an optimal variety of a crop at a site, given that different varieties exhibit different degrees of stability in yield as environmental conditions vary from period to period. Environmental factors are treated as inputs in the production functions of varieties, and the influence on yields (on average per growing period, season or crop) of fluctuations in environmental conditions from one growing period or crop to the next is studied. The focus is on the effect on average yields (in time) of variations in environmental conditions between growing periods or crops rather than within the growing period of each crop. Quantities of environmental inputs during the growing period for a crop might be approximated by such measures as the number of stress days encountered, amount of rainfall received in the period, number of days for which temperature is above (or below) a particular level and so on. While some writers, for example, Ryan and Perrin (1974), Rosegrant (1978) and Wickham *et al.*, (1978), have introduced environmental factors into agricultural production functions in this way, few have compared such functions for different crop varieties.

The comparison possibilities can be illustrated by the simplified hypothetical example in Figure 1. This shows the yield per growing period of a crop variety I by curve ABC as a function of the gradient of the relevant environmental condition (such as rainfall, number of stress days) during its growing period, other inputs, such as fertiliser, held constant. Curve DEF shows the yield for another crop variety II under similar environmental conditions. If environmental conditions fall in the range $x_1 < x < x_2$, variety I gives the greatest yield but if they fall outside this range variety II gives the greatest yield.

For simplicity, the analysis given here is for a single environmental variable but it can be generalised to take account of several environmental variables. The straightforward procedures for generalising for convex or concave functions are indicated by Karlin (1959). In the case where the yield function is quadratic and depends upon a number of environmental variables, expected yield depends non-linearly upon the mean values of these variables and linearly upon their variances and covariances. Thus, the single independent variable case generalises without difficulty.

If farmers happened to know in advance what environmental conditions would prevail during the growing of any crop, it would be easy in principle to select the variety giving the greatest yield under those environmental conditions. But, as a rule, a farmer cannot know these environmental conditions exactly in advance. His choice of variety is made under conditions

THE OPTIMAL CHOICE OF A VARIETY OF A SPECIES

Figure 1

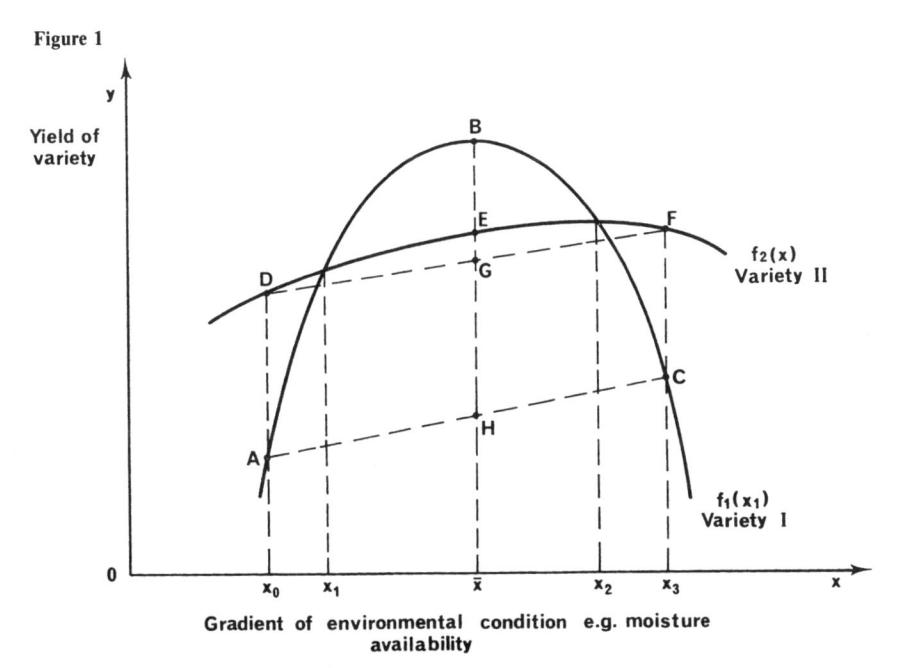

Figure 1 In the above example, in which yield functions are strictly concave, increased environmental variability leads to the choice of the more tolerant variety which is also the one with the lowest maximum yield.

of risk or uncertainty about environmental conditions. In this paper, I shall assume that a stationary objective probability or relative frequency distribution of environmental conditions applies and that the objective is to select the variety that gives the highest yield on average or expected yield in time. Other criteria could also be considered without great difficulty such as the minimax gain criterion and expected utility maximisation criterion (for instance, with utility expressed as some function of yield) in terms of the analysis outlined here.

Hudson (1971) suggests that the production functions of two varieties of cotton (Acala and Albar) in the Sudan *might* be like those in Figure 1 when 'weather' is taken as the independent variable. The yield function for Acala is like curve ABC and for Albar like curve DEF. He points out that 'present methods of testing varieties are specifically designed to identify "all weather" varieties rather than those that are likely to do well in one type of a season or another'. As will be seen from the discussion below, Albar tends to give the highest yields on average over time in an area where weather conditions between seasons fluctuate widely and, of course, this variety would also be the minimax choice *given* the yield functions shown in Figure 1. However, the work of Lomas and Levin (1979), considered later in this paper, suggests that the yield function for Acala may not be strictly concave.

If yields are non-linear functions of environmental conditions, average yields of crops over time depend not only on the averages of environmental conditions but upon the variability of these conditions from growing season to growing season, as for example summarised by the higher moments of the relative frequency distributions of environmental conditions. What implications does such variability have for the optimal choice of a variety?

This will be considered generally at first for a strictly concave yield function (reflecting diminishing marginal productivity from the environmental factor, as in Figure 1) and then for cases in which yields as a function of the environmental factor are quadratic. In the latter cases, average yields over time depend purely on the mean and variance of environmental conditions, other inputs held constant. Finally, strictly convex yield functions are considered. The biological law of tolerance, as for example outlined by Owen (1975, pp. 32-34) suggests that these may be relevant when a crop or species is being grown towards the limits of its environmental range.

Environmental Variability Favours the More Tolerant or Flexible Variety

Under environmental conditions known to fall always in the range $x_1 \leqslant x \leqslant x_2$, variety I would always be chosen in preference to variety II in order to maximise yields if the yield functions shown in Figure 1 apply. However, once environmental conditions begin to vary beyond this range, variety II may maximise yields on average. For example, if the environmental condition is always $\bar{x}$, variety I, with a yield corresponding to B, provides greater yields than variety II with a yield corresponding to E. But if the variability, say, of moisture availability should increase, so that x_0 occurs with a relative frequency of 0.5 and x_3 with the same relative frequency (and moisture conditions are the same on average, $0.5x_0 + 0.5x_3 = \bar{x}$), average yields from variety I then correspond to H and those from variety II correspond to G. The more tolerant or flexible variety, the one with least rate of change in marginal yields*, now gives the highest yield on average. Variety II is now preferable to variety I.

This can also be viewed differently. Letting $f_1(x)$ represent the yield function of variety I and $f_2(x)$ that of variety II, the difference between these functions is

$$g(x) = f_1(x) - f_2(x). \qquad (1)$$

Hence

$$g'' = f_1'' - f_2'' < 0. \qquad (2)$$

Consequently, function $g(x)$ is strictly concave as in Figure 2.

When $g(x)$ is positive, variety I, the less tolerant variety, gives the greatest yield. When $g(x)$ is negative, variety II, the more tolerant variety gives the highest yield. If environmental conditions vary over time, variety I (the less tolerant variety) gives the highest yield on average if $E[g(x)] > 0$. But if $E[g(x)]$ is negative, variety II, the more tolerant variety, gives the highest yield on average. Other things equal, the greater the variability of environmental conditions, the more likely is the tolerant variety to be the one that maximises yields on average. For example, in Figure 2, $E[g(x)] = NR$ if the environmental condition is x in every period. But if environmental condition x_0 occurs with a relative frequency of 0.5 as does x_3 (and so the average condition is unaltered), $E[g(x)] = NL$. Average yields are maximised by switching from the less tolerant variety to the more tolerant one.

---

* This definition is similar to that used by Stigler (1939) and Tisdell (1968) to define the flexibility of a technique.

In general, on the basis of the theorem outlined by Hardy, Littlewood and Polya (1934, pp. 74 - 75), it is known from the properties of concave functions that
$$E[g(x)] < g(E[x]) \qquad (3)$$

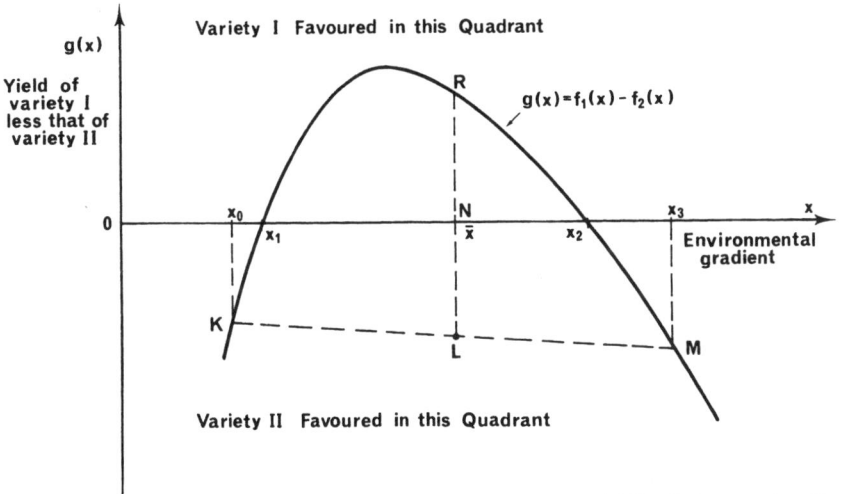

Figure 2

Figure 2 Another illustration of how increased environmental variability favours variety II, the more tolerant variety.

if not all x values are equal. Equality of these expressions only occurs if all x values are equal. Thus, environmental variability increases the likelihood of $E[g(x)]$ being negative, that is, of the more tolerant species maximising yields on average.

When we are, however, comparing one variable situation with another, ambiguity can arise as to whether variability is greater in one situation than in another. But there is no ambiguity in many cases. For example, if the relative frequency of the relevant environmental conditions, e.g., moisture availability, is symmetric about its mean and the mean remains unchanged, variability of the environmental condition increases

(a) if the size of one or more of the deviations below (above) the mean increases and is balanced by corresponding increases in the size of deviations above (below) the mean or

(b) if the probability of larger deviations increases at the expense of the probability of smaller ones.

In these circumstances, the strict concavity of $g(x)$ implies that the likelihood increases that the tolerant variety maximises yields on average. This is because $E[g(x)]$ falls, as indicated by a theorem of Karlin (1959, p. 406), with increased variability of environmental conditions generated in the above way.

Concave Quadratic Yield Functions and Choice of Variety

When the yield functions of varieties of species can be approximated by quadratic functions, this simplifies the task of choosing the variety that maximises yields on average under variable environmental conditions. This is because the average value of a quadratic function depends only on the mean and variance of its independent variable. Consequently, variability of the dependent variable, in this case the magnitude of the relevant environmental condition, can be measured by its variance alone.

Suppose that the per-period yield of variety I as a function of the environmental condition, x, can be expressed as

$$y_1 = -a_1 x^2 + b_1 x + c_1 \qquad (4)$$

and for variety II as

$$y_2 = -a_2 x^2 + b_2 x + c_2 \qquad (5)$$

Because variety II is the more tolerant species, $a_2 < a_1$; it exhibits less rate of change (of marginal product) than the function for variety I.

Now, for variable environmental conditions, the average yield using variety I can be expressed as

$$E[y_1] = -a_1 E[x^2] + b_1 E[x] + c_1 \qquad (6)$$

$$= -a_1 E[x]^2 - a_1 \operatorname{var} x + b_1 E[x] + c_1. \qquad (7)$$

The expression for $E[y_2]$, the average per period yield from variety II, is similar; subscript 1 is merely replaced by subscript 2.

One can of course directly substitute the mean and variance of the relevant environmental condition into the expressions for $E[y_1]$ and $E[y_2]$ so as to determine which of these is the largest and choose the variety accordingly. Or one can specify the difference function $E[g] = E[y_1] - E[y_2]$, carry out the same type of substitution, adopting variety I if $E[g]$ is positive and variety II if it is negative.

For some purposes, however, a simple graphical presentation such as that illustrated in Figure 3 may facilitate choice. For any given average value of the environmental condition, that is for any given $E[x]$ value, the average yield of a variety can be graphed as a linear function of the variance of the environmental condition. Taking the average value of x that concerns us we can substitute this for x in yield functions (4) and (5). Let the resultant value of yields be $\hat{y}_1$ and $\hat{y}_2$ respectively for an $E[x]$ value of $\hat{x}$. Then, given this average environmental condition, average yields per period from the two varieties, I and II, as a function of the variance of environmental conditions are respectively

$$E[y_1] = \hat{y}_1 - a_1 \operatorname{var} x \qquad (8)$$

and

$$E[y_2] = \hat{y}_2 - a_2 \operatorname{var} x. \qquad (9)$$

If average conditions fall in a region where, in the absence of environmental variation, the yield from variety I would exceed that from variety II, yield functions corresponding to (8) and (9) would be like those shown by lines STU and VTW respectively in Figure 3. The downward slope of STU, $-a_1$, exceeds that of VTW, $-a_2$, so the two lines intersect. The critical value of the variance, where they intersect, is σ^2. Given the prevailing average environmental conditions, the less tolerant variety maximises yields on average if the variance is less than σ_2 but the more tolerant variety, II, is needed to do this if the variance of environmental conditions exceeds σ^2.

THE OPTIMAL CHOICE OF A VARIETY OF A SPECIES

Figure 3

[Graph showing: y-axis "Average yield", x-axis "Variance of x (Var x)". Line from $\hat{y}_1$ at S sloping down through T to U, labeled "Variety I" and $E[y_1] = \hat{y}_1 - a_1 \text{ var } x$. Line from $\hat{y}_2$ at V sloping down through T to W, labeled "Variety II" and $E[y_2] = \hat{y}_2 - a_2 \text{ var } x$. Vertical dashed line at $\tilde{\sigma}^2$ labeled "Critical value of variance". Point σ^2 on x-axis.]

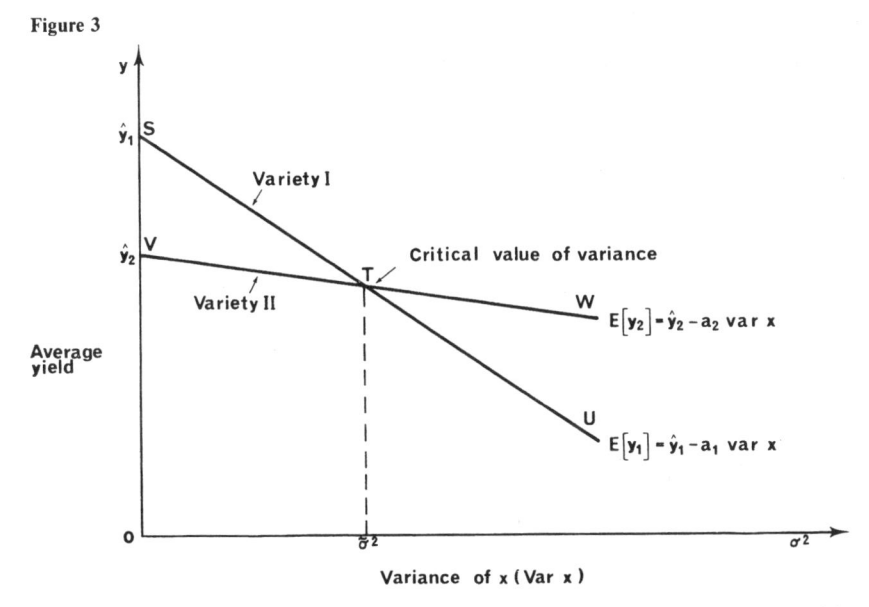

Figure 3 In the case of quadratic yield functions, average yields depend linearly on the size of the variance of the relevant environmental condition, the mean of this condition constant. In the above case, variety II, the more tolerant one is preferred, if var $x > \tilde{\sigma}^2$.

The yield functions shown in Figure 3 shift in a parallel fashion as the mean environmental condition (parameter) is altered. When mean environmental conditions are such as to maximise the excess of $\hat{y}_1$ over $\hat{y}_2$, the variance of the relevant environmental condition can be of its greatest value compatible with variety I's maximising yields on average. As the average environmental condition is varied, the difference between $\hat{y}_1$ and $\hat{y}_2$ alters and, for some values, $\hat{y}_1$ is less than $\hat{y}_2$ so that variety II is chosen no matter what is the value of the variance. The above graphical procedure provides a simple means for determining which variety provides greater yields on average given the relative frequency distribution of the environmental condition being confronted.

Environmental Variability in a Convex Yield Zone and Choice of Variety

Especially when species are growing towards the limits of their environmental range, their yield functions may be strictly convex in this relevant region rather than strictly concave, the condition assumed so far. In the relevant environmental region, yields as a function of the relevant environmental condition may vary at an increasing rate rather than at a decreasing rate as has been supposed so far. In fact, the biological law of tolerance, as for example outlined by Owen (1975), predicts a bell-shaped yield function. This implies that concave yield functions are approximations to the yield function towards the limit of the species environmental range, whereas concave functions provide an approximation to yields when the species is growing in favourable environmental conditions.

The same type of analysis can be applied to the convex case as to the concave one but the results may be a little surprising to some. In contrast to the concave case, increased environmental variability in the convex case favours the less tolerant variety, that is the variety having the greatest variation in the rate of change of its yield function. This can be illustrated from Figure 4.

Here variety I is shown as having the yield function $y_1 = f_1(x)$ represented by the curve ABC and the variety II is shown as having the yield function $y_2 = f_2(x)$ represented by curve DBE. The change in the marginal yield, $f_1''(x)$, from variety I is greater than the change in the marginal yield $f_2''(x)$, from variety II as environmental conditions alter. Variety I is less tolerant to changed environmental conditions than is variety II.

Figure 4

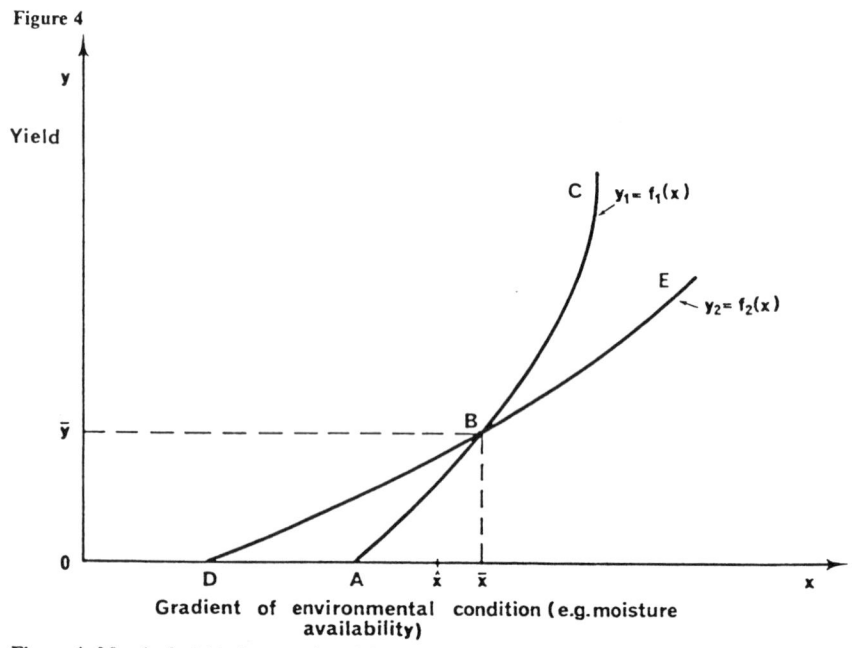

Figure 4 Marginal yields from variety I increase with improved environmental conditions at a faster rate than for variety II. Increased variability of the relevant environmental condition tends to favour variety I.

If the environmental condition as shown in Figure 4 happens to be $\bar{x}$, both varieties, give equal yields. However, if the environmental gradient is less than $\bar{x}$, variety II maximises yields and variety I does so if the gradient exceeds $\bar{x}$. But, if the mean of the environmental gradient remains at $\bar{x}$ and the variability of the gradient increases, variety I is favoured. This follows since if we consider the function $g(x)$ equal to $f_1(x) - f_2(x)$, it is strictly convex because

$$g''(x) = f_1''(x) - f_2''(x) > 0. \qquad (10)$$

Because the function $g(x)$ is strictly convex, it has the property, in the light of a theorem of Hardy, Littlewood and Polya (1934, pp. 74 - 75), that

$$E[g(x)] > g(E[x]) \qquad (11)$$

if not all values of x are equal. Furthermore, if the relative frequency distribution of x is symmetric, remains so and remains distributed around the same mean, $E[g(x)]$ increases

(a) as the size of one or more of the deviations (above or below the mean balanced by change on the other side) from the mean increase (other deviations unchanged) or

(b) the probability of larger deviations increases.

In such cases, variability unambiguously increases, and the likelihood increases that $E[g(x)]$ is positive. Thus, the likelihood increases that the less tolerant variety maximises yields on average.

An example of the yield relationships shown in Figure 4 can be obtained from the data given by Lomas and Levin (1979) for two varieties of cotton grown on the central coastal plain of Israel. Lomas and Levin report lint yields in kg/ha for two varieties (Acala and Deltapine) in relation to the seasonal evapotranspiration of the cotton expressed in mm. Evapotranspiration rates reflect amongst other things the availability of water to the plants. Seasonal evapotranspiration rates between 147 mm and 347 mm were observed. Within this range of observations, lint yields of both varieties of cotton appear to increase at an increasing rate with the level of seasonal evapotranspiration. Fitting logarithmic functions to the data, the least squares fit for the Acala variety is

$$\log y_a = 2.859387 + .00099377 x_e; R^2 = .545 \qquad (12)$$
$$(0.0686)$$

and for the Deltaphine variety

$$\log y_d = 2.84479 + .0010662 x_e; R^2 = .799 \qquad (13)$$
$$(0.0408)$$

where y_a represents the lint yield of Acala in kg/ha, y_d is the lint yield of Deltapine in kg/ha, and x_e is the seasonal evapotranspiration rate in mm. The figures shown in parenthesis are standard errors of the estimates. The t-value of coefficient of x_e in (12) is 2.682 which is significant at 2.5 per cent level. The t-value of the coefficient of x_e in (12) is 2.682 which is significant at 2.5 per cent level. The t-value of corresponding coefficient in (13) is 4.838 which is significant at 0.5 per cent level.

Lint yields of Deltaphine are less than Acala for seasonal evapotranspiration rates of less than 216 mm but exceed lint yields of Acala at higher seasonal evapotranspiration rates. Thus if a minimax approach is taken to risk, and seasonal evapotranspiration rates below 216 mm are possible, Acala will be preferred; Deltapine will be preferred if seasonal evapotranspiration rates are always above 216 mm, but within the observed range.

If seasonal evapotranspiration rates on average are in excess of 216 mm, Deltapine gives the greatest lint yield on average per season. If the average seasonal evapotranspiration is below 216 mm it is still possible for Deltapine to maximise yields on average if there is sufficient dispersion of the evapotranspiration between seasons. The greater the dispersion of evapotranspiration rates, the more likely is Deltapine (the less tolerant variety) to maximise lint yields on average.

In some cases, it may be possible to approximate the yield functions in the convex case by the upward sloping branches of quadratic functions. For example, the yield from variety I might be represented by the upward sloping branch of

$$y_1 = a_1 x^2 - b_1 x + c_1 \tag{14}$$

and that for variety II by

$$y_2 = a_2 x^2 - b_2 x + c_2 \tag{15}$$

As before, the expected value of these functions can be expressed as a function of the mean of the environmental condition x and the variance of x. Furthermore if the difference between these functions is represented by g,

$$\frac{\delta E[g(x)]}{\delta \operatorname{var} x} = (a_1 - a_2) > 0 \tag{16}$$

because $a_1 > a_2$. Hence, other things equal, an increase in the variance of the environmental condition makes it more likely that variety I, the less tolerant variety, will give the largest yield on average.

In this quadratic case, for any set of average environmental conditions, $E[y_1]$ and $E[y_2]$ can be expressed as a linear function of var x. Suppose that the mean value of x is $\hat{x}$, and let $\hat{y}_1$ represent the value of y_1 when the variance of x is zero and let $\hat{y}_2$ represent the value of y_2 when the variance of x is zero. Then,

$$E[y_1] = \hat{y}_1 + a_1 \sigma^2 \tag{17}$$

$$E[y_2] = \hat{y}_2 + a_2 \sigma^2 \tag{18}$$

where σ^2 represents the variance of x. The average yields of each of these varieties can be expressed as linear functions of their variances and represented by upward sloping straight lines shown as a function of σ^2. They can be diagramatically represented by a similar figure to Figure 3. Because $a_1 > a_2$, the straight-line representing (17) intersects that representing (18) from below. Letting $\bar{\sigma}^2$ represent the variance of x giving the solution for these equations (corresponding to the value at which the lines intersect) it is a critical value given the mean condition $\hat{x}$. Variety II maximises yield on average if the environmental variance is less than $\bar{\sigma}^2$ but variety I, the less tolerant variety, does this if the environmental variance exceeds $\bar{\sigma}^2$.

While the convex yield functions shown in Figure 4 are increasing ones, the argument also applies to strictly convex decreasing ones which, for example, might be encountered towards the top end of a bell-shaped tolerance function, suggested by Owen (1975).

Qualifications and Conclusions

It might be held that the objective of maximising average profit is more relevant than that of maximising yields on average. If this is so, all that needs to be done is to re-interpret the above yield functions as profit functions and the previous results hold, *mutatis mutandis*. It might even be considered desirable to go further than this and adopt the objective of maximising expected utility, either expressing utility as a function of the yield (not an unreasonable relationship in a subsistence economy) or as a function of profit. If diminishing marginal utility is the rule, there will be a premium on reducing the variance of yields or of profits as the case may be, and this will tend to weigh in favour of the more tolerant variety.

Nevertheless the above results are of interest in themselves. The result in the concave yield case that, other things unchanged, increased variability of environmental conditions tends to favour the more tolerant variety in terms of average yields was not unexpected. However, that the less tolerant species should be favoured by increased variability in the convex yield case was not intuitively clear, but in the light of the biological law of tolerance has important implications for the choice of varieties where species are growing towards the limits of their biological range. For quadratic cases, simple rules were outlined for choosing between varieties of species facing variable environmental conditions. Thus, some progress has been made with an important problem affecting the economic welfare of a large portion of mankind.

References

Evenson, R. E., O'Toole, J. C., Herdt, R. W., Coffman, W. R. and Kauffman, H. E. (1978). *Risk and Uncertainty as Factors in Crop Improvement Research.* The International Rice Research Institute, Manila, Research Paper Series No. 15.

Finlay, K. W. and Wilkinson, G. N. (1963). The Analysis of Adaptation in a Plant Breeding Programme, *Aust. J. agric. Research,* 14, 742 - 754.

Hardy, G. W., Littlewood, J. E. and Polya, G. (1934). *Inequalities.* Cambridge: Cambridge University Press.

Hudson, J. P. (1971). Horticulture in 2000 AD, in P. F. Wareing and J. P. Cooper (eds.), *Potential Crop Production.* London : Heinemann Educational.

Lomas, J. and Levin, J. (1979). Irrigation. In J. Seeman, Y. I. Chirkov, J. Lomas and B. Primault (eds.) *Agrometeorology.* Berlin : Springer-Verlag.

Karlin, S. (1959). *Mathematical Methods and Theory of Games, Programming and Economics,* Vol. 1. Reading : Addison-Wesley.

Owen, O. S. (1975). *Natural Resource Conservation: An Ecological Approach.* New York: Macmillan Publishing.

Rosegrant, M. W. (1977). *Risk and Farmer Decision-Making: A Model for Policy Analysis.* Agricultural Economics Department, The International Rice Research Institute (Manila) paper No. 77 - 5.

Ryan, J. G. and Perrin, R. K. (1974). Fertiliser Response Information and Income Gains: The Case of Potatoes in Peru, *American J. agric. Econ.,* 56, 337 - 343.

Stigler, G. J. (1939). Production and Distribution in the Short Run, *J. pol. Econ.,* 47, 305 - 328.

Tisdell, C. A. (1968). *The Theory of Price Uncertainty, Production and Profit.* Princeton: Princeton University Press.

Wickham, T., Barker, R. and Rosegrant, M. V. (1978). Complementarities Among Irrigation Fertiliser and Modern Rice Varieties, in *Economic Consequences of New Rice Technology.* Manila: The International Rice Research Institute.

ID: 9 : 99—109, 1983

THE BIOLOGICAL LAW OF TOLERANCE, AVERAGE BIOMASS AND PRODUCTION IN A VARIABLE UNCONTROLLED ENVIRONMENT

C. A. TISDELL

Department of Economics, University of Newcastle, N. S. W. 2308, Australia.

ABSTRACT

The biological law of tolerance holds that the tolerance function of a species can be represented by a bell-shaped curve showing the dendence of the population size or biomass of a species upon the gradient of (relevant) environmental conditions. The law implies (it is shown) that when environmental conditions are favourable to a species an increase in the variabiliıy of these conditions (average environmental conditions unchanged) reduces the average population level of the species or its biomass on average. The opposite the outcome if environmental conditions are unfavourable to the species. An increase in the variability of environmental conditions raises the average biomass or population of a species. Similar results are shown to hold for the average level of production and the average profit from harvesting a species influenced by uncontrolled variability of environmental conditions. In marginal areas (e.g. arid zones) great variability in environmental conditions, such as in the amount of rainfall annually, may be advantageous rather than as has sometimes been supposed, disadvantageous to the population of a species and to production. Convexity and concavity properties of tolerance functions are used to provide general conclusions.

ZUSSAMENFASSUNG

Das Gesetz biologischer Toleranz behauptet, dass die Toleranzfunktion einer Species durch eine glockenformige Kurve ausgedruckt werden kann, die die Abhängigkeit der biologischen Masse einer Species vom Gradient relevanter Umweltfaktoren zeigt. Wie hier gezeigt wird, stellt dieses Gesetz fest, dass, wenn die fur eine Species gunstigen Umweltbedingungen vorherrschen, durch die Zunahme in der Variationsbreite dieser gunstigen Umstände, die Durchschnittszahl der biologischen Masse einer Species reduziert wird, vorausgesetzt, dass die durchschnittlichen Umweltbedingungen unverändert bleiben. Das gegenteilige Ergebnis wird erzielt, wenn die fur eine Species ungunstigen Umweltbedingungen vorherrschen. Eine Zunahme der Variationsbreite der Unweltbedingungen erhoht die biologische Durchschnittsmasse einer Species. Ähnliche Resultate treffen auf den durchschnittlichen Produktionsstand und auf den durchschnittlichen Ertrag einer Species zu, die durch unkontrollierte Variationsbreiten der Umweltbedingungen beeinflusst werden. In Grenzgebieten (z.B. in Trockenzonen) kann sich eine grosse Variationsbreite in Umweltbedingungen, z.B. in der Hohe der jahrlichen Regenfalle, vorteilhaft und nicht, wie oft behauptet worden ist, nachiteilig auf die Anzahl der Species und auf ihren Ertrag auswirken. Konvexe und konkave Eigenschaften der Toleranzfunktion werden benutzt, um allgemeine Schlussfolgerungen zu ziehen.

INTRODUCTION

Despite the injunction of the eminent Cambridge economist, Alfred Marshall (1920, p. xii) that economists should pay greater attention to biological relationships and analogies, on the whole this has not become fashionable in economics. Rather interest in analogies from physics have tended to persist. But some biological relationships have direct economic application. The biological law of tolerance, for example, is of direct relevance to agricultural, forestry, fishing and hunting production based on the use of natural resources. Production in these rural industries is often dependent upon variable and uncontrolled environmental conditions, such as variations in rainfall, in temperatures or in the occurrence of pests. The law has various implications for the average levels of population and biomass of species, the average level of available harvest, the average level of production and profitability, that are well worth exploring.

The law implies, as argued in this paper, that where a species is being grown in marginal environmental conditions for it, variability of these conditions increases the population or biomass of the species on average. The average level and profitability of production may also rise in these circumstances. On the other hand, if production is occurring in a range of optimal environmental conditions, increased variation over time of environmental conditions in a locality may well reduce the average biomass, available harvest, the average level 'of production and profitability.

The purpose of this paper is to show that such relationships are a consequence of convexity and concavity properties of tolerance relationships (for the species) as a function of limiting environmental factors. The underlying ecological relationships have different implications for populations, yields, productivity and economic gain to that popularly imagined. For example, where a crop is being grown towards the limit of a region suitable for it, *i. e.* in a locality in which annual rainfall is much less than that required for the optimal (maximum) level of production or biomass per unit area from the crop, an increase in the variability of rainfall between years (the average rainfall over the years remaining unchanged) may actually increase total yield or biomass, (that is, yield on average per year) obtained from the crop even though variability in yield or biomass between years rises. Thus fluctuations in environmental conditions can be an advantage (as explained below) rather than a disadvantage.

The approach to modelling adopted here is similar to that of Levins (1968) who emphasizes that in ecology "there is no single', best all-purpose model. In particular, it is not possible to maximize simultaneously generality, realism and precision". (Levins 1968). Levins opts for generality in preference to precision and bases much of his theory on the properties of convexity or concavity of mathematical functions as is done here. However, Levins' (1968) main concern is with the theory of evolution in changing environments taking into account environmental heterogeneity, the adaptability of species and competition between species. This is not the focus of this paper. In this paper, the nature of the species is assumed to be fixed, and any competition which occurs with other species is assumed to be stationary. Furthermore, the paper does not deal with the dynamics of population sizes, with species interactions and impacts on gene frequencies (evolution) in random variable environments as does a considerable amount of recent literature in theoretical ecology (Turelli 1977). Much of this literature, especially that using stochastic differential equations, is very complicated.

The analysis used here is simplified and involves considerable abstraction from the complexities and interdependence between ecological factors. Initially the population of or biomass of a species in an area (and associated potential economic production from it) are assumed to depend on a single environmental variable[1] and this rela-

---

1. Other environmental factors having an influence on the biomass of the species are assumed to be constant.

tionship is supposed not to alter with the passage of time. Furthermore, the probability distribution of the value of the environmental variable is assumed to be stationary in time. In effect, the environmental variable in any period is a random variable drawn from a stationary probability distribution and the biomass or level of population of the species under consideration is assumed in each period to adjust 'perfectly' in accordance with the tolerance function (discussed below) to the value of the environmental variable experienced in each period. Thus time is implicity treated as a discrete rather than a continuous variable in the analysis[2]. The analysis can easily be extended to cover situations in which several environmental variables or limiting factors are variable, and its assumptions do not appear to be inconsistent with the principles of limiting factors as set out by Odum (1975, Ch. 5).

Once the biological law of tolerance and the implications of it for the biomass of a species is outlined, consequences for the average per period biomass or population (and possible harvest) of a species are analyzed relying on the convexity and concavity properties of the tolerance function. This in turn is shown to have consequences for the possible harvest, production and economic profit from a species.

THE BIOLOGICAL LAW OF TOLERANCE OUTLINED

The law of biological tolerance states that a species population or its biomass attains a peak under optimal environmental conditions and falls away under less favourable environmental conditions to form a bell-shaped curve[3] as in Figure 1 (Owen 1975). Towards extremes of its environmental range a species suffers from physiological stress

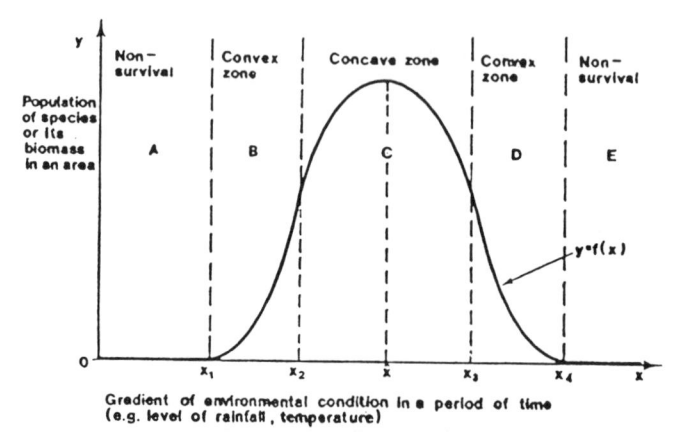

FIG. 1 Bell-shaped curve showing the relationship between the population of a species or its biomass as a function of the gradient of a relevant environmental condition, as suggested by the biological law of tolerance.

---

2. In essence the population or biomass of the species is assumed to adjust rapidly to the environmental condition in each period.
3. Although Owen deals with only one environmental variable, the bell-shaped curve can be generalized to take account of a number of environmental conditions subject to variation.

and in extreme environmental conditions the sepcies fails to survive. As environmental conditions within the survival range of a species alter from those that are less favourable to the species, its level of population (or biomass) increases at first at an increasing rate until the optimum condition is reached when it reaches a maximum population level.

Let us call the function $y=f(x)$ shown in Figure 1, the tolerance function. It can be divided into 5 zones, A, B, C, D and E. In zone A the environmental factor is so low that the species fails to survive in an area where such conditions prevail, and in zone E the relevant environmental factor is so great that the species also cannot exist. Within the range $(x_1 \leq x \leq x_4)$ where the species is able to survive, zone C corresponds to the strictly concave portion of the tolerance function and zones B and D correspond to its strictly convex functions.[4] In zone C, $f''<0$ and in zones B and D, $f''>0$. For example, $f(x)$ increases at an increasing rate in zone B.

When environmental conditions are variable, as they very often are, these zones have important implications for the average level of the population of a species (or the average of its biomass) and the average size of the available harvest of the species. In some areas, for instance a number of arid regions in the world, environmental conditions are extremely variable between years and the implications of this for the biomass and/or available harvest either of species introduced by man or natural species are important. We shall consider these implications shortly.

In principle, it is desirable to have concrete empirical support for the type of tolerance function portrayed by Own (1975) and to have estimates for it for particular species and variable environmental conditions. Unfortunately, my search of the literature and enquiries have not led to the discovery of completely specified tolerance function for a species. Empirical relationships for some of its range for some species have been found, for instance for yields (a biomass proxy?) from some varieties of cotton as a function of evapotranspiration rates (Lomas and Levin 1979, Tisdell 1983). Coefficients of adaptability of different crops to varying environmental conditions also have been calculated (Finlay and Wilkinson, 1963, Evanson et al., 1978) but these are not tolerance functions and they have a number of limitations for analytical work (cf. Tisdell 1983).

In considering quantitative evidence on limiting factors, Odum (1971) approvingly mentions empirical work by Klages (1942) on average barley yields (on a State-basis) and the coefficient of the variability of their yields in the United States as a result of alterations in environmental conditions between States. However, the data does not enable a relationship like that in Figure 1 to be derived. Several environmental and other factors vary across States as the cross-sectional data is subject to many influences. On the other hand, Odum appears to be critical of field experiments designed to test tolerance relationships and measure tolerance functions.[5]

Despite the difficulty of obtaining relevant data, the relationships between the biomass or level of population of a species need not be of a bell-shape for the analysis outlined below to be of general interest. It is likely to be of value provided that certain general properties, such as convexity or concavity of tolerance functions are

---
4. The shape of the tolerance function is similar to that often assumed by economists for the production function.
5. Odum correctly points out that the response of biomass to a particular level of an environmental variable depends upon the duration for which it is held at that level. In the model used here, this is abstracted from.

known to be empirically satisfied within the range of environmental conditions likely to be experienced.

THE AVERAGE BIOMASS OF SPECIES AND ENVIRONMENTAL VARIABILITY

The impact of increased environmental variability over time on the average biomass of a species depends upon whether the variation in the environmental variable is limited to the concave zone of the tolerance function or to one of its convex zones. If average environmental conditions over time remain the same and (a) if increased environmental variability is confined to the strictly concave zone C, of the tolerance function, the average biomass of the species per period falls, but (b) if the heightened environmental variability is restricted in a strictly convex zones B or D, of the tolerance function, the average biomass of the species per period rises. These propositions can be illustrated by Figure 2. In each period, the population of the species is assumed to adjust completely to the value of the environmental variable in the period.

Consider proposition (a). Suppose that the environmental variable is originally stable at h_5 in zone C. The average biomass (or population level) per period of the species is then Y_3. Now let conditions become more variable so that h_4 occurs with a relative frequency of 0.5 as does h_4. The average environmental condition is then $0.5h_4 + 0.5\ h_6 = h_5$. The increased environmental variability causes the average per period biomass (or population) of the species to fall from Y_4 to Y_2.

Now consider proposition (b). Suppose that environmental condition is initially stable in zone B at h_2. The per period biomass or population is then Y_1. However, suppose that variability of environmental conditions so that h_1 occurs 0.5 of the time as does h_3. The average value of these environmental occurrences is equal to h_2; $0.5\ h_1 + 0.5\ h_3 = h_2$. As a result of the variability, the average per period biomass or population of the species rises from Y_1 to Y_1. Under variable environmental conditions, biomass corresponds to K half of the time and to M half of the time so the average corresponds to N. This proposition can be similarly illustrated in zone D.

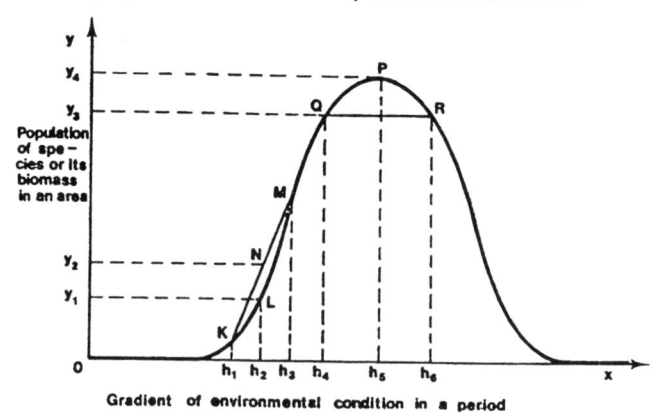

FIG. 2. Increased environmental variability tends to reduce the average population of or biomass of a species when environmental conditions are favourable to the species, but the opposite is the case when environmental conditions are unfavourable to the species.

If the available level of harvest of a species is linearly and positively related to its biomass, then the general propositions that have just been advanced about the average level of the biomass apply equally to the average level of the available harvest. For instance, variability of the environmental variable within a convex zone raises the average available harvest.

The above results are a reflection of the general rule that if not all values of x are the same, then accordingly as $f(x)$ is strictly convex or is strictly concave

$$E[f(x)] \stackrel{>}{<} f(E[x]) \qquad (1)$$

where E indicates the average or expected value of the variable in square parenthesis (Hardy et al. 1934, theorem 90). But this only refers to differences arising from a variable compared to a non-variable value of x, the environmental condition in this case.

The problem when we speak of increased variability is that it is not always clear how that is being or is to be measured. However, there are some circumstances in which variability unambiguously increases. If the relative frequency distribution of the variable x is symmetric about tthe mean of x and remains 2O, and if the mean value of x remains unchanged, an increase in the size of the deviations between the variable and its mean or in the probability of larger deviations greater variability of x. If the size of one or more deviations of x below its mean get larger or if the probability of larger deviations below the mean rises (with corresponding changes occurring symmetrically above the mean) the variability of x unambiguously increases. In this case the value of $E[f(x)]$, the average value of the biomass, increases with increased variability if $f(x)$ is strictly convex (if the tolerance function is strictly convex in the region of variation of x).[8] This average value falls if $f(x)$ is strictly concave, that is if the tolerance function is strictly concave in the region of variation of x.

In some circumstances, the variance of x is an entirely adequate measure of variability. This is so when the dependent variable, y, of the objective function is quadratic function of x. In that case the average value of y depends only on the mean of x and the variance of x.[9] For example if

$$y = ax^2 + bx + c, \qquad (2)$$
$$E(y) = aE(x)^2 + bE(x) + c \qquad (3)$$
$$= aE(x)^2 + a \text{ var } x + b \ E(x) + c \qquad (4)$$

If $a>0$, $f(x)$ is strictly convex and $E(y)$ increases with the variance of x, var x, If $a>0$, $f(x)$ is strictly concave and $E(y)$ falls with increases in the variance of x.

---

6. Relationship (1) generalizes to cover the case where a number of environmental factors are variable. If X represents the vector of these environmental variables it can be substituted in relationship (1) for x and the inequalities are satisfied respectively depending upon whether f (x) is strictly convex or strictly concave.
7. There are a number of possible measures of the variability of an environmental variable. It could be measured for instance by the range of the relevant variable, its variance or the coefficient of variation or by other measures. It is known however that the expected value of a polynomial function of the n-th power depends upon the first n moments of the probability (or relative frequency) distribution of the independent variable. Where a function is a quadratic the variance of the dependent variable is therefore the appropriate measure of variability.
8. For further discussion of relevant properties of convex functions, see Karlin (1959).
9. For further discussion and references, see Tisdell (1968, p. 26, 27).

It may be possible to approximate the tolerance function in the region in which the species is able to exist by three quadratic functions. The tolerance function in zone B might be approximated by the positively sloping branch of a strictly convex quadratic function, in zone C by a strictly concave quadratic function and in zone D by the negatively sloping branch of a strictly convex quadratic function. The rate of change of the biomass associated with such a composite function is illustrated in Figure 3.

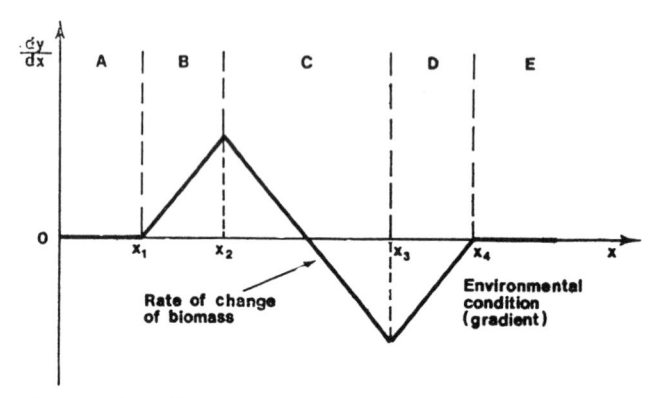

FIG. 3. The marginal population or biomass of a species as a function of the gradient of a relevant environmental condition when the tolerance function is approximated by a series of quadratic functions.

It follows in this case that if average environmental conditions are the same an increase in the variance of environmental conditions within regions B or D (marginal regions for the species) increases the average biomass per period or the average population level of the species. Variability can be considered to be an advantage. The average biomass (per period) increases with the variance of environmental conditions and rises by a larger amount the greater is the size of f'' (that is, the greater is the upward slope of the positively sloping segments in Figure 3).

On the other hand, if average environmental conditions are unaltered, an increase in the variance of environmental conditions in region C lowers the average biomass (per period). This decrease is larger the smaller (more negative) is f'' (that is, the greater is the downward slope of the negatively sloped segment in Figure 3) or the larger is the negative value of coefficient a. Within the region where optimal environmental conditions for the species tend to prevail, an increase in the variance of environmental conditions is undesirable from the point of view of maximising the average biomass or population level of the species.

Nothing has been said so far about the consequences of environmental conditions occurring in a 'non-survival' (zero-population) zone for the species. However, if a species can regenerate after such an experience, and the variability of environmental conditions straddle zones A and B or zones D and E, increased variability of these conditions (average conditions constant) raises the average per period biomass or population level of the species. This further supports the view that in marginal areas increased variability of environmental conditions can be advantageous.

This is illustrated in Figure 4 where a tolerance function for zones A and B is shown. Assume that originally environmental conditions are stable at g_2, so the average (and actual) harvest available in each period is Y_1. Now imagine that environmental condition g_1 occurs with a relative frequency 0.5 as does g_3 so that on average environmental conditions are equal to g_2. The average biomass then increases to Y_2. The furhter g_1 and g_2 deviate from one another the greater is y_2. This is because in the survival zone B, an increased addition is made to the biomass by an increased deviation in it, whereas in the non-survival zone the available harvest remains at zero.

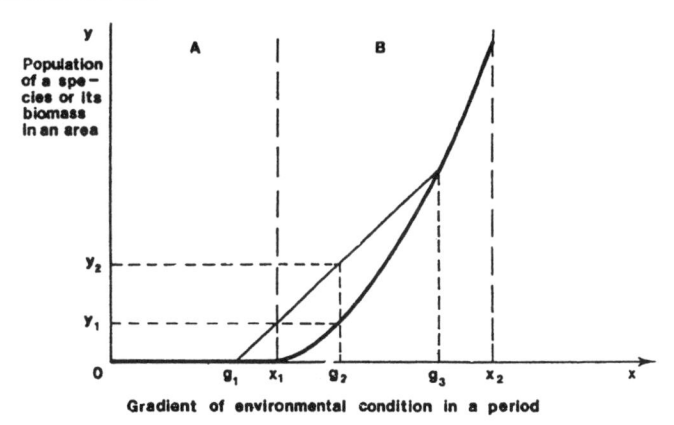

FIG. 4. Increased environmental variability in a marginal (unfavourable) zone for a species can increase its average biomass even when it periodically pushes the species into a 'non-survival' (zero-population) zone. This is so provided that the species is able to regenerate.

The population of a species in an area may be able to recover after it has been pushed to zero because regenerative material such as seeds or rhizomes remain in the area, or because immigration of new stock of the species or inward movement of seeds or other regenerative material occurs either naturally or under the husbandry of man. The possibility of regeneration of this type is well recognized in the ecological literature (Noy-Meir 1975).

It is, of course, possible for environmental variability to straddle both convex and concave regions of the tolerance function. In that case it can be difficult to generalize about the impact of increased environmental variability on the average biomass even though it can be evaluated for any particular case. Nevertheless the above discussion establishes some significant general relationships.

PRODUCTION AND PROFITABILITY

While the above information about the level of population of a species or its biomass (and the level of the available harvest) is of interest, the profitable pattern of the actual harvest may not be an exact replica of the tolerance function. It is possible that actual harvesting may be unprofitable, at low levels of population of the species, for example because the species is very scattered and high costs are involved in gathering it. It is also possible that the marginal productivity of harvesting a species may begin to decline before the end of zone B in Figure 1, before x_2 is reached. Thus the convex zone of the production function when harvesting is most profitably adjusted to environmental conditions might have a greater range than for the tolerance function. Nevertheless it is likely that this production (actual harvest) function can be divided into 5 zones like those in Figure 1 and that the above theorems will apply with the necessary changes.

As for the firm's maximum profit curve as a function of the gradient of the environmental condition, it might take the form shown in Figure 5. It is assumed that the firm can adjust factors of production for harvesting (or otherwise) so as to maximise profit for each environmental condition. For simplicity, it is assumed that all costs are escapable by not producing.[10]

The maximum profit function in Figure 5 has been divided into 5 zones, I, II, III, IV and V. Zones I and V correspond to environmental conditions in which production is unprofitable and are wider in range than zones A and E in Figure 1. Zones II and IV correspond to the strictly convex range of the profit function and zone III corresponds to its strictly concave range. It seems likely that the range of zone III will exceed that of zone C in Figure I, with the ranges of zones II and IV being smaller than those of B and D. It is also possible that the profit function could reach a maximum before x (which gives maximum abundance of the species) is reached. This could occur if beyond a point, abundance becomes an impediment to harvesting. However, in that case the function might have multiple local maxima. I shall assume that this complication does not occur.

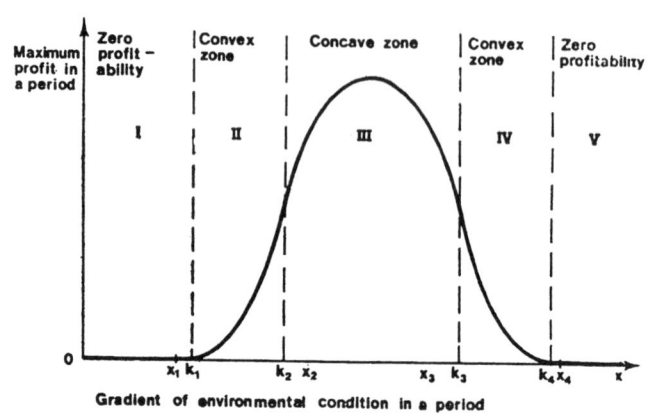

FIG. 5. Increased environmental variability may reduce a firm's average per period profit, when a harvested species is being grown in a range of favourable environmental conditions, but tends to raise this profit when a species is being grown under relatively unfavourable environmental conditions.

Using the same type of arguments as in the last section, if average environmental conditions are unaltered, increased variability of environmental conditions in the strictly concave zone III, of the profit function reduces average profit. Average profi-

---

10. Allowance can be made for inescapable costs. In any period in which production ceases, profit will have a negative value equal to inescapable costs.

tability is reduced by increased environmental variability when a species is growing in environmental conditions that tend to be optimal for it.

On the other hand, in zones II and IV where the (maximum) profit function is strictly convex, increased environmental variability increases average profit per period, if environmental conditions remain unaltered on average. When an organism to be harvested exists under marginal environmental conditions, increased variability of these conditions tends to be desirable from a profitability viewpoint. The arguments also extend to variations that straddle zones I and II or zones IV and V.

The argument can also be extended in a similar way to that in the last section to take advantage of quadratic functions. This can be done if the profit function in zone II is approximated by a quadratic function, and that in zones II and IV is approximated by the relevant branches of quadratic functions. The variance of the gradient of the environmental condition is then the relevant measure of environmental variability.

CONCLUSIONS

The biological law of tolerance implies that when a species is enjoying an optimal range of environmental conditions, that an increase in the variability of these conditions, their average value unaltered, reduces the average per period level of the population or biomass of the species. The average available harvest is likely to go down. However, when a species is subject to unfavourable environmental conditions from its biological point of view, the opposite consequence follows: its average available biomass tends to be increased by greater environmental variability. Similar results apply to the firm's average per period profits from harvesting the species.

It is frequently believed that greater variability of environmental conditions is undesirable for a species as well as for economic profitability for harvesting it. But this paper indicates that it may well enable a greater *average* population of the species to exist and that it may add to the level of average profits of firms utilizing the species. This is likely to be so where a species occurs towards the limits of its biological range. On the other hand, where the species occurs well within its optimal biological range such variability does reduce its average per period biomass and population level and average profits from utilizing it. At least this appears to be so if the biological law of tolerance of the form suggested by Owen (1975) applies. Even if all cases do not accord with the bell-shaped relationship outlined by Owen, provided that the tolerance function is known to be strictly convex or strictly concave within the range of environmental variation, the general theory outlined is applicable and useful in the way suggested by Levins (1968).

ACKNOWLEDGEMENTS

I wish to thank Dr. M. G. Young, of the Institute of Biological Resources, CSIRO, for his comments on the earlier draft of this paper, the reviewers for their constructive suggestions. The usual *caveat* applies.

REFERENCES

EVENSON, R. E., J. G. O'TOOLE, R. W. HERDT, W. R. COFFMAN, AND H. E. KAUFFMAN. 1978. *Risk and Uncertainty as Factors in Crop Improvement Research.* Research Paper Series No. 15, The International Rice Research Institute, Manila.

FINLAY, K.W., AND G.N. WILKINSON. 1963. The analysis of adaptation in a plant breeding programme. *Aust. J. Agric. Research* **14** : 742-754.

HARDY, G.H., LITTLEWOOD, AND G. POLYA. 1934. *Inequalities*. Cambridge University Press, Cambridge.

KARLIN, S. 1959. *Mathematical Methods, Theory of Games, Programming and Economics*. Addison-Wesley, Reading, Mass.

KLAGES, K.H.W. 1942. *Ecological Crop Geography*. Macmillan Co., New York.

LEVINS, RICHARD. 1968. *Evolution in Changing Environments : Some Theoretical Explorations*. Princeton University Press, Princeton, N. J.

LOMAS, J., AND, J. LEVIN. 1979. Irrigation, in : J. SEEMAN, Y. I. CHIRKOV, J. LOMAS AND B. PRIMAULT (Eds). *Agrometerology*. Springer-Verlag, Berlin.

MARSHALL, ALFRED. 1920. *Principles of Economics*. Macmillan, London.

NOY-MEIR, I. 1975. Stability of grazing systems : in application of predator-prey graphs. *J. Ecol.* **63** : 459-481.

ODUM, EUGENE P. 1971. *Fundamentals of Ecology*. Third Edition, Saunders, Philadelphia.

OWEN, O.S. 1975. *Natural Resource Conservation : An Ecological Approach*. Macmillan Company, New York.

TISDELL, C.A. 1968. *The Theory of Price Uncertainty, Production and Profit*. Princeton University Press, Princeton, N. J.

TISDELL, C.A. 1983. The optimal choice of a variety of a species for variable environmental conditions. *J.Agric. Econ.* **34** : 175-185.

TURELLI, M. 1977. Random environments and stochastic calculus. *Theo. Pop. Biol.* **12** : 140-178.

[4]
NEW CROP VARIETIES: IMPACT ON DIVERSIFICATION AND STABILITY OF YIELDS*

CLEM TISDELL and MOHAMMAD ALAUDDIN

University of Queensland　　　*University of Melbourne*

I. INTRODUCTION

It is frequently argued that high yielding varieties (HYVs) of crops create greater uncertainty about and instability of farm yields and production. While earlier empirical evidence (*e.g.* Hazell, 1982, 1985; Mehra, 1981) pointed towards greater relative instability of crop yields as a result of the introduction of HYVs, more recent research including our own (Alauddin and Tisdell, 1988a, 1988d) indicates the opposite tendency.

In earlier publications, we have isolated increased multiple cropping (Aluaddin and Tisdell, 1986b) and increased control over agricultural micro-environments due to greater use of HYV associated techniques (Alauddin and Tisdell, 1988d) as significant factors contributing to reduced relative variability of crop yields with the adoption of HYVs. However, we have not given in-depth attention to the possibility that increased crop diversification, including greater diversity in varieties of crop season, may also contribute to this result. This paper extends our earlier research by concentrating on the diversification aspect. In so doing, it applies portfolio diversification analysis to this issue, makes use of lower semi-variance and Cherbychev's inequality and for the first time derives from Bangladeshi data, trends in semi-variances of foodgrain yields and probabilities of disaster yield levels as well as shifts in the efficiency locus for the mean yield/semi-variance of foodgrain yields. But even if this efficiency locus shifts outward, it may not in fact be associated with greater crop diversification. We provide evidence from two Bangladeshi villages which indicates that the range of rice crop varieties grown is becoming less diverse. From a risk point of view this may not be a major problem unless existing varieties disappear. If all existing varieties remain in existence, farmers always have the option of choosing their old portfolio of cropping combinations and if they do not choose them then on the surface at least, they regard the new combination as superior. But if existing varieties disappear then the available portfolio can change irreversibility and raise problems that may be overlooked in short-period diversification analysis. Let us consider these issues after first of all briefly reviewing findings in the literature about the relationship between instability of crop yields and the introduction of HYVs.

---

*Revised version of a contributed paper presented at the Thirty Second Annual Conference of the Australian Agricultural Economics Society, La Trobe University, Bundoora, Victoria 3083, February 9-11, 1988. Subject to the usual *caveats* we wish to thank various participants in the conference especially Dan Etherington, John Brennan and Kailash Sharma for their comments. We also gratefully acknowledge useful comments on an earlier draft of the paper by two anonymous referees and an editor of this journal and Janet Piper for computational assistance.

II. Relationship Between HYV Introduction and Instability of Foodgrain Yields: Brief Literature Review

Empirical studies dealing with the impact of the introduction of HYVs on instability of foodgrain yields and production have reported conflicting results. However, evidence is tending to support the hypothesis that the introduction of HYVs is associated with a reduction in the relative instability of crop yields, especially of rice.

Mehra (1981) and Hazell (1982, 1985) have found evidence of increased instability in Indian agricultural production following the introduction of modern HYV agricultural technology. This was supported by Parthasarathy (1984, p.A74) who indicates yield instability to be positively associated with districts experiencing higher agricultural growth rates in the Indian state of Andhra Pradesh. Hazell (1982, p. 10) goes so far as to conclude that "production instability is an inevitable consequence of rapid agricultural growth and there is little that can be effectively done about it".

One needs to be reminded of course that despite the apparent similarity of their conclusions, Mehra (1981) and Hazell (1982) differ in an important respect: Mehra (1981) hypothesizes a causal link between the new technology and increased production and greater yield instability. Thus while Mehra (1981) attributes most of the production variation to yield instability, Hazell (1982, 1985) attributes rising production variability to greater yield variability as well as a reduction in the offsetting patterns of yield variation (a rise in covariation of yields) between crops and regions.

While the studies by Hazell (1982) and Mehra (1981) are substantial, in our view, they are subject to two methodological limitations: (1) arbitrary cut-off points, and (2) inconsistency in deletion of certain observations. These have been discussed elsewhere by Alauddin and Tisdell (1988a, 1988d) so we do not wish to repeat the discussion here. However, it is pertinent to mention that while Hazell (1982) and Mehra (1981) found increased variability in foodgrain production and yield following the introduction of the 'Green Revolution' technologies, other studies indicate a contrary relationship. In an earlier study, Sarma and Roy (1979) found that the coefficient of variation of Indian foodgrain production declined from 14 per cent in the pre-'Green Revolution' (1949-50 to 1964-65) to eight per cent in the period following it *i.e.* 1967-68 to 1976-77 (Dantwala, 1985, pp. 112, 123; see also Government of India, 1982).

A recent study (Jain *et al.*, 1986) extends the Hazell (1982) analysis for India to 1983-84. Without dropping any observations from either period, they find that the period of new technology (1967-68 to 1983-84) is associated with a lower production and yield variability compared to the earlier period (1949-50 to 1966-67). One further point that emerges from recent studies is that while Hazell (1986, p. 16) shows that the probability of a five per cent fall below the trend in world cereal production may have doubled in recent years, there are wide variations between regions and commodities. For instance, the coefficients of variation of both rice and wheat production have declined in recent years (Hazell, 1986, p. 18; Evans, 1986, p. 2). Hazell (1986, p. 18) also claims that "the least risky countries are those that predominantly grow rice, presumably because much of the crop is irrigated. These countries include Indonesia, Thailand, Bangladesh and Japan". Thus more recent evidence seems to cast some doubt on the validity of the earlier Hazell contention of greater instability being an inevitable concomitant of HYV penetration.

Recent studies (Alauddin and Tisdell, 1988a; 1988d) have also addressed the issue of foodgrain (rice and wheat) production and yield variability in the context of Bangladeshi 'Green Revolution'. Aluaddin and Tisdell (1988a) measured the degree of production and yield variability (standard deviation and coefficient of variation) in terms of deviations from a moving average of a five-year period. The variance of a variable for any year was calculated as its observed variance for the five-year period up to and including the year under consideration. So the variance itself was a moving value. Using this approach and applying it to Bangladeshi time series data for the period 1947-48 to 1984-85 the changing behaviour of absolute (standard deviation) and relative (coefficient of variation) measures of variability in production and yield was examined. The evidence suggested no increase in production and yield variability in the post-'Green Revolution' period. On the basis of an in-depth and analysis of aggregate time series data it was suggested that the 'Green Revolution' has had a stabilising impact on the relative variability of production and yield.

In a separate study (Alauddin and Tisdell, 1988d) employed the Hazell (1982) approach of fitting trend lines to time series data and measured foodgrain production and yield variability in terms of deviations from the trend values for Bangladeshi national and district level data both before and after the introduction of the new agricultural technology. The results indicated that the 'Green Revolution' may (in contrast to Hazell's findings for India) have reduced relative variability of foodgrain production and yield. Districts experiencing greater penetration of HYVs and associated techniques seemed to have lower relative variability. The probability of production and yield falling five per cent below the trend (Hazell, 1985, p. 149) was found by Alauddin and Tisdell (1988d) to be lower for districts with higher rates of HYV adoption. For Bangladesh as a whole, intertemporal analysis indicated falling relative variability of foodgrain production and yield, and this is also true for most districts in Bangladesh.[1]

Neither of the two sets of results emerging from the employment of the alternative methodology (Alauddin and Tisdell, 1988a) and Hazell's own method (Alauddin and Tisdell, 1988d) provide any evidence to support Hazell's findings for India of rising relative instability of foodgrain production and yields in Bangladesh following the introduction of the 'Green Revolution' technologies. The evidence points towards a decline in such instability.[2]

III. Factors Contributing to Declining Instability of Foodgrain Yields

There is increasing evidence that the introduction of HYVs is associated with a fall in relative variability of foodgrain yields and has been widely observed in relation to rice. This appears to be so in Bangladesh. Three factors may help to explain this: (i) increased

---

[1] Absolute variability is measured in terms of standard error of estimate (SE) while relative variability is measured by the coefficient of variation (CV). Thus $CV = (SE/Mean) \times 100$. It should be noted that the definition of coefficient of variation employed by Hazell (1982) and Alauddin and Tisdell (1988d) does not conform to the standard definition used in statistical literature. As deviations are measured around the trend values instead of the actual mean values, it is perhaps more appropriate to term it coefficient of "unexplained" variation rather than coefficient of variation.

[2] A similar evidence is provided by Murshid (1986) who examined variability of Bangladeshi foodgrain production for the twenty year period 1960-61 to 1979-80. There did not appear to be any trend in instability even though it showed some tendency to increase in the early 1970s (Murshid, 1986, p. 66).

incidence of multiple cropping; (ii) greater use of techniques such as irrigation which exert more control over crop environment; and (iii) greater scope for crop diversification as new varieties are added to the new stock. We have previously provided evidence to show the significance of the two factors first mentioned (Alauddin and Tisdell, 1986b, 1988a, 1988d). We wish here to concentrate on the possible significance of the diversification aspect, to apply portfolio analysis to the subject and in particular to consider the impact of HYV adoption on the mean yield/risk efficiency locus for foodgrains, paying particular attention to the lower semi-variance as a measure of risk or variability. To that end this section deals with conceptual issues whereas the next examines trends and shifts in the lower semi-variance, coefficient of lower semi-variation and shifts in the efficiency locus, mean foodgrain yields and lower semi-variance of these yields.

The 'Green Revolution' has had a strong effect in raising average or expected yields of crops (Herdt and Capule, 1983; Alauddin and Tisdell, 1986b). This is because in some cases HYVs raise within season yields when they replace traditional varieties as well as increasing the scope for multiple cropping and so on an annual basis also add to expected yields. If a single crop of a HYV has a lower yield than a single crop of a traditional variety and if the former permits multiple cropping but the latter does not, annual expected yield may be much higher with HYV introduction. To the extent also that yields between seasons are not perfectly correlated, this will tend to reduce risks by, for example, lowering the coefficient of variation of annual yields, even though as we have discussed elsewhere (Alauddin and Tisdell, 1987; see also Boyce, 1987) production sustainability problems may emerge in the long term. Multiple cropping is likely to reduce the probability of annual farm income falling below a disaster level, if we leave secular problem to one side (Anderson *et al.*, 1977, p. 211).

While HYVs may have a higher expected yield and greater risk or yield variability than traditional varieties, in some cases HYVs may involve higher yields and less risks to individual farmers than traditional varieties. In the latter case, they dominate traditional varieties and can be expected to replace them in due course. As explained in the next section, techniques associated with HYVs, by ensuring greater environmental control, may help to establish this dominance.

As is well known, there is no simple shorthand way of measuring uncertainty and instability. But for the purpose of this exercise let us take the variance or the standard

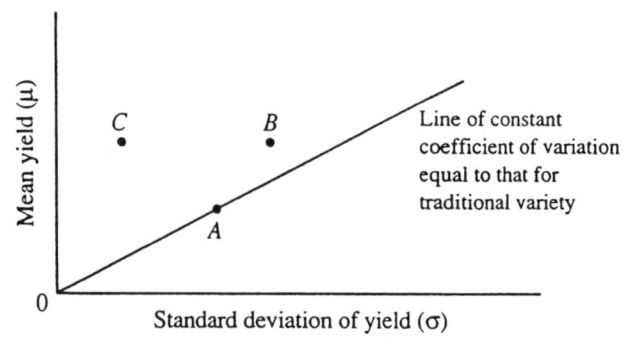

Figure 1

deviation as an indicator. In Figure 1, the mean income for a farm and its standard deviation are shown. The combination at A may indicate the situation of the farm using a traditional variety. The advent of an HYV may if it replaces the traditional variety make B or even C possible. Aggregate evidence from Bangladesh indicates a rise in absolute variability but a reduction in relative variability following the 'Green Revolution' as well as a rise in mean yields (Alauddin and Tisdell, 1988a; 1988d). So the overall situation is depicted by a point to the right of A located above line 0A. The slope of 0A is determined by the mean yield of the traditional variety divided by its standard deviation. In the case illustrated, the yield characteristics of HYV at B lie above the line 0A and reduce the coefficient of variation even though the HYV has a much greater variance than the traditional variety.

Where the variability of yields from HYVs are greater than that for traditional varieties, diversification of varieties grown can be used as a strategy to reduce the risk of growing some HYV provided yields of varieties are not perfectly correlated. If the yields of the varieties are perfectly correlated, and if A and B are the unmixed crop variety possibilities, the efficiency locus (Markowitz, 1959) is indicated by the line AB in Figure 2 (Anderson et al., 1977, p. 193). Lack of perfect correlation between returns from the different crop varieties results in the efficiency locus or curve joining A and B bulging to the left so that a curve like AKB in Figure 2 may result (for details on the nature of this curve see Anderson et al., 1977, p. 193). Thus when returns are not perfectly correlated, variety diversification can reduce risk and lower the coefficient of variation.

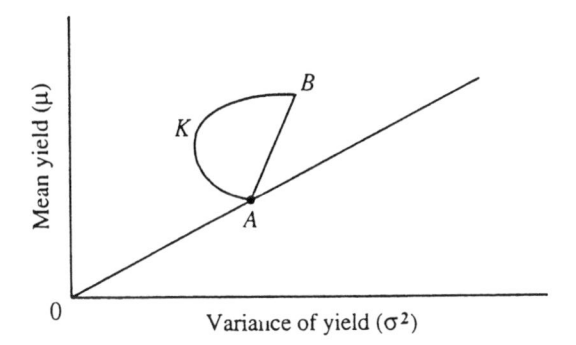

Cherbychev's inequality and variations on it can be used to explore this matter further (Anderson et al., 1977, p. 211). According to this inequality the probability that a random variable, x, deviates from its expected value, μ, by more than an amount k is equal to or less than its variance divided by k, that is

$$\Pr(|x-\mu| \geq k) \leq \sigma^2/k \qquad (1)$$

According to the inequality, if an income of x_1 or less than x_1 is to be avoided (would be a disaster), the probability of not avoiding this is given by the following inequality

$$\Pr(x \leq x_1) \leq \sigma^2/(\mu-x_1)^2 \qquad (2)$$

A farmer may require the probability to be less than a particular value, say k. This will be satisfied if

$$\sigma^2/(\mu-x_1)^2 \leq k \tag{3}$$

Thus if σ^2 is assumed to be the dependent variable, all combinations of (μ, σ^2) on or below the parabola

$$\sigma^2 = k(\mu-x_1)^2 \tag{4}$$

satisfy this constraint, except for all combinations involving a value of $\mu < x_1$, x_1 being assumed to be less than μ. Thus for the case illustrated in Figure 3 combinations to the left of the *DEF*, the positive branch of relevant parabola, may not satisfy the safety first constraint whereas those to the right of this branch or on it do satisfy the constraint.

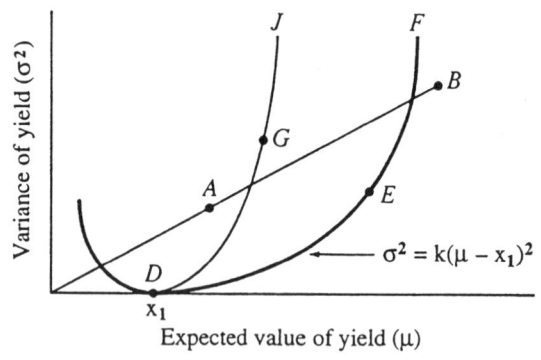

Figure 3

Several points may be noted:

(1) Where returns from different varieties are not correlated and on the basis of the argument illustrated in Figure 2, it may be possible to meet the safety first constraint by diversification of varieties grown.

(2) If the variance and mean level of income increase in the same proportion, the likelihood of the safety first constraint being satisfied rises. For example, if in Figure 3, A is the combination of (μ, σ^2) for the traditional variety and B corresponds to that for a HYV, the HYV meets the constraint but the traditional variety does not. As one moves out further along the line OA, the likelihood of the constraint being satisfied rises and the probability declines of incomes falling below x_1. This can even happen, up to a point, when μ rises and σ^2 increases more than proportionately.

It should be noted Cherbychev's inequality is not very powerful. Because of this, there *may* be combinations to the left of curve *DEF* in Figure 3 which also satisfy the probability constraint. In that respect a modification of this inequality so that it is based on the lower semi-variance is more powerful (Tisdell, 1962). If the probability distribution of income is symmetric about its mean the lower semi-variance is $0.5\sigma^2$ and it follows that

$$\Pr(x<x_1) \leq 0.5\sigma^2/(\mu-x_1)^2 \tag{5}$$

The relevant safety first parabola now is

$$\sigma^2 = 2k(\mu - x_1)^2 \tag{6}$$

and the relevant branch might be as indicated by curve *DGJ* in Figure 3. Clearly similar consequences follow to those mentioned earlier.

IV. Lower Semi-Variance Estimates and Shifts in Mean/Risk Efficiency Locus for Foodgrain Yields in Bangladesh

The lower semi-variance and the modified Cherbychev inequality as set out in Expression (5) provides a more relevant measure of the risk from yield variability faced by the farmers than does the variance and Cherbychev's inequality. We used the variance in our earlier analysis (Alauddin and Tisdell, 1988a) but we wish in this section to consider changes in the lower semi-variance using Bangladeshi data for foodgrain yields and also consider trends in the probability of a disaster yield in Bangladesh using the modified Cherbychev inequality. This will lead on to estimates of shifts in the mean/risk foodgrain yield efficiency frontier for Bangladesh, a matter which does not seem to have been previously considered.

Let us now consider empirical estimates using Bangladeshi foodgrain yield data. First of all annual indices of overall foodgrain yield were derived using the average of the triennium ending 1977-78 as the base. In measuring the degree of variability the present paper follows the approach used in one of our earlier studies (Alauddin and Tisdell, 1988a). As the primary concern of this paper lies in the deviations below the mean, we choose a period of longer duration (than five years employed by Alauddin and Tisdell, 1988a). This is because a five yearly period is likely to reduce significantly the number of observations for negative deviations and is unlikely to provide a robust basis of anlysis. Using this approach and applying it to Bangladeshi time-series data for the period 1947-48 to 1984-85 we specify the changing behaviour over time of yield variability below the mean. Moving values for 9 and 11 years were tried and the results were similar. The results presented in this paper are based on the 11-year period. The relevant data are set out in Table I.

A few observations seem pertinent. First, there is a steady increase in foodgrain yields. Secondly, there seems to be little *overall* time trend in lower semi-variance, (lower) coefficient of semi-variation and probability of yield falling 50 per cent or more below the initial 11 year average yield. However, these measures of variability seem to increase initially and then decline before showing a rising tendency once again at a slower rate than initially. Indeed at about the mid-1960s there appears to be a fundamental change in relative variability of yields. A large and significant downward shift occurs in the trend of relative variability and it increases at a slower rate than prior to the mid-1960s.

These aspects come into sharper focus when one plots the relevant observations against time. Figure 4 plots (lower) semi-variance (*SEMVAR*). Once can visually identify two distinct phases (the first phase 1952-63, that is 1952-53 to 1963-64 and the second phase 1964-79, that is, 1964-65 to 1979-80). Up to the early 1960s (lower) semi-variance of yield shows a strong tendency to increase. In order to make quantitative comparison of change in its behaviour three regression lines with time (*T*) as the explanatory variable were estimated. These are presented as Equations (7), (8) and (9) for the corresponding periods. Figure 4 illustrates Equations (7) and (8) for the first and second phases respectively.

TABLE I
Mean Values, Lower Semi-Variance, Coefficient of Semi Variation of Yield of all Foodgrains, and Probability of Yield Falling below a 'Disaster Level': Bangladesh, 1947-48 to 1984-85

Year	Index	MEANYLD	LAVERAG	SEMVAR	SEMCOV	PROB50
1947	70.74					
1948	78.79					
1949	75.44					
1950	73.29					
1951	69.23					
1952	70.50	73.231	69.386	5.441	3.362	0.00406
1953	74.80	73.195	69.305	5.203	3.291	0.00389
1954	71.06	73.304	69.970	6.807	3.729	0.00506
1955	65.40	74.332	69.970	6.807	3.729	0.00478
1956	81.34	75.853	70.899	10.874	4.651	0.00706
1957	74.96	76.925	71.178	12.396	4.946	0.00763
1958	70.34	79.032	71.314	15.357	5.495	0.00854
1959	79.98	80.452	72.349	29.751	7.539	0.01550
1960	86.75	82.096	75.507	42.678	8.652	0.02060
1961	90.02	83.783	77.527	22.724	6.149	0.01020
1962	81.02	84.761	78.053	29.112	6.913	0.01260
1963	93.68	86.492	78.826	34.836	7.488	0.01400
1964	90.42	88.512	82.928	9.345	3.686	0.00347
1965	89.14	89.374	85.218	12.327	4.120	0.00443
1966	83.96	89.222	84.800	11.317	3.967	0.00409
1967	92.09	88.599	86.606	2.946	2.060	0.00109
1968	94.00	89.939	86.163	8.686	3.420	0.00305
1969	92.57	89.787	86.163	8.686	3.420	0.00307
1970	89.46	90.540	86.163	8.686	3.420	0.00299
1971	85.08	91.098	86.418	7.882	3.287	0.00266
1972	83.17	93.104	89.063	16.185	4.517	0.00507
1973	95.75	94.233	89.681	19.108	4.891	0.00576
1974	92.01	94.958	88.467	17.482	4.727	0.00514
1975	98.70	96.493	90.125	27.217	5.789	0.00759
1976	95.28	97.977	90.267	33.891	6.450	0.00900
1977	106.03	100.230	92.982	35.737	6.429	0.00883
1978	104.52	102.953	96.742	14.167	3.891	0.00322
1979	101.97	104.944	98.495	25.196	5.096	0.00540
1980	109.46					
1981	105.79					
1982	109.85					
1983	113.12					
1984	117.66					

Notes: 1947 means 1947-48 (July 1947 to June 1948) etc. Index numbers are constructed with the average of the triennium ending 1977-78 = 100. Mean values are 11 yearly moving

averages of the index numbers. SEMVAR is lower semivariance. LAVERAG is average of indices below their mean. SEMCOV=[$\sqrt{SEMVAR/LAVERAG}$]x100. PROB50 is probability of yield falling 50 per cent or more below the first of the 11 yearly moving mean yield (i.e., 73.231).
Source: Based on Alauddin and Tisdell (1988a, pp. 145-146).

Equation (7) clearly indicates a strong time trend in (lower) semi-variance in terms of both explanatory power and statistical significance. Equation (8) shows a similar trend in the later period. However, the relative rate of change is much lower in the second phase compared to the one in the first as indicated by their respective slopes. More importantly, there is a change in trend apparent with the relative variability indicated by Equation (8) being much lower than that for (9) as a function of time.

Phase 1: $SEMVAR = 1.4045 + 3.1080T$, $R^2 = 0.7529$, t-value $= 5.52$ (7)

Phase 2: $SEMVAR = -13.1311 + 1.5031T$, $R^2 = 0.5443$, t-value $= 4.09$ (8)

Entire: $SEMVAR = 11.7585 + 0.4011T$, $R^2 = 0.0895$, t-value $= 1.60$ (9)

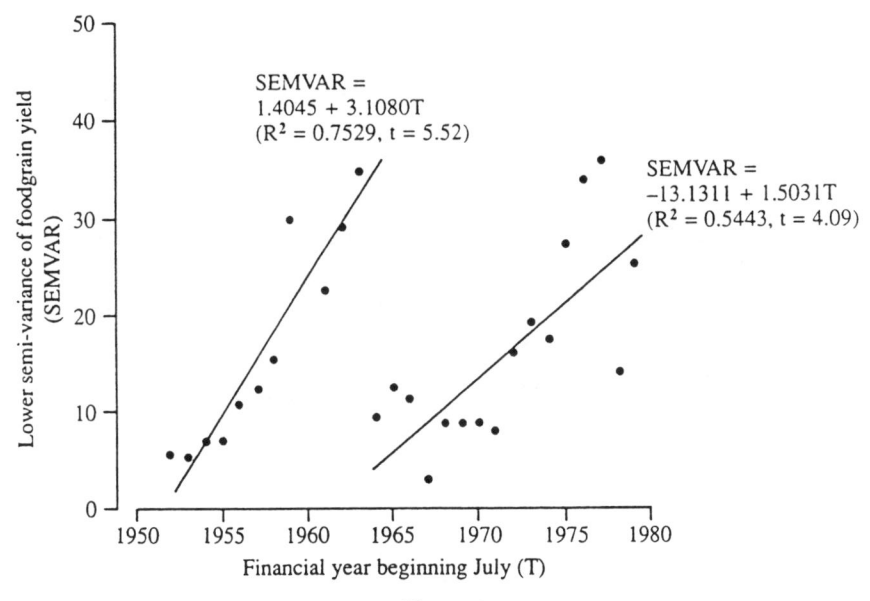

Figure 4

Figure 5 plots coefficients of (lower) semi-variation (*SEMCOV*) against time. One can identify two similar phases in its behaviour to those for (lower) semi-variance. To facilitate quantitative comparisons, Equations (10), (11) and (12) were estimated by least squares linear regression. Equations (10) and (11) have been illustrated in Figure 5. Equation (10) corresponds to the first phase and shows that the coefficient of (lower) semi-variation had

Figure 5

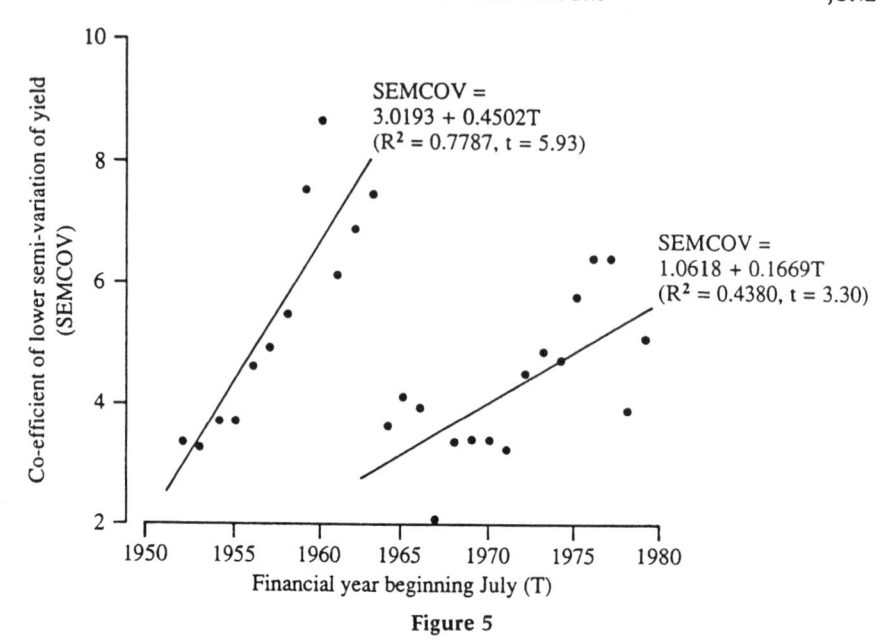

Plot with y-axis "Co-efficient of lower semi-variation of yield (SEMCOV)" from 2 to 10, and x-axis "Financial year beginning July (T)" from 1950 to 1980.

Phase 1 line: SEMCOV = 3.0193 + 0.4502T (R^2 = 0.7787, t = 5.93)

Phase 2 line: SEMCOV = 1.0618 + 0.1669T (R^2 = 0.4380, t = 3.30)

a strong tendency to increase during Phase 1. The strong explanatory power and high t-value lend clear support to this claim. However, there is an important change in its behaviour when one considers Equation (11) which relates to the second phase and Equation (12) which relates to the entire period. A comparison of Equations (10) and (11) clearly shows a much slower rate of increase in the coefficient of (lower) semi variation in the second phase. Once again a strong downward shift in relative variability is suggested by the comparison of Equations (10) and (11) and their lines shown in Figure 5.

Phase 1: $SEMCOV = 3.0193 + 0.4502T$, $R^2 = 0.7787$, t-value = 5.93 (10)

Phase 2: $SEMCOV = 1.0681 + 0.1669T$, $R^2 = 0.4380$, t-value = 3.30 (11)

Entire: $SEMCOV = 4.7619 + 0.0047T$, $R^2 = 0.0006$, t-value = 0.12 (12)

Figure 6 depicts the behaviour of probability of yield falling 50 or more below the average of the initial 11 year period (PROB50). Overall no time trend can be established. But two distinct phases can be identified. It increases during the period up to the early 1960s (Phase 1) and falls to lower values on average in the subsequent period (Phase 2). The behaviour of PROB50 can be placed into a pattern by considering the estimated regression equations in Figure 3 and Equations (13), (14) and (15) which relate respectively to the first and second phases and the entire period. A comparison between Equations (13) and (14) and the corresponding lines shown in Figure 7 suggests that the trend in probability of a disaster level of yield had undergone a significant downward shift around the mid-1960s.

Phase 1: $PROB50 = 0.00320 + 0.00115T$, $R^2 = 0.6236$, t-value = 4.07 (13)

Phase 2: $PROB50 = -0.00075 + 0.00028T$, $R^2 = 0.3491$, t-value = 2.74 (14)

Entire: $PROB50 = 0.00824 + 0.00011T$, $R^2 = 0.0420$, t-value = 1.07 (15)

Figure 6

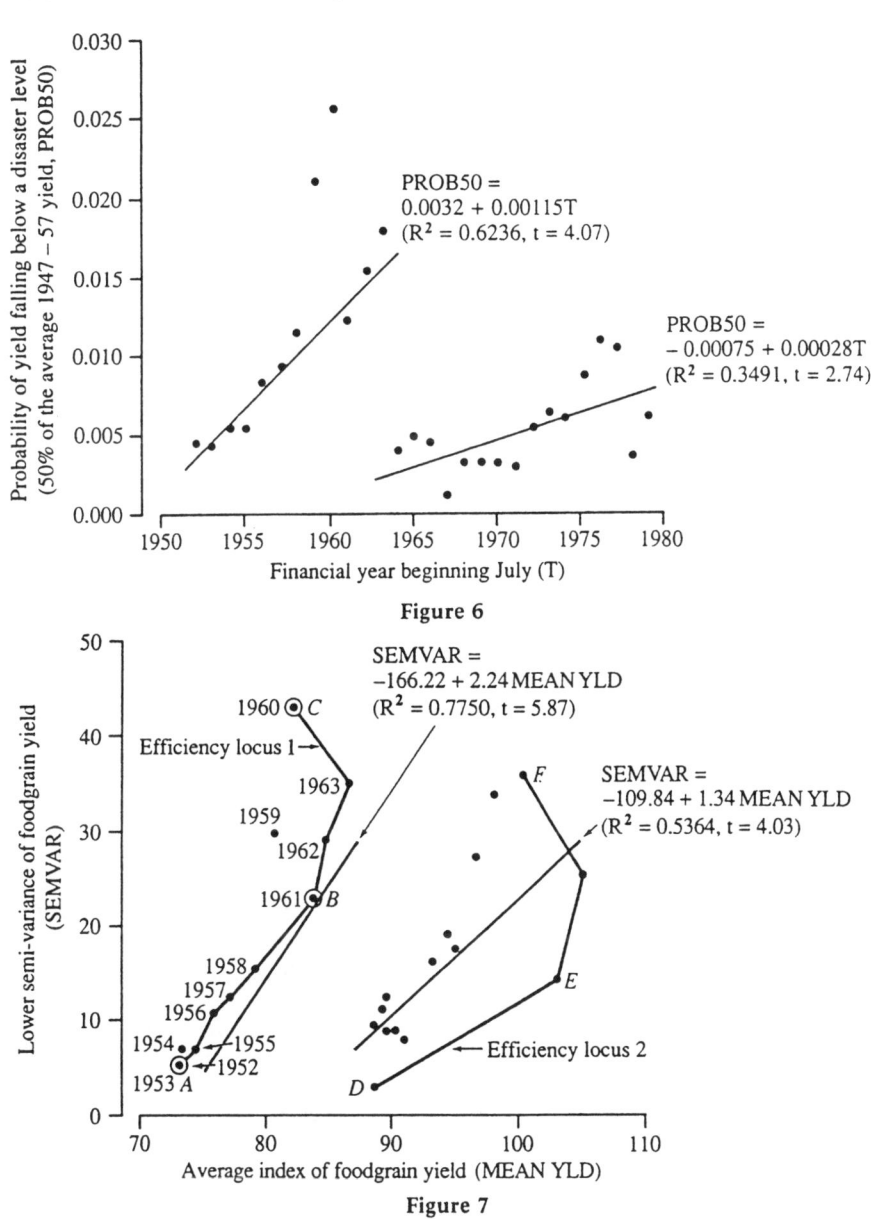

Figure 7

Figure 7 plots (lower) semi-variance against mean yield (*MEANYLD*). It follows a similar pattern to those in Figures (4)-(6).

Note that the scatter falls into two distinct clusters and sets, with observations centred on 1963 or earlier being well to the left of those for the later period. If a linear least squares approximation is made to the two clusters, the following equations are obtained:

CLUSTER 1: $SEMVAR = -166.2208 + 2.2395 MEANYLD$,
$R^2 = 0.7750$, t-value = 5.87 (16)

CLUSTER 2: $SEMVAR = -109.8443 + 1.3425 MEANYLD$,
$R^2 = 0.5364$, t-value = 4.03 (17)

These indicate a significant shift downward in the (lower) semi-variance in relation to mean foodgrain yield about the mid-1960s and that the (lower) semi-variance is increasing at a slower rate in relation to the mean level of foodgrain yield in the second phase than in the first. In addition, *if it is assumed* that Cluster 1 is drawn from a distinct population, the efficiency locus in Phase 1 is as indicated by the heavy line segments passing through the points such as ABC whereas in Phase 2 it is indicated by heavy line segments passing through the points such as *DEF*. This would imply that there has been a considerable outward shift (*i.e.* to the right) in the efficiency locus. All the available evidence strongly points towards this conclusion. Note, however, the efficiency loci as shown are rough approximations and more than two loci are likely to have applied in the period under consideration. Also such loci, unlike Locus 1, should be non-reentrant. But the order of change is such as to override such considerations.

From the available data, it seems that relative variability of Bangladeshi foodgrain production shifted downwards significantly around the mid-1960s. While the relative variability below the mean is still continuing to rise it appears to be doing so at a much slower rate than prior to the early 1960s. Similar shifts have occurred in relation to the probability of a 'disaster' level of yield, meaning the shift was downward around the mid-1960s.

Why does the apparent break occur in trends in variability? The second period, 1964-79 corresponds to the commencement of the introduction of the new technology *to control* agricultural production. Increased irrigation and fertilizer use followed later by the introduction of HYVs occurred during the second phase, as distinguished here. Note, however, that in both phases there is a tendency for relative variability of yield below the mean to increase with time. The reason for this trend is unclear at present but is worthy of future investigation.

It should, however, be pointed out that this whole analysis depends on mathematically *expected* values (averaging procedures) even when it takes account of higher moments. While there may be a reduction in the *relative frequency* or *probability* of disaster, disaster when 'disaster' comes may be more *catastrophic*. This is not captured by averaging procedures. Thus while use of HYVs may reduce the probability of a disaster level of income (or less) occurring, a catastrophically lower income may occur when disaster strikes. More research into the probability of this is needed.

V. Evidence from Bangladesh about the Extent of Crop Diversification Following the Green Revolution

As observed in the last section, the mean/lower semi-variance of foodgrain yield in Bangladesh appears to have shifted to the right with the introduction of HYVs. While at least in the short run the introduction of new varieties makes greater crop diversification possible, there is no guarantee that that increased diversification will occur. Indeed, if the expected return/risk combination of the new varieties are sufficiently favourable, less diversification both of crops and varieties of crops can occur. In fact, the limited field evidence which we have points towards less diversification following the 'Green Revolution'.

Before considering that evidence note that in Bangladesh the 'Green Revolution' is confined to cereals (rice and wheat) and partly to jute and sugarcane. Even though some success in research in other crops like potato, summer pulses and oilseeds is believed to have been achieved (Alauddin and Tisdell, 1986a; Gill, 1983; Pray and Anderson, 1985) it is yet to take-off in any real sense. Recent evidence (*e.g.* Alauddin and Tisdell, 1988c) indicates that the area under rice and wheat has expanded at the expense of non-cereals. This substitution of cereals for non-cereals has resulted from, among other things, the improvement in the technology specific to the former relative to the latter. The output of non-cereals as a whole has declined as a result of decline in yield and hectareage. Thus there is a comparative 'crowding out' of non-cereal crop cereal production following the 'Green Revolution'. On the other hand, cultivated land that was once left fallow for a significant part of each year (for example, during the dry season) is now used for crops such as wheat and dry season rice varieties. Thus the 'Green Revolution' has induced greater *monocultural* multiple cropping.

We recently conducted a field survey on the 1985-86 crop year in two Bangladeshi villages: Ekdala in the north-western district of (greater) Rajshahi and South Rampur in the south-eastern district of Comilla.[3] Both villages have a relatively long history of HYV technology adoption. South Rampur being in the laboratory area of the Bangladesh Academy of Rural Development (BARD), was one of the earliest to adopt HYVs. Ekdala was also one of the early adopters of the new technology in the region. However, the villages differ ecologically. Ekdala is located in the low rainfall area and is considered drought-prone while South Rampur is in the high rainfall area and is flood-prone. The two villages differ in respect of access to irrigation. While South Rampur is completely irrigated Ekadala is only partially irrigated.

In all 58 landowning farmers were interviewed from each of the two villages on the basis of stratified random sampling. Each category of farmers (small, medium and large) had approximately the same proportionate representation as their respective total numbers in the aggregate number of farming households in the villages.

An analysis of the farm-level data indicates that the two villages exhibit very different cropping patterns. Ekdala has a more traditional cropping pattern. Apart from growing cereals (rice and wheat), a number of other crops including sugarcane, pulses, jute and oilseeds and fruit crops such as banana and watermelon are also grown. Cereals occupy 65 per cent of gross cropped area. Overall cropping intensity[4] is quite high (over 178 per cent) and is considerably in excess of the national figure of 150 per cent in recent years (Alauddin and Tisdell, 1987).

---

[3] A detailed description of the survey area and sampling procedure has been presented in Alauddin (1988). A brief description also appears in Alauddin and Tisdell (1987; 1988b).

[4] Cropping intensity is defined as the percentage of gross cropped area (*i.e.* including multiple cropping) over net cropped area (*i.e.* actual area cultivated) on an annual basis.

South Rampur on the other hand has a much more specialised cropping pattern in that rice is the only crop grown. Almost every plot of land is double cropped with rice. Intensity of cropping is significantly higher (nearly 200 per cent) than that in Ekdala. Furthermore, it is observed from the Ekdala data that farmers with access to irrigation allocate a significantly higher percentage of gross cropped area to rice and wheat than those without access to irrigation. The latter category of farmers allocate a significantly higher percentage of gross cropped area to non-cereals like sugarcane, pulses, oilseeds, watermelon. Irrigated farms cultivate land much more intensively than non-irrigated farms where cropping intensity is much lower (about 116 per cent) than the overall intensity of cropping for Bangladesh.

Thus the availability of irrigation has considerable impact on cropping pattern and intensity of cropping. A plot of land under irrigation is normally double cropped with rice (*boro* HYV rice followed by local or HYV *aman* rice). This, however, involves (rice) monoculture multiple cropping. This seems to be consistent with the changes in cropping pattern for Bangladesh as a whole. This trend is likely to be a cause for concern.

Firstly, monocultural multiple cropping to cereals may prove to be ecologically unsustainable or less sustainable than in the past and there are already signs that it is becoming more costly and difficult to maintain productivity using such cultural practices (Hamid *et al.*, 1978, p. 40; Alauddin and Tisdell, 1987).

Secondly, mere production of increased quantities of rice at the expense of other non-cereal food crops *e.g.*, pulses, vegetables is unlikely to solve food and nutritional problems. Recent evidence[5] indicates that overall cereal production has been absorbed by rising population in Bangladesh with per capita cereal availability remaining roughly constant. On the other hand, per capita availability of pulses, fruits and spices has fallen markedly in the post-'Green Revolution' period and on average the availability of vegetables per head has fallen. Furthermore, there has been a significant fall in the per capita protein content of the Bangladeshi diet. The average Bangladeshi diet now appears to be less varied and balanced, and *a priori* is less nutritious with adverse welfare implications.

It would be useful to have data for Bangladesh indicating whether the number of strains or varieties for crops such as rice grown on farms is increasing or decreasing and have some information about trends in the total number of varieties available. It is possible that with the introduction of HYVs the number of available varieties at first increases and then decline. At the same time, the *fundamental* gene bank may be declining while the number of available strains or varieties is at first increasing. So a difficult measurement problem in relation to genetic diversity exists.

Only limited evidence is available from our survey areas of Ekdala and South Rampur. The farm-level data from Ekdala indicate that farmer primarily rely on two or three varieties of HYV rice. During the 1985-86 crop year, BR11 and China varieties constituted 82 per cent of the gross area planted to HYV rice. In South Rampur, farmers were found to allocate over 70 per cent of HYV rice area to four rice varieties: BR3 (21 per cent), BR11 (26

---

[5] A detailed analysis of the nutritional and welfare implications has been provided in Alauddin and Tisdell (1988c).

per cent), *Paijam* (13 per cent) and Taipei (10 per cent). Furthermore, the South Rampur Survey indicated that *the number of rice varieties in use has fallen from about 12-15 in the pre-'Green Revolution' period to 7-10 in the post-'Green Revolution' phase.* This also seems to be supported by the Ekdala evidence.

VI. Problems of Reduced Genetic Diversity of Crops: A Secular Decline in Available Varieties and Strains

Our limited evidence from Bangladesh indicates that the introduction of HYVs has been accompanied by less crop diversification on farms even though the relative variability of foodgrain yield has fallen its mean/risk efficiency locus has become more favourable. The latter trends thus cannot be attributed to any significant extent to greater diversification of crops.

While a favourable reduction in yield risks appears to have accompanied the 'Green Revolution', these trends may disguise long-term rising risk. In the long-term, the introduction of HYVs may paradoxically reduce the scope for diversification. This problem would not arise if varieties in existence prior to the introduction of HYVs and the development of other improved varieties continued to be available. But improved varieties quite frequently drive existing varieties out of usage and existence so that in the long-term, less choice of varieties may be available than prior to the introduction of HYVs (*cf.* Plucknett *et al.*, 1986, esp. pp. 8-12). Furthermore, the varieties which disappear may be those which provide the most valuable genetic building blocks for development of new varieties. Thus crop productivity may become dependent on a limited number of varieties and the risk of production being unsustainable may rise considerably (*cf.* Plucknett *et al.*, 1986).

New varieties of crops appear to have a limited life on average. The World Conservation Strategy (WCS) document (IUCN, 1980) suggests that wheat and other cereal varieties in Europe and North America have a life-time of only 5-15 years. "This is because pests and diseases evolve new strains and overcome resistance; climates alter, soils vary; consumer demand change. Farmers and other crop-producers, therefore, cannot do without the reservoir of still-evolving possibilities available in the range of varieties of crops, domesticated animals, and their wild relations. The continued existence of wild and primitive varieties of the world's crop plants provides humanity's chief insurance against their destruction by the equivalents for those crops of chestnut blight and Dutch elm disease" (IUCN, 1980, sec 3.3).

The WCS document goes on to point out how the bulk of Canadian wheat production now depends on four varieties and so does most of US potato production and provides other examples of growing agricultural dependence on narrower range of varieties. Furthermore, there appears to have been a rapid disappearance of primitive cultivars, for instance, the percentage of primitive cultivars in the Greek wheat crop fell from over 80 per cent in the 1930s to under ten per cent by the 1970s and the absolute number of these declined quite considerably. (IUCN, 1980, sec 3.4; Allen, 1980, p. 41). Such declines are claimed to be "typical of most crops in most countries" (IUCN, 1980, sec 3.4). Thus while in the shorter-term new varieties may become available and increase the scope for reducing instability of production and income, for instance via diversification, in the long-term the opposite tendency may be present.

This raises a number of questions. For instance, if varieties disappear is some market failure present and will the disappearance lead to a socially sub-optimal outcome? Is government intervention required to correct the situation? If so, what guidelines should be adopted in determining which crop varieties to preserve? What social mechanisms should be adopted to bring about the range of conservation of varieties required?

These complex questions cannot be addressed in detail here. However, there is reason to believe that market failure and transaction costs can lead to the socially sub-optimal disappearance of species (see our Conference paper mentioned in the first footnote). Furthermore, a variety of criteria have been suggested for determining which array of species and varieties to conserve. These range from approaches using cost-benefit analysis (CBA) to those advocating a safe minimum standard (SMS) (Bishop, 1978; Chisholm, 1988; Ciriacy-Wantrup, 1968; Quiggin, 1982; Randall, 1986; Smith and Krutilla, 1979).

VII. Concluding Comments

The available evidence from Bangladesh points to a major decline in the relative variability of foodgrain yields following the introduction of new crop technology associated with the 'Green Revolution'. An analysis of mean yields and (lower) semi-variance data for Bangladesh indicates a strong shift downwards in the trend line of relative instability of crop yields from about the mid-1960s and a major break in trends with rates of increase in yield instability being lower after the mid-1960s. Factors which may be responsible for this phenomenon were discussed and particular attention was given to the diversification of crops and varieties of particular crops grown as a possible contributor. Our data from Bangladesh are inadequate at present to determine whether there is a more (or less) crop diversification following the 'Green Revolution'. While initially the advent of the new varieties would seem to expand the available choice of varieties and possibilities for crop diversification for farmers, this may not be so in the long-term. Some traditional varieties may be dominated *initially* by some favourable characteristics of new varieties. Consequently, these traditional varieties may disappear thereby reducing available genetic diversity and raising risks in the long run (*cf.* Plucknett *et al.* 1986, Ch. 1). This raises a possible dilemma and a basic issue for policy — namely, what is the responsibility of governments to preserve genetic diversity and how should governments decide which varieties to preserve? Economists are divided in their advice to governments about the appropriate decision-making model to apply to such choice problems (*cf.* Randall, 1986).

References

Alauddin, M. (1988). "The 'Green Revolution' in Bangladesh: Implications for Growth, Distribution, Stability and Sustainability", unpublished Ph.D. thesis (Newcastle, Australia: University of Newcastle).

Alauddin, M. and Tisdell, C. A. (1986a). "Bangladeshi and International Agricultural Research: Administrative and Economic Issues", *Agricultural Administration*, vol. 21, no. 1.

Alauddin, M. and Tisdell, C. A. (1986b). "Decomposition Methods, Agricultural Productivity Growth and Technological Change: A Critique Supported by Bangladeshi Data", *Oxford Bulletin of Economics and Statistics*, vol. 48, no. 4.

Alauddin, M. and Tisdell, C. A. (1987). "Trends and Projections for Bangladeshi Food Production: An Alternative Viewpoint", *Food Policy*, vol. 12, no. 4.

Alauddin, M. and Tisdell, C. A. (1988a), "Has the 'Green Revolution' Destabilized Food Production? Some Evidence from Bangladesh", *Developing Economies*, vol. 26, no. 2.

Alauddin, M. and Tisdell, C. A. (1988b), "Patterns and Determinants of Adoption of High Yielding Varieties: Farm-Level Evidence from Bangladesh", *Pakistan Development Review*, vol. 27, no. 2.

Alauddin, M. and Tisdell, C. A. (1988c), "Bangladeshi Crop Production and Food Supply: Growth Rates and Changing Composition in Response to New Technology", Research Paper No. 210, Department of Economics, University of Melbourne.

Alauddin, M. and Tisdell, C. A. (1988d), "Impact of New Agricultural Technology: Data Analysis for Bangladesh and Its Districts", *Journal of Development Economics*, vol. 29.

Allen, R. (1980), *How to Save the World* (London: Kogan Page).

Anderson, J. R., Dillon, J. L. and Hardakar, J. B. (1977), *Agricultural Decision Analysis* (Ames, Iowa: Iowa University Press).

Bishop, R. (1978), "Endangered Species Uncertainty: The Economics of a Safe Minimum Standard", *American Journal of Agricultural Economics*, vol. 60, no. 1.

Boyce, J. K. (1987), "Trends and Projections for Bangladeshi Food Production: Rejoinder to M. Alauddin and C. Tisdell", *Food Policy*, vol. 12, no. 4.

Chisholm, A. H. (1988), "Uncertainty, Irreversibility and Rational Choice", in C. A. Tisdell and P. Maitra (eds) *Technological Change, Development and the Environment: Socio-Economic Perspectives* (London: Routledge).

Ciriacy-Wantrup, S. V. (1968), "Resource Conservation: Economics and Politics", Division of Agricultural Services (Berkeley and Los Angeles: University of California).

Conway, G. (1986), *Agroecosystem Analysis for Research and Development* (Bangkok: Winrock International).

Dantwala, M. L. (1985), "Technology, Growth and Equity in Agriculture" in J. W. Mellor and G. M. Desai (eds) *Agricultural Change and Rural Poverty: Variations of a Theme by Dharm Narain* (Baltimore, Md.: Johns and Hopkins University).

Gill, G. J. (1983), "Agricultural Research in Bangladesh: Costs and Returns" (Dhaka: Bangladesh Agricultural Research Council) (mimeo.).

Government of India (1982), *Economic Survey 1981-82* (New Delhi: Government of India, Ministry of Finance).

Hamid, M. A., Saha, S. K., Rahman, M. A. and Khan, A. J. (1978), *Irrigation Technologies in Bangladesh: A Study in Some Selected Areas* (Department of Economics: Rajshahi University, Bangladesh).

Hazell, P. B. R. (1982), *Instability in Indian Foodgrain Production*, Research Report No. 30 (Washington, D.C.: International Food Policy Research Institute).

Hazell, P. B. R. (1985), "Sources of Increased Variability in World Cereal Production Since the 1960s", *Journal of Agricultural Economics*, vol. 36, no. 2.

Hazell, P. B. R. (1986), "Introduction", P. B. R. Hazell (ed.) Summary Proceedings of a Workshop on Cereal Yield Variability (Washington, D.C.: International Food Policy Research Institute).

Herdt, R. W. and Capule, C. (1983), *Adoption, Spread and Production Impact of Modern Rice Varieties in Asia* (Los Banos, Philippines: International Rice Research Institute).

IUCN (1980), *World Conservation Strategy: Living Resources Conservation for Sustainable Development* (Glands, Switzerland: International Union for the Conservation of Nature and Natural Resources).

Jain, H. K., Dagg, M. and Taylor, T. A. (1986), "Yield Variability and the Transition of the New Technology" in P. B. R. Hazell (ed.) Summary Proceedings of a Workshop on Cereal Yield Variability (Washington, D.C.: International Food Policy Research Institute).

Krutilla, J. V. (1967), "Conservation Reconsidered", *American Economic Review*, vol. 57.

Markowitz, H. M. (1959), "Portfolio Selection – Efficient Diversification of Investments", Cowles Foundation Monograph No. 16 (New York: John Wiley and Sons Inc.).

Mehra, S. (1981), "Instability in the Indian Agriculture in the Context of the New Technology", Research Report No. 25 (Washington, D.C.: International Food Policy Research Institute).

Murshid, K. A. S. (1986), "Instability in Foodgrain Production in Bangladesh: Nature, Levels and Trends", *Bangladesh Development Studies*, vol. 14, no. 2.

Parthasarathy, G. (1984), "Growth Rates and Fluctuations of Agricultural Production: A District-Wise Analysis in Andhra Pradesh", *Economic and Political Weekly*, vol. 19, no. 26.

Pray, C. E. and Anderson, J. R. (1985), "Bangladesh and the CGIAR Centers: A Study of their Collaboration in Agricultural Research", CGIAR Study Paper No. 8 (Washington, D.C.: World Bank).

Plucknett, D. P., Smith, N. J. H., Williams, J. J. and Anishety, N. M. (1986), *Gene Bank and the World's Food Supply* (Princeton, N.J.: Princeton University Press).

Quiggin, J. (1982), "A Theory of Anticipated Utility", *Journal of Economic Behaviour and Organization*, vol. 3.

Randall, A. (1986), "Human Preferences, Economics and Preferences and the Preservation of Species", in B. G. Norton (ed.) *The Preservation of Species: The Value of Biological Diversity* (Princeton, N.J.: Princeton University Press).

Smith, V. K. and Krutilla, J. V. (1979), "Endangered Species, Irreversibility and Uncertainty: A Comment", *American Journal of Agricultural Economics*, vol. 61.

Tisdell, C. A. (1962), "Decision Making and the Probability of Loss", *Australian Economic Papers*, vol. 1, no. 1.

[5]

The economic spacing of trees and other crops

C. A. TISDELL*
Department of Economics, University of Newcastle, Australia
N. T. M. H. DE SILVA**
Economic Research Division, Coconut Development Authority, Colombo, Sri Lanka

(first received April 1983, final version received August 1983)

Summary

Density related yield functions are used in this paper to consider the conditions that determine the profitablity of increasing or reducing the density of crops. The optimal density of the crop when densities need to be increased to raise profit is shown to be usually less than the profit-maximising density when densities need to be reduced to increase profit. It is shown that when actual densities fall in a particular density band or range (which straddles the density for which yield per hectare is at a maximum) profit is maximised by leaving the existing density unchanged. This range or band is described as an 'inert' area. Determinants of the width of this band are discussed.

The existence of this 'inert' area has not been recognised in the earlier literature. Its agricultural policy implications have been ignored by policy-makers, many of whom also fail to take account of other economic considerations discussed in this paper.

1. Introduction

The amount of spacing between individual plants (the density of plants) significantly influences the productivity of tree crops and other crops (Manthriratna and Abeywardena, 1979; Willey and Heath, 1969) and many agricultural extension services provide advice to farmers about the optimal spacing of crops. How-

---

[*] Professor of Economics at the University of Newcastle, Australia.
[**] Currently Research Scholar at above address. On leave as Deputy Director, Division of Economic Research, Coconut Development Authority, Sri Lanka.
  The authors would like to thank two anonymous reviewers for their constructive suggestions on an earlier draft of this paper. The usual *caveat* applies.

Euro. R. agr. Eco. 10 (1983), 281-293   0165-1587/83/0010-0281 $2.00
                                          © Mouton Publishers, Amsterdam

ever, the economics of spacing is much more complicated than is commonly realised, and, indeed, advice on this subject is often given without taking economic factors into account.

Although the optimal economic density of crops has been discussed in the agricultural economics literature (see bibliography in Dillon, 1977), optimal densities have obtained greater coverage in the literature dealing with forestry (Clark, 1976; Gregory, 1954; Pearse, 1967; Scott, 1972; Smith, 1962) and in the literature on agronomy and biology (Christiansen and Fenchel, 1977; Downey, 1971; May, 1973; Willey and Heath, 1969). Yet the possibility, identified in this paper, that density adjustment is uneconomical within certain density ranges or bands, the so-called 'inert' adjustment *areas* (Leibenstein, 1978), does not appear to have been canvassed in the literature. Furthermore, the density recommended by policy-advisers and agricultural extension services is often either that which maximises yield per plant or tree, or, alternatively, that which maximises yield per hectare. In fact, as pointed out in this article, both of these densities are rarely the most economic ones.

The importance of the problem can be illustrated by the cultivation of coconuts in Sri Lanka, where they are a staple crop. The Coconut Research Institute of Sri Lanka recommends to coconut growers an optimal density of 64 mature palms per acre. This is the density for which yield per acre is believed to be maximised. Economic factors are not taken into account in this recommendation. The desirability of achieving optimal densities is considered to be of sufficient importance for the Government of Sri Lanka to provide subsidies (as do governments in other countries) to growers to encourage them to reduce crowded coconut stands to recommended densities. As mentioned below, there are other specific tree crops (many providing food) for which the problem is important, but its importance is not limited to perennials.

2. Density-related yield functions

Other things being equal, the yield of a crop per unit of area depends on the density of the planting. If the biological law of tolerance (Owen, 1975: 3) applies to this case, the relationship for many crops may form a 'bell-shaped' curve. However, the conclusions from this analysis do not depend upon the precise shape of the yield-density curve, which, as we shall see below, has different empirical forms depending upon the crop under consideration and the relevant yield, that is, whether the relevant yield is, for example, grain, fruit, total drymatter, useable volume of timber, and so on. In the discussion a monocrop is assumed.

At low crop densities, competition from weeds and competitive plants may be severe (Whitehead and Smith, 1968), pollination may be uncertain and so on, but as the density increases yield per plant or tree rises. Nevertheless, beyond a certain point, plants of the same type begin to compete with one another for available sunlight and nutrients, many show less resistance to pests and diseases and eventually crowding may become so great that the productivity of the whole crop declines. A typical relationship per unit of time might be like that illustrated in Figure 1 by the solid line *OABCD*. In this case, maximum yield per tree (or plant) occurs for a spacing between trees corresponding to a density of x_1 but a greater density, x_2 (closer spacing) of trees is required to maximise yield per hectare per unit of time.

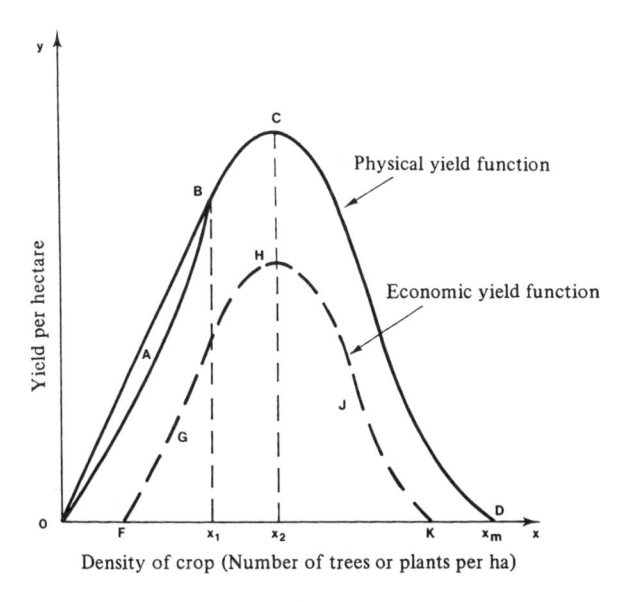

Figure 1. *The physical yield of a crop appears to be related to its density and in some cases, may be 'bell-shaped' like curve ABCD. The economic harvest may however fall below the physical yield as indicated by the broken curve*

Curve *OABCD* is the physical yield function per period. It is very often assumed that if harvesting occurs, then the whole physical yield is harvested, at least in a long-term bioeconomic equilibrium. In practice, however, it is not always economic or practical to harvest the *whole* available physical harvest – some of the crop is likely to be left in the field because usually the more thorough the harvesting, the greater the cost. Economic harvesting curves are liable to fall below the physical yield curves. For example, given prevailing harvesting costs and prices for the harvest, an economic harvesting curve such as *FGHJK* may apply.

In the foregoing analysis, it is assumed that there is no significant difference between the physical yield functions and the economic harvesting function. It is worth observing too that if the difference between the two functions is a constant, they reach their relative maxima for the same densities, and the distinction between the curves would not be of great importance from an optimisation point of view.

Turning to empirical evidence on yield function: yields of all tree and field crops appear to be influenced by the density of planting because of interplant competition (Donald, 1968). Downey (1971) showed that maize (*grain*) yields per unit of area rise and then fall as density of planting is increased and the yield function follows a 'bell-like' shape. One important factor that reduces the yield of maize grain at high densities is the increasing frequency of barrenness of plants in a stand as density is increased. Similar types of yield functions have been suggested for perennial crops. See, for example, Eden (1965) for tea, Child (1974) for coconuts, and Cremer, Cromer and Florence (1978) for eucalyptuses. Gregory (1954, 1955), using experimental data, showed that the gross yield per hectare and grade of bananas depends upon the density of the planting of bananas. He observed that the gross weight-yield of bananas, their quality and the production of ratoons increases at first with the density of the banana planting and then declines. Harvey (1983) finds that the value of a stand of patula pine (*Pinus patula Schiede et Deppe*) and its yield are both influenced by the density of planting.

While some progress has been made in estimating coconut yields as a function of the density of palms, agricultural experimental findings are still restricted to a limited number and range of observations (Coomans, 1974; Manthriratna and Abeywardena, 1979). While results show that nuts per palm and copra per palm decline with increased densities in the range of densities observed, insufficient data is available to fit the complete relationship indicated in Figure 1. This is disappointing, but it points to an area of agricultural research which is needed to provide a basis for rational economic advice. It is relevant, however, to observe that Manthriratna and Abeywardena find that density, rather than the spatial arrangement of palms, is the most important determinant of coconut yields.

In practice, the nature of yield-density functions depends on the type of crop under consideration and the biological phenomenon that is regarded as yield. The yield function can be expected to be different if we are interested in the total biomass or population of a species (or a characteristic of it) as a yield, to that if we are interested in the rate of change of the biomass or population (or a characteristic of it) as a yield. Thus, if the relevant biomass is a logistic function of density, as it is in some cases (Christiansen and Fenchel, 1977; Clark, 1976), the rate of change of the population (this is the harvestable yield in some cases) may be of an inverted U-shape or even 'bell-shaped'. Yield-density functions for crop dry matter also appear to differ from those for reproductive types of yields such as grains and seeds.

For crops where the final yield of dry matter is relevant, review of empirical evidence indicates that asymptotic yield-density relationships (logistic, power function and the like) provide reasonable approximations to observed data (Willey and Heath, 1969). The Langsaeter relationship for growth of timber as a function of the stock of timber per unit area (Smith, 1962: 43) indicates that an asymptotic relationship is likely to hold for total timber yield as a function of density.

In contrast, reproductive yields per hectare (grains, seeds) appear to first increase with crop density and then decline. These relationships can reasonably be approximated in many cases by parabolic equations, as is amply demonstrated in the literature (Willey and Heath, 1969). Downey (1971) has shown for maize that its yield of dry matter per hectare varies asymptotically with its density, whereas the grain yield follows a polynomial form, rising at first with density and then declining.

It is apparent that, depending upon the problem at hand, various types of yield-density functions can be empirically relevant. In the following analysis, we shall follow the common assumption in economics that the per-unit productivity functions are of an inverted U-shape. This implies a 'bell-like' yield-density function. However, the analysis discussed here can be extended to any type of continuous and differentiable density-related yield function.

3. Optimality of increasing densities of trees or plants

The most economic density of trees or plants differs depending upon whether their *existing* density is above or below that which maximises yield per unit of area. When the existing density is below the level that gives this maximum yield, it *may* be economic to increase densities, but this is not always so. Where the density is above the level that maximises yield per unit of area, it *may* be economic to thin the trees or plants, but this is not always the case.

Let us suppose that it is desired to maximise profit (or the surplus) from the crop in a *single* period of time, and that only the density of the crop is to be varied to achieve this. Furthermore, assume that no crop exists so that the entire crop has to be planted. Let us also make the following assumptions: (a) each unit of resource (labour, or labour capital and raw material in fixed proportions) can plant b trees or plants so that, where L represents the quantity of resources employed in planting per hectare, $x = bL$ is the number of trees planted per hectare; (b) each unit of resources used in planting costs w, and the production unit has adequate supplies at this price. Therefore the total cost of planting is $C = wL$ or $C = (w/b)x$; (c) the production unit is unable to influence the price received from selling the produce from the trees and receives a price per unit of p for the produce.

Given the above assumptions, the profit, Π, received by the productive unit in the period of time is $\Pi = py - C = pf(x) - (w/b)x$ where w/b represents the resources used per tree planted. The necessary condition for maximum profit is that $pf'(x) = w/b$, that is, the value of increase in yield from planting an extra tree be equal to the additional cost of planting it. Rearranging this equation, this condition is satisfied when $f'(x) = (w/b)/p$.

Assuming the same type of yield function as shown in Figure 1, Figure 2 can be used to illustrate the economic optimum. The curve *LMN* represents the yield per tree or plant, and the curve *RMS* represents the rate of change of total yield of the plants, that is, the marginal total yield, as the density of the crop is raised. Curve *LMN* represents $f(x)/x$, the average product, and curve *RMS* represents $f'(x)$, the marginal product. Line *UV* represents the real cost (relative cost, not to be confused with inflation-adjusted cost) of planting each tree, that is $(w/b)/p$. In the case illustrated, profits (or the surplus) are maximised by planting $\bar{x}$ trees per hectare.

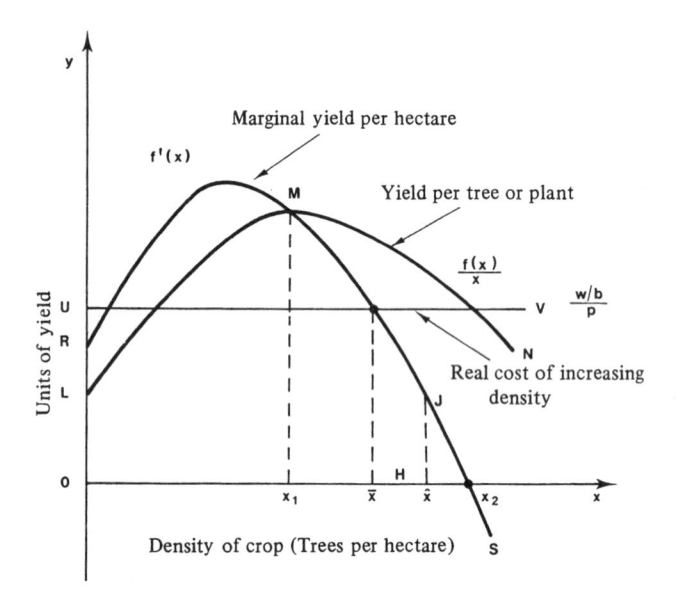

Figure 2. *When profit or the surplus can be raised by increasing the density of the crop, it is optimal to raise the density until the real additional cost of raising the density equals its addition to total yield*

Note that, other things being equal, the lower the cost per tree planted (w/b) or the greater the price, p, obtained per unit of yield, the lower is the line *UV* and the higher is the optimal density of trees per hectare. (The scarcer land is as a factor, the higher the price of produce from it is likely to be). But only in the limiting case (never reached in practice), where the cost of increased planting is zero or the value per unit of yield is infinite, would profit be maximised by planting the number of trees per hectare that maximise yield per hectare. The optimal economic density in this model is either equal to or lies between that which maximises yield per tree, x_1, and that which maximizes yield per hectare, x_2. Usually, the optimal economic density falls between these two extremes.

In many instances, the existing densities of trees or crops are not zero. Natural densities of trees or historically determined densities may be positive. The question arises in such cases of the extent to which it is economical to alter densities. The additional cost of varying densities needs to be balanced against the extra revenue obtained from this variation.

Assuming that the marginal cost per tree planted remains constant at w/b, and that the real marginal cost is constant at $(w/b)/p$, it is optimal if the existing density, x^*, is less than $\bar{x}$ *to increase* this density to $\bar{x}$, the density for which the marginal yield equals the real marginal cost of increasing the density of the trees. Profit is then equal to

$$\Pi = pf(\bar{x}) - \int_{x^*}^{\bar{x}} (w/b)\, dx \text{ and the surplus is } \xi = \Pi/p.$$

If the existing density, x^*, exceeds $\bar{x}$ but is x_2 or less (that is, equal to or less than that which maximises yield per hectare), *profit (or the surplus) is maximised by not altering the density of the crop*. Any increase in the density would add more to cost than to revenue. Profit is $\Pi = pf(x^*)$ and the surplus is $\xi = f(x^*)$ in this case.

To give examples of the above two cases: if $x^* = x_1$ in Figure 2, profit or the surplus is maximised by increasing the density (or plantings) by $\bar{x} - x_1$. The surplus is then equal to the area under curve *RMS* between 0 and x_1 plus the area between this curve and *UV* between x_1 and $\bar{x}$.

In the second case, if $x^* = \hat{x}$, the additional real cost of expanding plantings exceeds the marginal contribution to yield of increased plantings, that is $OU > HJ$. Thus, profit is maximised by leaving the existing density unaltered, and the surplus amounts to the area under curve *RMS* between 0 and $\hat{x}$.

The existing density of trees or plants may exceed that which maximises yield per hectare. Coconut palms, oil palms, natural forests, are sometimes too crowded to give maximum yield per hectare. In other words, the existing density x^* often exceeds x_2. To what extent should the density be then reduced to increase the profit or the surplus? Let us consider this matter.

4. Optimality of reducing densities of trees or plants

When existing densities of trees or plants exceed that which maximises yield per hectare, it *may* be possible to increase profit or the surplus by reducing these densities. But just as costs are involved in increasing densities, so are costs and effort involved in reducing densities, for example, by thinning out trees.

The effect of reduced densities on yields can be seen from Figure 1. We can consider the effect on total yield as the density is reduced from a maximum value x_m in Figure 1 towards x_2. Given the type of total yield function shown in Figure 1, the marginal productivity from reduced densities are like those shown in Figure 3 by curve ABC. Where curve RS indicates the real marginal cost of reducing density, the surplus (or profit) is maximised when $\tilde{T}$ trees are eliminated (assuming that the existing density is x_m) or when the density is reduced to $\tilde{x}$.

The problem can be posed specifically as follows:

Let $T = kL$ represent the relationship between trees 'removed' and the amount of labour used (or labour and other resources used in fixed proportions), and assume that each unit of the resources, used to thin the crop, costs w. Then the total cost in the period is $C = wL = (w/k)T$. Let $y = h(T)$ represent the relationship between total yield in the period and the number of trees removed assuming that the existing density is x_m. Total profit can then be expressed as: $\Pi = py - C = p\,h(T) - (w/k)T$. Examining the necessary condition for a maximum of this expression, profit is maximised in the single period under consideration when sufficient trees or plants are removed to equate the value of the marginal product from removing trees with the cost of removing each tree. The surplus is maximised when the increase in yield from removing trees equals the real cost of removing each tree. In Figure 3, this is satisfied for $T = \tilde{T}$.

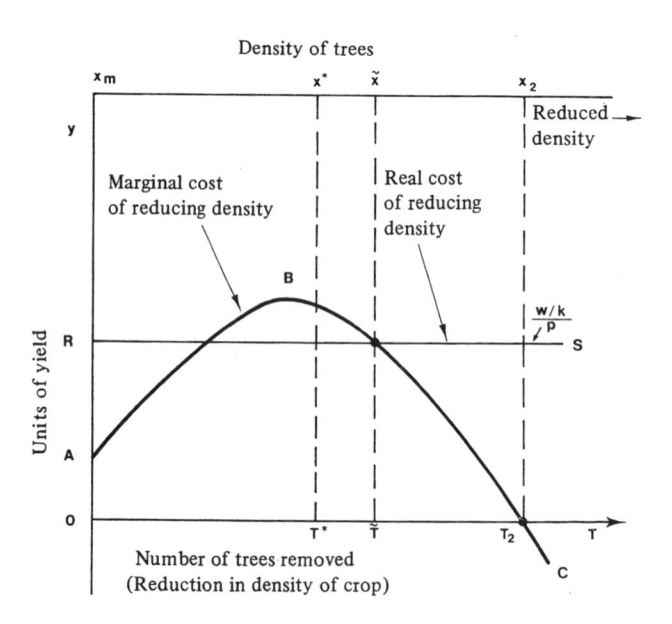

Figure 3. *When profit can be increased by reducing the density of the crop, it is optimal to reduce density until the extra real cost of lowering density equals the extra contribution to total yield*

In practice, actual densities can be expected to be less than x_m, which is the greatest density likely to be encountered. Nevertheless, if the actual density is beyond that which maximises yield per hectare and is one for which the value of the increase in yield from thinning exceeds the marginal cost of thinning (the cost of removing each tree), the basic condition for an economic optimum still needs to be satisfied. For example, if the actual density of the crop is x^*, as indicated in Figure 3, it might be supposed that 'nature' has removed T^* trees. Nevertheless, the surplus is maximised by removing a further $\tilde{T} - T^*$ trees. Compared to a situation in which x_m trees exist, the surplus here is larger by $OR \cdot T^*$. The actual surplus will be

$$S = f(x^*) + \int_{T^*}^{\tilde{T}} h(T)dT - \frac{w/k}{p}(\tilde{T} - T^*)$$

where the last two terms indicate the net increase in surplus as a result of the thinning of the 'natural' tree cover.

It might be observed from Figure 3, that the optimal amount of thinning is greater, other things being equal, (i) the higher the value of p, the price received for each unit of the yield; (ii) the lower the value of w/k, the cost per tree removed (which in turn is lower the smaller the cost of each unit of resource used in tree removal, w, and the greater the productivity, k, of each unit of resources in removing trees); and (iii) the slower the decline is in the marginal productivity curve ABC. However, only in extreme circumstances would it pay to thin the tree cover sufficiently to obtain a density that maximises yield per hectare. This extreme occurs if the cost of removing trees is zero, or the value of the yield is infinite — two unlikely circumstances.

If the existing density of trees is less than x_2, (the level maximising yield per hectare) but greater than $\tilde{x}$ (the density for which the real marginal cost of reducing density is equal to the marginal yield from reducing density), the profit or the surplus is maximised by leaving the existing density unaltered. This is so because the real marginal cost of reducing density exceeds the increase in yield from lowering the density. Other things being equal, the greater the real marginal cost of reducing density, the more likely it is to be unprofitable to alter the existing density. The real marginal cost of lowering the density is greater, the lower the price per unit of the produce and the higher the value of w/k, the per unit cost of removing trees.

5. The combined model

The above discussion can be brought together by means of Figure 4. In Figure 4, the curve $HJKLM$ indicates marginal yield (the rate of change of total yield) as a function of the density of the crop. The line UV shows the marginal real cost of increasing the density of the crop and the line RS the marginal real cost of reducing the density of the crop.

If the actual density is less than $\bar{x}$, profit is maximised by increasing the density to $\bar{x}$; if the actual density is greater than $\tilde{x}$ profit is maximised by reducing the density to $\tilde{x}$, assuming in both cases that total revenue covers total costs. However, if the actual density falls in the range $\bar{x} \leqslant x \leqslant \tilde{x}$, profit (or the surplus) is maximised by not changing the existing density. This range may be called an 'inert' area and shows that it can be uneconomic to alter a wide range of densities.

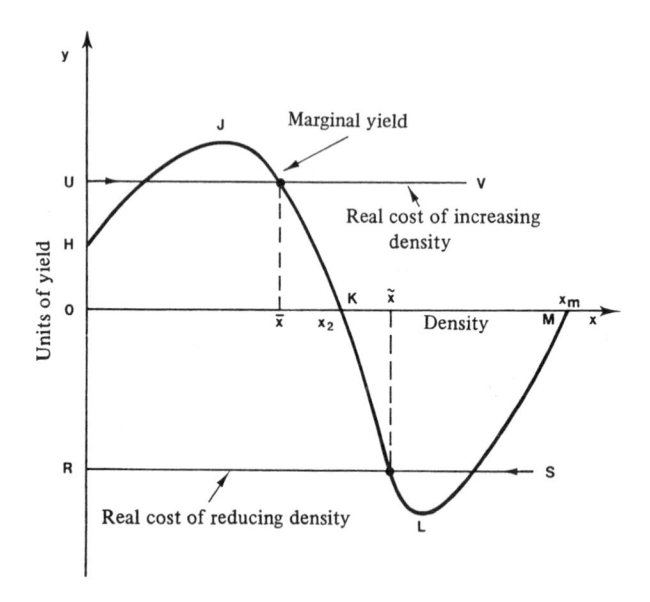

Figure 4. If the actual density of plants or trees falls in the range $\bar{x} \leqslant x \leqslant \tilde{x}$ it is unprofitable to alter their density. The higher the extra cost of increasing density or of lowering density, the wider (other things being equal) is this range

The range $\bar{x} \leqslant x \leqslant \tilde{x}$ in which it does not pay to alter the existing density of trees or crops is likely to be greater, other things being equal:
(a) the lower the value of p, the price obtained per unit of yield;
(b) the higher the marginal cost of increasing the density of the crop, for example by planting trees (if a tree crop is involved), which, in turn, depends upon the cost of the resources used to increase density and their productivity in tree planting or increasing density;
(c) the greater the marginal cost of reducing the density of the crop (the cost of removing each plant in this case), which, in turn, depends on the cost of resources used to reduce density and their productivity in lowering density; and
(d) the more slowly marginal yield declines (when it does decline) as the density of the crop is increased.
There may be a wide density-range (inert range) for which it does not pay

to alter the existing density of a crop, such as a tree crop, and this needs to be taken into account in giving advice to farmers. As a rule, this inert range *includes* the density that maximises yields per hectare. If this is the existing density, it is uneconomic to alter it. However, if this is not the existing density, it is normally not economic to try to attain such a density, nor to advise this density as optimal.

6. Time as a Variable

In solving many problems, the passage of time needs to be taken specifically into account. While the main results are not altered by taking time into account, the optimization problem is more complicated. However, in the following simple case the preceding results are very easily adapted. Assume that the productive unit has a planning interval covering n periods. Suppose that in the initial period the unit can adjust the density of its crop but that adjusted density remains unchanged throughout the planning interval. Furthermore, suppose that the earlier conditions apply and that the productive unit aims to adjust the density so as to maximise the net present value of the crop (Tisdell, 1982). Letting p_t^* represent the discounted present price of the product in period t, and subscript 1 refer to the initial period in the interval, the necessary condition for maximum profit or surplus is $f'(x_1) = (w_1/b)/\Sigma p_t^*$. Thus the earlier theorems can be re-interpreted by replacing p by $\sum_{t=1}^{n} p_t^*$, the sum of the present discounted prices of the produce during the planning interval. Similar conclusions apply if the density of the crop needs to be reduced rather than increased in order to raise profits.

Clearly, when the planning period extends beyond a single period a greater amount of adjustment of densities tends to be profitable. However, the higher the rate of interest, the smaller is the value of Σp_t^* and hence the smaller is the amount of optimal adjustment of densities. Also the higher the rate of interest, other things being equal, the greater is the range in which it does not pay to alter densities.

Note that this time-model involves a once and for all adjustment of crop density at the beginning of the planning period. Thus it differs from models discussed in the forestry literature which allow for thinning and density reduction at various times throughout the life of the crop (Clark, 1976; Pearse, 1967; Smith, 1962). The main principle (namely, the possibility of an 'inert' adjustment area in which it neither pays to thin nor to increase densities, for example, by infilling) can be expected to apply in this more general case too.

7. Concluding comments

The above discussion abstracts considerably from complications encountered in the real world in order to establish a general principle. For example, it assumes that densities are uniformly altered, but such variation is not always practical or profitable. Furthermore, yields depend not only on average densities but on variability of densities on a plot of land, and may alter in an uncontrolled way with the passage of time. Again, optimal densities vary with the age of some crops. For some crops, a higher density is optimal when they are younger, followed by thinning when they are older. The optimal spacing of crops may also be influenced by the desirability of allowing access to the crop for cultivation purposes. Nevertheless, some general principles can be established.

There is widespread empirical evidence that the yield of crops depends on their density (Willey and Heath (1969), and other sources previously noted), as well as other factors, but more agricultural information is needed to calculate these functions for specific crops. However, given general evidence about the nature of these functions, this paper brings attention to optimality principles, that seem not to have been appreciated in theory and in policy advice. It points out that there may not be a unique optimal economic density for crops. The optimal economic density when existing densities are below those in the inert area (identified in the paper) is likely to be less than that when existing densities are above the inert area. Furthermore, the inert area identifies a range of densities (which may eventuate as a result of natural events) that a farmer does not find it profitable to alter.

Applying this economic principle to the cultivation of coconuts as an example, advice needs to be of the following form: In sparse stands, the density of palms should be brought up to the minimum density of the inert area, and in crowded stands densities should be thinned to the upper density of the inert area. Stands with densities falling within the inert area should be left unchanged. Present policy advice does not take this form.

A number of economic influences on the size of the inert area have been noted. For example, the size of the inert area is smaller, other things being equal, the higher the price per unit obtained for the yield of the monocrop. Consequently, optimal economic densities approach more closely to densities maximising yields per hectare the higher the price of the produce. The price of the produce will tend to be high if land suitable for the crop is in relatively short supply. Similarly, other implications of the analysis can be spelled out to provide an abstract (but, in some ways, more satisfactory) overview of the optimal density problem, than has been the case in the past.

REFERENCES

Child, R. (1974). *Coconuts*, 2nd ed. London: Longman.
Christiansen, F. B. and Fenchel, T. M. (1977). *Theories of Populations in Biological Communities*. Heidelberg: Springer-Verlag.
Clark, C. W. (1976). *Mathematical Bioeconomics*. New York: Wiley.
Coomans, P. (1974). Densitées de plantation pour le cocotier. *Oleagineux* 29: 409-414.
Cremer, K. W., Cromer, R. N. and Florence, R. G. (1978). Stand establishment. In W. C. Hillis

and R. G. Florence (eds.), *Eucalypts for Wood Production*, pp. 81-138. Canberra: Commonwealth Scientific and Industrial Research Organization.

Dillon, J. L. (1977). *The Analysis of Response in Crop and Livestock Production*, 2nd ed. Oxford: Pergamon Press.

Donald, C. M. (1968). Competition among crop and pasture plants. *Advances in Agronomy* 15: 1-114.

Downey, L. A. (1971). Plant density-yield relations in maize. *Journal of the Australian Institute of Agricultural Science* 37: 138-146.

Eden, T. (1965). *Tea*. London: Longman.

Gregory, E. J. (1954). Investigations. *Department of Agriculture Jamaica Bulletin* 53: 163-165.

– (1955). Investigations. *Department of Agriculture Jamaica Bulletin* 54: 130-132.

Harvey, A. M. (1983). *Growth, Volume and Value of Production of Patula Pine in a Free Growth Spacing Trial*. Technical paper no. 33. Brisbane: Department of Forestry.

Leibenstein, H. (1978). *General X-Efficiency Theory and Economic Development*. New York: Oxford University Press.

Manthriratna, M. A. P. P. and Abeywardena, V. (1979). Planting densities and planting systems for coconut, *Cocos Nucifera L*: Study of yield characteristics and the economics of planting at different densities. *Ceylon Coconut Quarterly* 30: 107-115.

May, R. M. (1973). *Stability and Complexity in Model Ecosystems*. Princeton: Princeton University Press.

Owen, O. S. (1975). *Natural Resource Conservation: An Ecological Approach*, 2nd ed. New York: Macmillan.

Pearse, P. H. (1967). Optimum forest rotation. *Forestry Chronicle* 43: 178-195.

Scott, A. D. (1972). *Natural Resources: The Economics of Conservation*. Toronto: McLelland-Stewart.

Smith, D. M. (1962). *The Practice of Silviculture*, 7th ed. New York: Wiley.

Tisdell, C. A. (1982). *Microeconomics of Markets*. Brisbane: Wiley.

Whitehead, R. A. S. and Smith, R. W. (1968). Results of a coconut spacing trial in Jamaica. *Tropical Agriculture, Trinidad* 45: 127-133.

Willey, R. W. and Heath, S. B. (1969). The quantitative relationships between plant population and crop yield. *Advances in Agronomy* 21: 282-321.

Professor C. A. Tisdell
Department of Economics
University of Newcastle
Newcastle 2308
Australia

N. T. M. H. De Silva
Deputy Director
Division of Economic Research
Coconut Development Authority
Colombo 10
Sri Lanka

SUPPLY-MAXIMISING AND VARIATION-MINIMISING REPLACEMENT CYCLES OF PERENNIAL CROPS AND SIMILAR ASSETS: THEORY ILLUSTRATED BY COCONUT CULTIVATION

C. A. Tisdell and N. T. M. H. De Silva*

University of Newcastle†

> *The paper argues that maximum sustainable yields rather than net present values often need to be considered for policy and identifies the length of the replacement cycle required for maximum sustainable yield or supply in multi-point or interval output time-phased production conditions, such as apply to perennial crops and to similar productive processes. It enables point-input point-output models which have been well explored (for instance, for forestry) to be treated as special cases. A virtue of the analysis is its use of a simple technical relationship to determine the replacement rate of a perennial crop to maximise sustainable yield. The analysis is illustrated by an empirical example drawn for the cultivation of the Sri Lankan 'tall' variety of coconut.*

Introduction

Optimal replacement cycles for forests have been discussed extensively in the economic literature (Faris, 1961; Chisholm, 1966; Perrin, 1972; Samuelson, 1976; Wan, 1976; Kemp and Long, 1983). Generally point-input point-output time-phased models have been used for this purpose and it is usually assumed that the net present value (NPV) of the forest in relation to a given land area is to be maximised by appropriate husbandry including an appropriate length for replacement cycles for the forest. Economists have been rather critical of the maximum sustained yield (MSY) approach proposed by many foresters (Samuelson, 1976). The optimal length of replacement cycle based upon the MSY usually will differ from that indicated by the NPV approach (Samuelson, 1976, pp.478-479).

While the NPV approach is valuable, it requires a considerable amount of information about prices and interest rates and more calculation than the MSY approach. This may complicate the analysis when either a point-input multi-point (interval) output time-phased model is used or when inputs themselves are multi-point or spread over an interval. Such patterns are usually the case for perennial crops, but also apply to produce from other types of biological resources, for example, milk from cattle. The purpose of this article is to explore the MSY strategy for such productive processes, paying particular attention to the following two questions and illustrating the analysis by coconut

---

* We would like to thank an anonymous referee for suggestions that were useful to us in revising the draft of this paper.

† Department of Economics, University of Newcastle, N.S.W., 2308, Australia. N. De Silva is on leave from the Division of Economic Research, Coconut Development Authority, Colombo, Sri Lanka.

cultivation. Under stationary conditions, what should be the length of the replacement cycle of the crop (productive asset) in order to maximise output per unit of time, that is to obtain MSY? What phasing or pattern of replacement is necessary in a steady state to minimise variation in supply over time from the crop or productive asset as its replacement occurs? In other words, what is the optimum age-profile to maintain for the crop so as to achieve this minimisation?

Without wishing to denigrate the NPV approach, it is pertinent to note that there are circumstances, particularly in LDCs, where the above MSY considerations seem to be of greater policy relevance. This would seem to be so for coconut production in Sri Lanka where the government desires to maximise the supply of coconuts from the area and resources currently allocated to coconut cultivation. In general, there is a desire also to avoid increased variability of supply when this not compensated for by a rise in overall supply. Furthermore, it needs to be remembered that many markets including capital markets are extremely imperfect or limited in their operation in LDCs and it may be quite inappropriate to assume, as is done in the usual NPV approach, that everyone can borrow and lend in unlimited amounts at the market-given competitive rate of interest.

A supply-orientated strategy may be relevant to subsistence growers if resources used in growing the crop have little opportunity cost. This may be the case for a number of perennial tree crops grown in the tropics. Again, a possible reaction of growers to uncertainty about prices and interest rates may be to avoid 'fine-tuning' of their production strategies to these variables. There seems therefore to be a case for exploring the implications of MSY strategies in the context of perennial crops. Let us do this first by considering some simple theory and then an empirical case involving the cultivation of coconuts, and follow this by further general discussion.

Some Theory

Suppose that the yield from a crop of uniform age and of given density on a given area of land can be specified by the function

$$Y = f(t) \qquad (1)$$

where Y represents the yield and t indicates the age of the crop, for instance the age of coconut palms on the land. The length of the replacement cycle needed to maximise yield averaged over time is that maximising the following expression with respect to t:

$$h(t) = t^{-1} \int f(t)\,dt \qquad (2)$$

For many perennial crops or similar productive assets this relationship is unimodal. An example of such a single-peaked curve is shown in Figure 1 by OAFCG.

In Figure 1, curve OABD represents the continuous yield of the crop as a function of its age. The crop has a pre-bearing period of OA, and its maximum yield occurs at age $\hat{t}$. The cumulative yield averaged over the age of the crop reaches a maximum when the crop is $\bar{t}$ years old. Hence, $\bar{t}$ is the optimal length of the replacement cycle for maximising supply on average per unit of time. Since f(t) can be regarded as a marginal curve and h(t) can be considered to be an average, the usual relationships between these curves apply. For example, when h(t) is increasing f(t) is above it and f(t) falls below h(t) when h(t) is declining (Chiang, 1967, p.174). In the single-peaked case, f(t) reaches a maximum before h(t) does. This implies that the replacement cycle needed to maximise supply on average per unit of time exceeds in length the period required for a crop to reach its maximum yield.

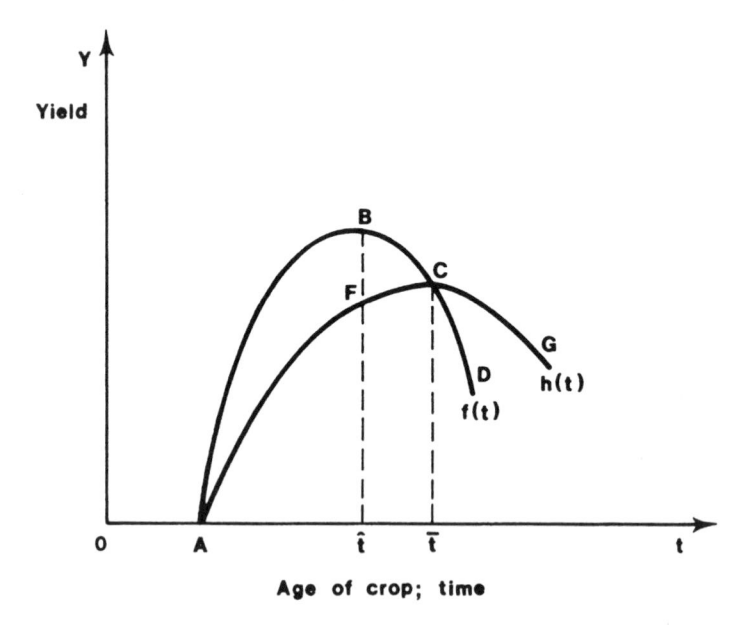

Figure 1. Hypothetical yield and average cumulative yield functions for a perennial crop as a function of its age. These curves can be respectively interpreted as marginal and average functions in relation to time.

This has determined the optimal length of the replacement cycle, but what in a steady state is the optimal age profile of the trees to minimise fluctuations in supply through time? The rectangular distribution where crops of each age between o and t occur with a relative frequency of t^{-1} does this. For instance, let us suppose that time consists of t discrete intervals, for example, years with f(t) representing yield in period t. In a steady state, the variation-minimising age profile of the crop will be that in which t^{-1} of the crop is of each of the ages 1, ..., t. This will provide an annual stationary yield or supply of produce from the land area area devoted to the crop of

$$R = t^{-1} \sum_{t=1}^{\bar{t}} f(t). \tag{3}$$

This yield can be maintained in perpetuity by replanting each section of the area planted with the crop when it reaches t in age. This pattern maximises supply over time since each segment of the crop is held for an optimal length of time and avoids any variability in supply because the replacement pattern is such that the age profile of the crop on the available land never changes.

However, if the age profile differs from the above, variation in supplies occur with the passage of time. Assuming that each segment of the crop is held for t years before replacement, the greater is the age between segments of the crop, the greater is the amplitude of fluctuations in available supplies but the lower is the frequency of such fluctuations over time. If the whole area is devoted to a crop of the same age the range and (amplitude) of fluctuations in supply is maximised over time. The time pattern of supply from different replacement profiles for the crop (age distribution of it) can be comparatively easily determined assuming steady state conditions. Let us illustrate the above general approach by the cultivation of coconuts.

Optimal Replacement Cycles and Age-Profiles for Coconuts

Coconuts are a perennial crop of considerable importance in tropical and semi-tropical areas. There are a number of varieties of coconut with different yield characteristics. For illustrative purposes, we concentrate on the Sri Lankan tall variety of coconut which has a pre-bearing period of approximately 8 years. Sri Lankan authorities recommend that palms be replaced at 60 years (Nathaniel, 1968). Such recommendations are based on experience of planters and administrators rather than objective evidence (cf. Etherington, 1978, concerning Malaysian rubber). The theory can, of course, be applied to other varieties of coconut and other crops such as rubber, oil palm, tea, cocoa, coffee and so on.

The yield of a perennial crop (or a similar asset) generally depends upon a variety of factors. In the case of coconut, the density of coconuts and their age are important influences on yield. Using a random sample of 59 holdings from the Matara district of Sri Lanka (for more details see De Silva and Tisdell, 1985), we have estimated the following yield function for coconuts in the area:

$$\text{Log } Y = 1.8304 + 2.04 \log t + 0.536 \, t/\log D$$
$$(5.48) \quad\quad (4.31)$$

$$-.0051 \, D - .0243 \, t \log D - 42.96 \log t/D$$
$$(4.22) \quad\quad (5.58) \quad\quad (8.94) \tag{4}$$

where Y is the yield per acre, D is the density of palms (palms/acre) and t is the age of the palms. For this function $R^2 = .8159$ and the t-statistics given in parentheses are significant at the 1% level.

This logarithmic function (cf. Shannon, 1978) was decided on after experimentally fitting to the data all alternative functional forms possible with package P9R of the Biomedical Computer Programs (Dixon and Brown, 1979). This functional form gave the best fit to the data. Furthermore, this functional form yields a curve with the properties suggested by earlier research on the relationship between yield of coconut palms and their age (Burgess, 1981).

As for the quality of the data, data was collected from field interviews with growers by trained coconut extension officers and checked by their observations. Since records of the age of palms are not kept by growers, the possibility exists of errors in recall or identification of the age of palms. However, growers appear to be rather accurate in identifying age of plantings by reference of the time of planting to major family events such as family births, marriages and so on. Furthermore, most growers and extension officers can readily identify the age of a group of palms by their morphological characteristics such as leaf spread, tapering of the palm, the shape of the bole and height. The close relationship of these factors to the age of palms has been established by a number of researchers (Child, 1974; Santhirasegaram, 1966) and is particularly important in estimating the age of older palms.

Equation (4) only has two variables, namely, age of palms and their density. While other variables such as fertiliser use and agro-climatic conditions influenced the yield of palms (cf. De Silva and Tisdell, 1985), their inclusion in this case has a masking effect and the two variables used appear to possess good explanatory powers in terms of this sample. If required, however, the relationship can be expanded using additional variables.

In order to illustrate the theory let us consider the age-dependent yield function of palms of a given density, namely 76 palms per acre. A density of about 76 palms to the acre appears to maximise yields from an acre (De Silva and Tisdell, 1985). In this case, for D=76, equation (4) reduces to

$$\text{Log } Y = 1.4428 + 1.4748 \log t - .0173t \tag{5}$$

This yield-age relationship for an acre of land planted to Sri Lankan tall is illustrated by curve OABCD in figure 2 using an arithmetical scale and assuming a pre-bearing period of 8 years for palms. Yield reaches a maximum when a palm is 36 years old, that is, $t = 36$.

The corresponding cumulative yield curve averaged over the age of the stand is shown by curve OACG. It reaches a maximum at 66 years. In this case, $t = 66$. The optimal yield-maximising replacement cycle of 66 years is longer than the 60 year period recommended by Sri Lankan authorities. Furthermore, length of the optimal replacement cycle is almost double the age at which palms reach their maximum annual yield.

If the age of palms in the area (on the acre of land) is not uniform but such that it has an age profile in which each $1/66$ of the area is planted with palms of uniform age ranging from 1 to 66 years old, annual supply is 991 nuts per acre and, if the profile can be maintained, is represented by the horizontal line HJ in Figure 3. Supply shows no variation in time. If, on the other hand, all the area is planted with palms of uniform age and replanted after these reach 66 years of age, yield follows the repeating pattern shown by OABCK. Yield varies from a low of zero in some years to a maximum of 1,323 nuts at the peak-yield age. For comparison, the heavy line shows the yield stream on the assumption that the coconut growing area is divided into three sections each with palms of equally spaced ages in the 66 years replacement cycle. Fluctuations are more frequent but of smaller amplitudes than in the previous case. Nut supplies vary between a low of 815 nuts and a maximum of 1,134 nuts. These results accord with our previous theoretical propositions. In the case of Sri Lanka, if the above result for Matara were to apply throughout the island it would be optimal to replace $1/66$ of the area under coconut palms each year, at least once a steady state is reached.

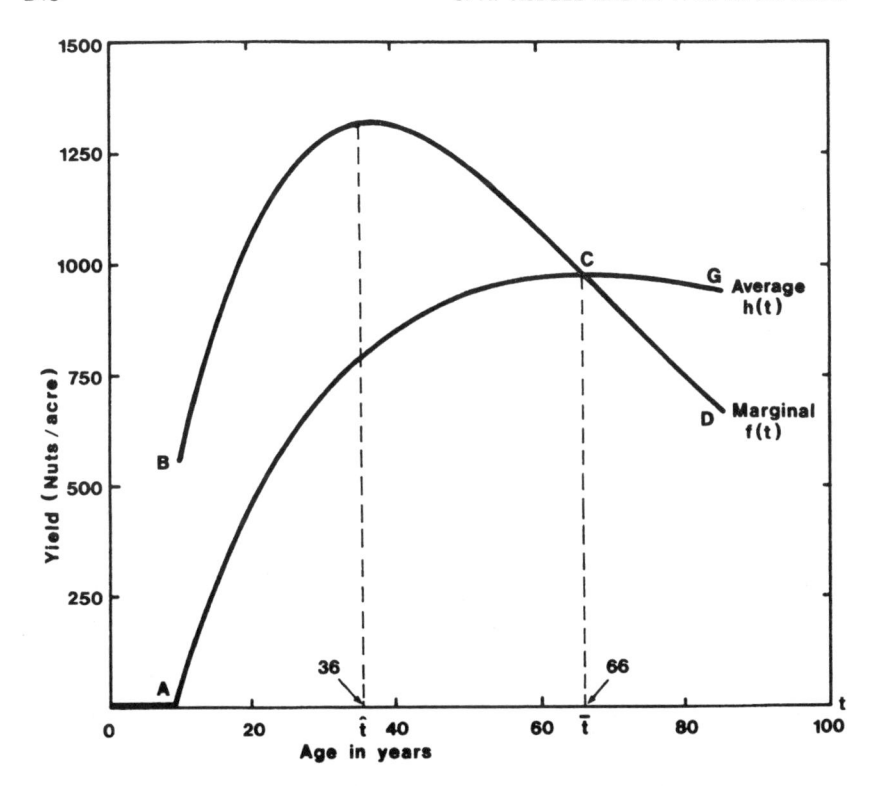

Figure 2. Empirically estimated yield function (marginal function) for the Sri Lankan tall variety of coconut and average cumulative yield as a function of the age of the palms.

Discussion

Yield functions may in practice be subject to uncertainty or variations with the passage of time. This has not been allowed for in the above analysis. Furthermore, yield functions may vary between localities. This can be easily taken into account. The optimal replacement cycle for a variety of a crop will vary from locality to locality depending upon differences in the yield curve between localities. The analysis can be applied depending upon differences in the yield curve between localities. The analysis can be applied *mutatis mutandis* to each locality.

The problem of how optimally to achieve the steady state age-distribution of the crop needed to minimise fluctuations in annual yield has not been discussed. There would be a number of alternative paths to this steady state path. The question of whether to phase such a path in over one cycle or several cycles is likely to depend upon the existing age composition of the crop.

It may at this stage be worthwhile considering some further aspects of the earlier simple theory and relating it to the point-input point-output case. For this purpose it is useful to consider the *cumulative* yield curve for a crop of uniform age on an area of land, that is, $\int f(t) \, dt$. This is shown in Figure 4 by curve OAPRQ. The length of cycle needed to obtain MSY is, as before, $\hat{t}$. If, however, costs are involved in replanting the crop and net sustainable yield is to

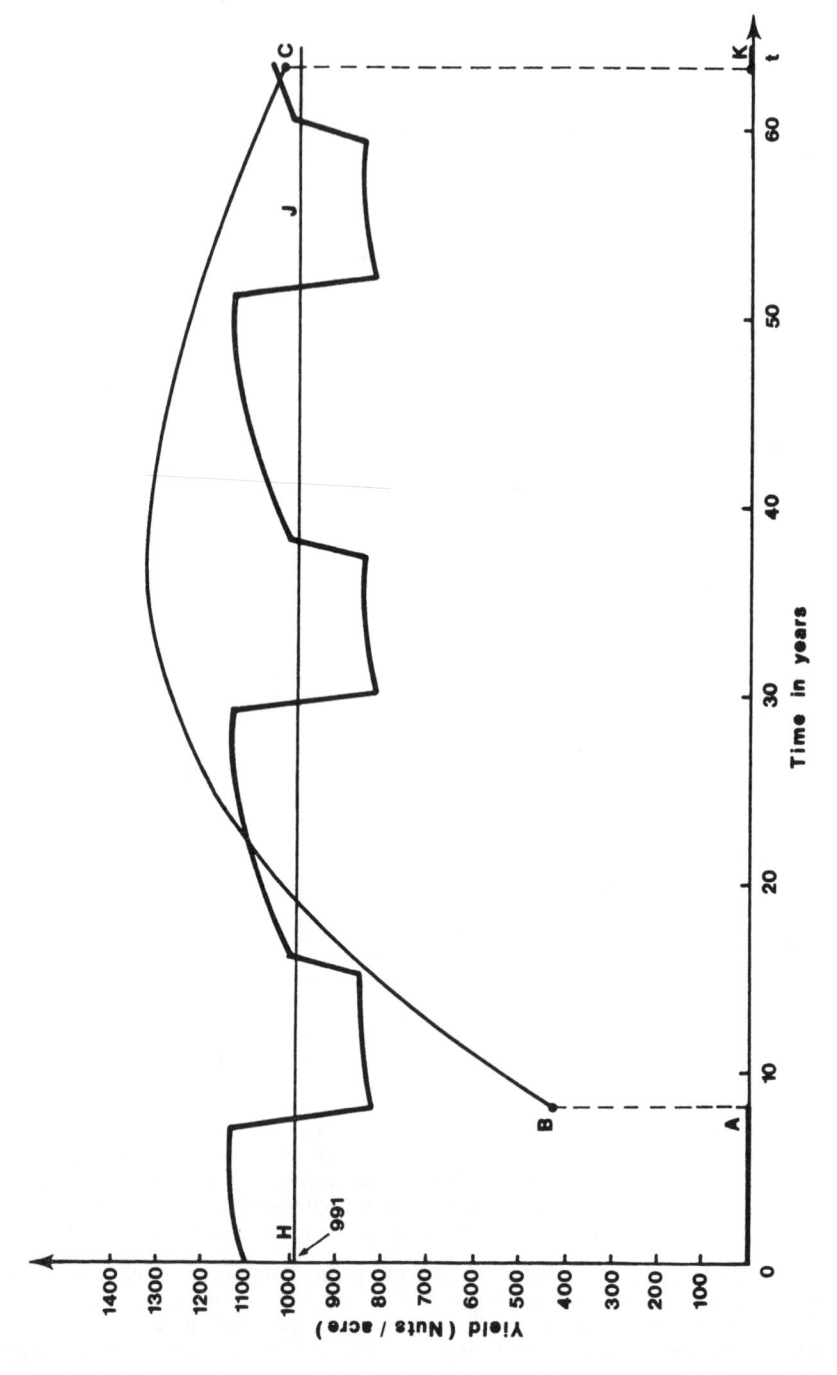

Figure 3. Yield stream of Sri Lankan tall coconuts for three different age-profiles (or phasings) of replacement assuming a replacement cycle of 66 years.

be maximised, the length of the optimal replacement cycle is altered. It is longer than when no such costs are involved and is greater the larger are these costs. In the case shown, costs of OR involved in replacement lengthen the optimal replacement cycle to t_1. Other things equal, an increase in the pre-bearing period for the crop (the distance OA in Figure 4) also lengthens the opimal replacement cycle. In this hypothetical case, segment AP RQ of the cumulative yield curve is displaced to the right by a constant amount.

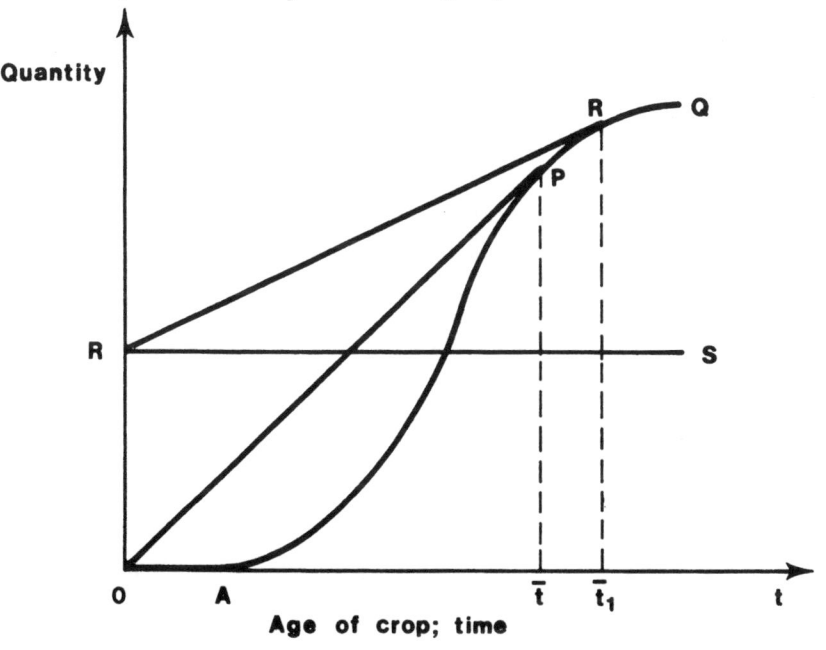

Figure 4. 'Typical' cumulative yield pattern for a perennial crop showing optimal length of replacement cycles for gross yield and for net yield.

In the point-input point-output case, the 'cumulative' yield from a crop or forest is the actual yield which one gets by felling it at particular time. In that sense it is not really cumulative at all. It does not consist of the summation of a stream of yields as in the multi-point output case. The same simple MSY theory however applies to this case (cf. Samuelson, 1976, p.477). Note cumulative yield never declines with the passage of time in the multiple-point or interval output case. However, it may do so in the point-output case for the output depends upon when it is decided to realize the *potential* yield.

In conclusion, it does seem worthwhile exploring alternatives to the NPV model for perennial crops and similar productive systems dependent upon multiple productive units, where the productivity of each unit varies with its age. Examples arise in agricultural production and other biologically-based productive systems as well as in some manufacturing industries where the productivity of machines is age-dependent (Kibria and Tisdell, 1985). At least for the conditions facing LDCs, it seems worthwhile exploring the maximum sustainable yield strategies for such productive processes. This paper has identified and illustrated replacement strategies that will maximise yield and minimise the variability of supply over time. The analysis itself may be applied to perennial crops in developed as well as less developed countries.

References

Burgess, R. J. (1981). *The Intercropping of Smallholder Coconuts in Western Samoa—An Analysis Using Multi-Stage Linear Programming*, MADE Research Series, Canberra: Development Studies Centre, Australian National University.

Chang, A.C.C. (1967). *Fundamental Methods of Mathematical Economics*, New York: McGraw-Hill Book Company.

Child, R. (1974). *Coconuts*, London: Longmans.

Chisholm, A. H. (1966). Criteria for Determining the Optimum Replacement Pattern, *Journal of Farm Economics*, **48(1)**, 107-112.

De Silva, N.T.M.H. and Tisdell, C.A. (1985). Density Related Yield Functions for Coconuts, *(Cocos nucifera L)*: An Empirical Estimation Procedure, *Experimental Agriculture*, **21(3)**, 259-269.

Dixon, W. J. and Brown, M.B. (1979). *Biomedical Computer Programs: P-Series*, Los. Angeles: University of California Press.

Etherington, D. M. (1977). A Stochastic Model for the Optimal Replacement of Rubber Trees, *Australian Journal of Agricultural Economics*, **11(1)**, 40-58.

Faris, J. E. (1960), Analytical Techniques used in Determining the Optimal Replacement Pattern, *Journal of Farm Economics*, **42(4)**, 755-766.

Kemp, M. C. and Long, N. V. (1983). The Economics of Forests, *International Economic Review*, **24(1)**, 371-395.

Kibria, M. G. and Tisdell, C.A. (1985). Operating Capital and Productivity in Jute Weaving in Bangladesh, *Journal of Development Economics*, **18**, 135-152.

Nathanael, W. R. N. (1968). The Coconut Industry and Planting Progress, *Ceylon Coconut Quarterly*, **20(3)**, 145-160.

Perrin, R. K. (1972). Asset Replacement Problems, *American Journal of Agricultural Economics*, **58(4)**, 888-894.

Samuelson, P. A. (1976). Economics of Forestry in an Evolving Society, *Economic Inquiry*, **14(3)**, 466-492.

Santhirasegaram, K. (1966). Intercropping with Coconuts, *Journal of National Agricultural Society of Ceylon*, **2(1)**, 1-11.

Shannon, R. E. (1978). *Systems Simulation—the Art and Science*. Englewood Cliffs, New Jersey: Prentice Hall.

Wan, H. Y. Jr., (1966). A Generalized Wicksellian Capital Model: An Application to Forestry, in Institute of Economics Academia, Taipei, *Economic Papers, Selected English Series*, **1**, 1-13.

Forum

Conserving and Planting Trees on Farms: Lessons from Australian Cases

C. A. Tisdell*

Background

Trees on farms can be used for a variety of purposes. These include shelter for stock, crops or pastures with consequent possible increases in productivity, the control of erosion (both gully and wind erosion), aesthetic appeal, the provision of farm wood and timber, the supply of commercial timber as in the case of agroforestry and sometimes farm trees may provide a fodder reserve for drought. They may also be a source of fruit and nuts, honey, eucalyptus, cork and pharmaceutical and chemical supplies and a habitat for wildlife. In some areas, evapotranspiration from deep-rooted trees may keep ground watertables from rising and so keep salting of soils and streams at bay.

One can of course, add to this list of possible benefits of trees on farms but it would be wrong to conclude that trees on farms are always beneficial from an economic point of view. Tree densities can be so high that they reduce the value of the land for grazing or agriculture. Or trees on the property might not be optimally grouped from an agricultural productivity point of view—for instance, they may be scattered and not provide suitable windbreaks or they may interfere with the operation of machinery used in cultivation. The latter factor has been important in decisions to remove hedges in Britain (Helliwell 1969). Very often if net benefits are to be realized from trees they must exist or be planted in a particular pattern.

In assessing the costs and benefits of the provision of trees on farms, it is useful to keep in mind the distinction between private cost-benefit analysis (CBA) and social cost-benefit analysis (Hufschmidt et al. 1983). The aim of private CBA is to maximize private net benefits (private gains less private costs) whereas the aim of social CBA is to maximize social net benefits (community-wide gains less community-wide costs). It is, of course, socially optimal to maximize social net benefits of tree provision on farms but private decisions are likely to be based on private CBA and this may result in socially inadequate provision of trees on farms.

In many cases, the community's benefits from tree provision on a farm will exceed the gains to the individual farmer. For example, runoff may be less rapid after rain where there is tree cover and this may mean less silting and likelihood of flooding in drainage areas. Other spillover benefits may include wildlife conservation and aesthetic appeal for passers by.

If social net benefits exceed private net benefits from tree provision, farmers may not retain or provide enough trees from a social point of view (such a divergence provides a rationale for the National Tree Program). This type of market failure is illustrated in Figure 1 for a hypothetical case. There curve ABC represents the marginal private net benefit of tree provision on a property. Hence, private net benefits are maximised when x_2 trees are on the property. However, marginal social net benefits

---

* University of Newcastle. This is a revised and extended version of a paper originally presented to a National Workshop on Benefits of Trees on Farms. I wish to thank the editors of this *Review* for making various suggestions for improving my original paper.

as shown by curve *DEF* exceed marginal private net benefits, and thus the socially optimal number of trees on the property x_3, exceeds the number provided by the farmer.

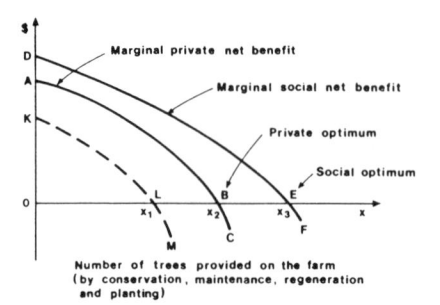

Figure 1: Possible divergence between the private and social optimum in tree provision on farms.

Some farmers may fail to realize the actual extent of private benefits from trees on their farms. For example, a farmer may believe that his marginal private benefits are like those shown by curve *KLM* and provide x_1 trees on his property whereas from his own point of view x_2 are optimal. In such a case it is possible to improve the farmer's choice by providing him with improved information. In this case, the farmer is better off and he also makes a better decision from the point of view of society when given greater information. It is always a happy situation when by providing improved information one can increase private and social gains from environmental decisions. To some extent the strategy (discussed later) of the Victorian Garden State Committee (V.G.S.C.) and the Victorian Farmers and Graziers Association (V.F.G.A.) in promoting farm trees is intended to take advantage of this parallelism.

In providing farm trees, a farmer has many alternatives to consider: Should he concentrate on conserving and maintaining existing trees? Should he consider replacing these trees by other varieties and if so which variety? Would he be better to concentrate on trying to regenerate local trees or to replant with seedlings? What pattern of tree cover should he attempt to establish? There are several complications to be taken into account. Take for instance regeneration. This may require areas to be fenced off or protected from stock while regeneration takes place. Similar costs are also likely to be incurred if planting seedlings. However, regeneration is uncertain—it depends on the suitability of the site and the availability of tree seeds in the area. Weeds, especially near cattle camps (Cross 1984), may interfere with regeneration and there is a risk of rubbish or non-climax trees or shrubs being propagated. Delays in self-establishment, the need for a particular tree pattern and type of tree may all tend to favour planting of tree seedlings. Nevertheless in some cases judicious stocking rates and appropriate spelling of paddocks may do much to encourage recovery of natural vegetation where this is required on large grazing properties, and in the long-term this may be the most effective method of sustaining production levels from grazing (cf. van Rensburg 1983, pp. 51–2).

Australia is a large country with varied environments and several types of agriculture. One therefore, has to be wary in generalizing from case studies carried out in one or a few localities and in applying results obtained in one State to make recommendations for another State in the Commonwealth. Nevertheless, if this proviso is borne in mind, it is useful to consider some case studies that have been done in Australia (*see* also Howes and Rummery 1978; Moore 1983). Four cases are outlined in turn: (1) shelter-belts and an economic study of trees on four Victorian farms, (2) the economic benefits of saving trees from dieback on the New England Tablelands as evaluated by Sinden and others, (3) the economics of agroforestry on the Southern Tablelands of N.S.W. as evaluated by Gisz and Sar, and (4) the economic aspects of dryland salting.

Case (1) is designed to indicate the net private benefits of trees on farms through their complementary impact on pure agricultural production. Case (2) examines the value of trees on farms in view of pure public good or collective good characteristics associated with them. Case (3) considers the private net benefits in particular cases of a system of mixed production involving agriculture and forestry, and Case (4) illustrates possible favourable production externalities on farms from the retention or provision of trees. Cases (2) and (4) concentrate on social dimensions in dealing with possible sources of market failure in tree provision on farms. The other two cases focus on private net benefits and *may* also provide information which improves the decisions of farmers from a social point of view. In the latter respect, two types of social benefits are possible: (a) farmers' decisions may be improved from a private point of view and resource-use improved, and (b) these improved decisions may give side benefits beyond the benefits to the individual farmers, for example, through public-good benefits from tree provisions or favourable production externalities. Improved information may have a social net benefit even if only possibility (a) is present, that is if there are no side or spillover benefits.

Shelter Belts and Trees on Four Victorian Farms

The value of shelter belts in raising agricultural productivity has been demonstrated in many countries. In Jutland (Denmark) and in northern Germany, the sole reason for planting and managing shelter belts and tall hedges is for increased agricultural production (Baltaxe 1961; quoted in Helliwell 1969). Bird (1981) points out that "there is ample evidence from studies in U.S.S.R., U.S.A., U.K., Germany, France, Switzerland, *etc.*, that shelter can improve crop yields by at least 25 per cent, pasture yields by 20-30 per cent, dairy milk production by 10-20 per cent". Bird provides American data (*see* also U.S.D.A. 1957) indicating approximately a 20 per cent increase in crop yields as a result of shelter belts and says that these results are consistent with those obtained at the Frankston Vegetable Research Station, Victoria. He also notes research at Hamilton (Victoria), which indicates that lamb mortality may be reduced by up to 50 per cent, and the survival rate of shorn sheep is raised, by provision of shelter. When stocking rates are approximately adjusted, total wool production can increase by more than 30 per cent and sheep liveweights by 20 per cent as a result of the provision of shelter in cold areas such as Armidale (Lynch, *et al* 1980).

Nevertheless, economic evaluation of shelter belts requires a number of productivity elements to be taken into account which are neglected in the above studies. Possibly the brief economic study of trees on four Victorian farms reported by V.F.G.A. and V.G.S.C. (1984) should also be put on a sounder economic basis than at present. Reported claims appear to exaggerate the private benefit of trees on farms.

Variations in yields of crops and pastures as a result of the provision of shelter belts are typically estimated from the inside *edge* of the shelter belt (Bird, 1981; Seipen, 1983; Sturroch, 1981; Breckwoldt, 1983; Forest Commission, Victoria, 1972). While crop/pasture yields near the shelter belt decline due to shading and competition, yields further out increase considerably due to the protection afforded from wind by the shelter belt. A typical pattern might be like that shown in Figure 2 (*cf.* Breckwoldt, 1983). On the leeward side of the shelter belt, crop/pasture production may on balance increase by 20 per cent. Increased crop/pasture production may occur for a distance of approximately 15 times the height of the shelter. However, realistically potential production on the areas occupied by the shelter must be subtracted from total production. In many cases, this is equal to the height of the shelter. If this is so the net increase in yield over the areas subject to the shelter belt amounts to 12.5 per cent rather than 20 per cent. However, if the increase in yield on the leeward side is only 10 per cent the net overall gain in crop yield is only a fraction over 3 per cent. If yield on the leeward side increases by less than 6.67 per cent (on balance), overall crop/pasture yield falls in the area modified by the shelter.

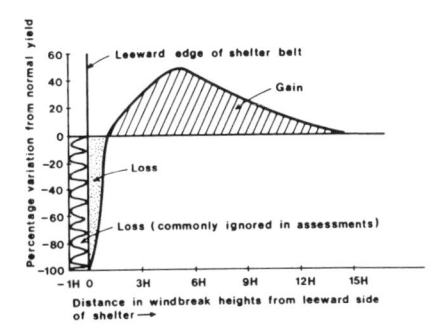

Figure 2: Yield variation of a crop or pasture as a function of the provision of a windbreak or shelter belt.

In order to obtain the long-run private benefits to the farmer of a shelter belt, it is necessary to take into account the change in his/her *profit*. Increases, say, in total revenue from greater production may overstate gains. This will be the case if higher costs are involved in greater production. Most studies only consider the increase in total production as a result of the provision of a shelter belt but this is inadequate from an economic point of view.

If a shelter belt has to be established, costs are involved in establishing it. Table 1 sets out the costs which the V.F.G.A. and the V.G.S.C. suggest are involved in establishing a three-row shelter belt. The actual cost, however, can be expected to be much higher because a number of years has to elapse before the shelter belt grows and provides effective shelter. During this period, interest is foregone on the money invested in establishing the belt and this should be taken into consideration as part of the cost. Furthermore, if the area in which the shelter belt is being established has to be fenced off from stock and/or withdrawn from cropping, the *net* value of production foregone on this area during the period of establishment has to be added to the cost of establishing the belt. When account is taken of such considerations the cost of establishment is much increased.

Table 1: Cost per 100 m of a Typical Three-row Shelter Belt

Cost Item	$
Fencing	180
Cultivation	40
Trees	50
Planting	40
Replacements (10 per cent)	5
Follow-up weeding	25
Weedicide	10
Vermin control (home-made wire guards)—	
materials	40
labour	30
Total (per 100 m)	$420

Therefore, to establish a shelter belt on one side of a 20 ha square paddock (450 m) costs $1,890.

Source: V.F.G.A. and V.G.S.C. (1984).

Table 1 indicates that the costs of establishing a shelter belt on *one* side of a 450m square paddock are $1,890. But one shelter belt is unlikely to shelter the whole paddock. If the shelter belt is 10m high and provides protection for 15 times its height, three shelter belts would be required to shelter the whole paddock. On this basis the cost could be of the order of $6,000. When account is taken of foregone production and interest during the establishment period, this cost could well increase to $12,000 to $15,000 per 20 ha paddock.

Furthermore, the estimates used by the V.F.G.A. and V.G.S.C. may paint an inadequate picture in another respect. Benefits have not been estimated as increased private net profits in these case studies. For example, one of the benefits to farmer C from shelter is said to be a 15 per cent increase in carrying capacity of ewes on a 20 ha paddock. It is claimed that since the *market value* of a ewe is $25 and an extra 26 ewes can be carried on a 20 ha paddock on farm C, the annual benefit to the farmer is $650. But this exaggerates private net benefits, if the net *income* per year per ewe is less than $25 as seems likely. Values of assets and sizes of income flows should not be confused in cost-benefit analysis.

Calculations of private economic benefits from shelter belts on Australian farms need to be based upon acceptable economic principles which often differ from accounting practices. It could well be (in fact, it is likely) that the provision of shelter belts on many farms is still privately profitable but the picture appears to be less rosy than that painted in this case study. There is also a need to supplement adequate economic analysis by further scientific and experimental work in Australia so that we have more knowledge about possible variations in yields for different crops in different localities as a result of the provision of shelter.

From a social point of view, other benefits from the provision of shelter belts need to be taken into account. For example, belts may reduce water run-off (thus reducing water-erosion, siltation and flooding), reduce wind-erosion, provide a suitable habitat for wildlife which is valued, and over a wide enough area may result in favourable climatic effects. In addition they may have aesthetic appeal. These possible benefits need to be measured and to be carefully calculated from an economic point of view. Sinden and collaborators have examined these in relation to Eucalyptus dieback in the New England region.

Eucalypt Dieback and Farm Income: A New England Case Study

An economic evaluation of the dieback problem in the New England area has been undertaken by Sinden, Jones and Fleming (1983). Their approach is based on willingness to pay principles and takes account of income benefits to farmers from a reduction in tree cover plus the willingness of the community at large to pay for the retention of tree cover on farms (Sinden and Jones 1983). The private farm income benefits from reduced tree cover were estimated from a cross sectional sample of farms in the area. The willingness of householders to pay for the retention of tree cover was *inferred* from a sample survey designed to elicit the maximum amount which households would be prepared to contribute towards research to prevent dieback in the region.

The results of Sinden *et al.* (1983) indicate that dieback in the New England area has been a net benefit to the community. Dieback has permitted stocking rates of livestock to be increased (for example, due to greater pasture cover) and this has meant, on average, a $2,000 increase in annual income per property in the area. The total increase in net farm income in the shires of Dumaresq, Uralla and Walcha as a result of woodland decline is conservatively estimated at $1,666,000 by Sinden and Jones (1983). They estimated that each household in the area (most households were in urban centres such as Armidale) is prepared to pay (on average) $6.84 per year to preserve trees in the area and that the aggregate amount they would pay is $79,000 per year. On this basis, they concluded that dieback to the extent that it has occurred in New England has been socially beneficial.

However, their calculations take no account of the possibility that residents outside the Tablelands area might be prepared to pay something for tree preservation in the area. The effect of trees on run-off, pest control (Davidson 1982; Tisdell 1983), and soil erosion is also neglected. They do suggest, however, that if dieback should proceed further, it could impose a net social cost on the community. Nevertheless, their conclusion is that it would not have been economic to save the trees that have so far disappeared in the New England area due to dieback. This of course does not show that it would be uneconomic to plant shelter belts in the New England region or even engage in agroforestry there—even though there is independent evidence to suggest that widespread agroforestry is not likely to be economic in the New England region at present (Sinden, pers. comm.).

The research by Sinden and collaborators represents the first attempt in Australia to take specific economic account of public good or collective good characteristics associated with tree cover. Their case illustrates the point that public good characteristics are not in themselves *sufficient* to justify a policy of tree conservation. Their analysis is also based upon the premise that tree cover on farms constitutes a mixed good (Barkley and Seckler 1972) that is a good with private characteristics for the landholder and with pure public characteristics for the community at large.

Mixed good characteristics of tree cover on farms can be illustrated by Figure 3. This represents a simplified example. Suppose AB represents the marginal cost of altering tree cover (in practice, as pointed out by Tisdell and De Silva (1983) this is likely to be asymmetric) and let curve CEF represent the marginal private on-farm benefits of altering tree cover. If farmers are well informed one would expect tree cover to be adjusted to x_1. However, there may be a collective demand for tree cover from urban dwellers and others for instance to enhance the scenic beauty of the countryside. If the sum of the social marginal evaluation of the characteristics of tree cover that can be collectively enjoyed is as shown by curve D_1D_1, the marginal social value of providing tree cover is represented by curve GEF. Ignoring the possibility of production externalities, well informed farmers will provide a socially optimal amount of trees. The collective demand element is irrelevant in this case from a Kaldor-Hicks point of view.

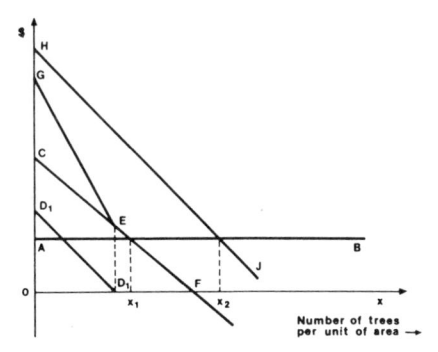

Figure 3: Illustration of the mixed good characteristic of trees on farms.

However, if the social marginal evaluation curve of public characteristics of trees is positive at x_1, farmers will not provide a socially optimal amount of tree cover. For example, if the demand for public characteristics amounts to the difference between curves *HJ* and *CF*, the marginal social value of tree provision is represented by curve *HJ*. In that case, the socially optimal amount of tree cover is x_2 and farmers will undersupply tree cover by $x_2 - x_1$ from a Kaldor-Hicks viewpoint. The public good characteristics of trees on farms are Paretian relevant in this case whereas they are Paretian irrelevant in the previous case. The problem parallels that for Paretian relevant and irrelevant externalities (Tisdell 1970; Walsh and Tisdell 1973). It should be noted that production-type externalities from tree provision are not taken into account in the above exposition but could also be allowed for. They are the main focus of attention in the dryland salting case considered below.

Agroforestry: A Southern Tablelands (N.S.W.) Study

Gisz and Sar (1980) and Gisz (1982) have considered the economics of agroforestry on a property near Tarago on the Southern Tablelands of N.S.W. They consider the alternatives of operating a purely sheep-grazing enterprise involved in wool production with the alternative of an integrated enterprise of widely spaced radiata pine and sheep grazing. It is assumed that the radiata pine is managed for saw logs and the sheep are managed basically for wool.

In their 1980 analysis they found that the "sheep only" alternative was more profitable than agroforestry at a rate of interest of 9 per cent or more. In 1982, however, Gisz was able to revise these estimates in the light of actual experience with costs and productivity of the agroforestry enterprise. When this was done, the economics of the agroforestry enterprise from the point of view of the farmer was more favourable. The new estimates indicate that agroforestry at the Tarago property is more profitable than the sheep only alternative at a discount rate of less than 12.75 per cent.

Gisz (1982), p. 10) concludes his study as follows:

> For the case-study farm, the analysis shows that agroforestry compares favourably with the sheep-only alternative when evaluated in terms of N.D.R. (net discounted return) per hectare at discount rates of less than 12.75 per cent. To generalize from this to other sites is hazardous because of the specific assumptions and limitations underlying the analysis.
>
> As a potential land use alternative for farms suitably located in relation to processing facilities, factors which should be considered include: availability of labour and management skills needed to produce high-quality saw logs; the relative profitability of alternative enterprises; the availability of capital to establish the agroforestry enterprise; the investor's assessment of risk and attitude to long-term investment; and potential benefits derived from agroforestry in mitigating soil erosion problems and buffering against periodic slumps in markets for agricultural products.

One needs to keep in mind that the economics of land-use alternatives is subject to variation so updating of calculations is required on a continuing basis. One may also need to consider from a social point of view the spillover or externality qualities associated with particular species of trees and types of agroforestry, for example, roles of these in preserving wildlife. There has been some criticism of the use of *Pinus radiata* in Australian plantations on the grounds that it does not provide suitable habitats for native wildlife.

Production possibility curves can in principle be used to analyze the economics of agroforestry or mixed productive systems involving trees as explained by Filius (1982). However, since the usual rendition of these concepts is static, time is not adequately accounted for. From a practical viewpoint one has to consider alternative time-paths of production strategies as done by Gisz. Nevertheless, some useful conceptions do follow from a consideration of production possibilities.

Five alternative types of production possibility sets are indicated in Figure 4 for a landholder. Assuming that the landholder does not influence the price of the products, it is only in cases (4) and (5) that mixed production (agroforestry) will always be optimal for profit maximization. In these cases forest production is complementary to agriculture. For the production possibility sets marked (2) and (3), agroforestry may maximize profit. It does not, however, do so when a corner solution at A or B prevails as will arise when the price of agricultural products is high relative to forest products. In case (1), there are increasing returns to specialization in production and agroforestry never maximizes profit—it pays to specialize either in agriculture or in forestry depending upon relative prices. Note that in cases (4) and (5) forest or tree products may as such have no commercial value yet it would be privately profitable to grow trees on farms. These cases would be in accord with the Victorian case studies mentioned earlier.

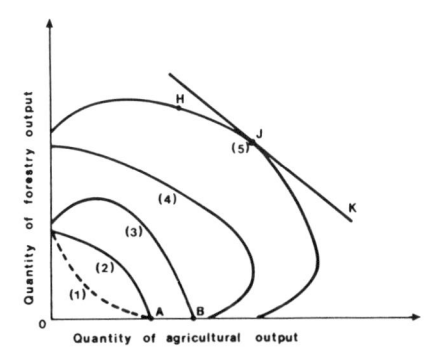

Figure 4: Implications of some different types of production possibilities curves for agroforestry.

Naturally, agroforestry or a system involving trees on farms need not always be socially optimal. For example, if the price line indicated by KJ in Figure 4 prevails and the production possibility curve is as indicated by (5), the combination at J is privately most profitable. However, if trees on farms have positive productive spillovers or public good characteristics (relevant for the quantity of trees associated with J), a greater quantity of trees may be socially optimal. For example, the combination at point H may be socially optimal from a Kaldor–Hicks viewpoint.

Dryland Salting

The removal of trees from farms can have adverse production externalities. When this occurs the private net benefits of tree removal are liable to diverge from the social net benefits. In some areas of Australia, the removal of trees gives rise to dryland salting. The process of dryland salting as the result of the removal of deep rooted trees in parts of Western Australia and Victoria is well-known (Bennett and Thomas 1982; Hodge 1982; Batini 1982; Mulcahy 1983; Nulsen and Baxter 1982; Greig and Devonshire 1981; Trotman 1974). Approximately 0.82 per cent of the cultivated land area of Australia has been reported to have been affected adversely by dryland salting. The clearing of land in areas subject to salting has unfavourable externalities on agricultural production on other properties and on the water quality in streams. The salting of the soil may result in bare patches of ground without pasture or with sparse or unpalatable pasture and limit the range of crops that can be grown. The salting of streams may result in the quality of water being unfit for irrigation, for human consumption or for animals. In addition, fish may be killed and the scenic qualities of streams impaired.

Hodge (1982) illustrates the type of market failure that occurs when unfavourable agricultural production externalities arise from tree removal which results in salting, and I shall not repeat his diagram here. Hodge (1982, p. 199) also outlines economic factors that should be taken into account in government regulation of the clearing of land in areas subject to dryland-seepage salinity. He concludes however, that:

> the issue of whether restrictions over clearance of land would show a positive economic return has not been considered. The evidence suggests this to be the case (*e.g.*, Lumley 1982), and it appears that landholders face incentives to clear an excessive land area. The value of restrictions over clearance has been assumed in this paper but would benefit from further empirical testing.

Dumsday *et al.* (1983) suggest that as far as dryland salinity control in Victoria is concerned the problem could be alleviated "by relatively minor changes in existing farming systems. More drastic approaches such as reforestation may only be required in limited areas." They suggest that greater use of lucerne (a deep-rooted salt-tolerant plant) in combination with deep-rooted perennial pastures would go far in controlling the salting problem and could be privately more profitable than present farming patterns as well as socially more satisfactory (see also Dumsday 1983). However, this approach cannot be used in all areas subject to dryland salting, e.g., environmental conditions for the successful cultivation of lucerne are not satisfied in all areas.

In Queensland, salting of non-irrigated land has been observed in areas subject to higher rainfall (Hughes 1984) and this has been a contributing factor to rural tree dieback (Johnston and Wylie 1984) which is on the increase in Queensland. Several other factors as listed by Wylie and Johnston (1984) have also contributed to this, including stock damage from rubbing and soil compaction. Feral animals such as the feral goat and the feral pig have also played a role in this process (Tisdell 1982).

Conclusion

More scientific data and knowledge are required to help us evaluate the value of trees on Australian farms. The case studies mentioned in the text illustrate the diversity of issues that need to be taken into account in determining the costs and benefits of tree conservation, maintenance and planting on farms. The appropriate economic estimation of the private costs and benefits of trees on farms is more complicated than is commonly recognized and social cost-benefit analysis is even more complex. Nevertheless, considerable progress has been made and can be made in measuring the costs and benefits of trees on farms in Australia. A moderate amount of determination and a small amount of resources for research if used systematically could quickly improve our knowledge about the social net benefits of farm trees in this country. It could enable us to decide the extent to which we should act on a recent editorial comment, namely,

a plan should already be formulating similar to the treeing of American farm land during the depression or the present taming of deserts in China. It is an enormous task and would not begin to show fruit for another 20 years. Australia certainly has the labour force and the need (Ball 1983).

Enthusiasm needs to be tempered by knowledge and logic. Much more economic expertise needs to go into the assessment of the value of farm trees in Australia and more scientific facts need to be gathered in specific Australian environments. Such information as we do have should also be used wisely in giving advice.

It may be pertinent to note that in searching for suitable case studies of the conservation of trees on farms, I found no case studies for the coastal and wetter regions of N.S.W. including the eastern slopes of the Great Divide. Some studies are however, being done of similar areas in Queensland (see, Johnston and Wylie 1984; Wylie and Johnston 1984). Particular studies of the higher rainfall areas are needed because tree cover in many of these areas is still substantial, the areas are favourable for tree growth and problems such as landslides or slumps are more acute in such areas. Such areas are also close to major population centres so the amenity value of trees is likely to be most important in the eastern coastal region.

In conclusion, the case studies discussed in this article illustrate the complexity of assessing the economic value of the provision of trees on farms. Apart from the need to more accurately assess private net benefits from tree conservation and planting on Australian farms, considerable scope exists for estimating the public economic dimensions of tree provision. These include the public good characteristics associated with trees and the various externality consequences of their provision or retention.

References

BALL, J. (1983). "The need for an evolutionary U-turn", *Australasian Farming* 4 (1) 1.

BALTAXE, R. (1961), "Shelter planting practice", *Scottish Forestry* 15 (1).

BARKLEY, P. W. and SECKLER, D. W. (1972), *Economic Growth and Environmental Decay*, Harcourt Brace Jovanich, New York.

BATINI, F. E. (1982), "Regional management—a case study. The Wellington Dam Catchment in Western Australia", pp. 77-83, in N. M. OATES *et al.* (eds.), *Focus on Farm Trees: The Decline of Trees in the Rural Landscape*, Natural Resources Conservation League, Springvale, Victoria.

BENNETT, D. and THOMAS, J. F., (eds.) (1982), *On Rational Grounds: Systems Analysis in Catchment Land Use Planning*, Elsevier, New York.

BIRD, R. (1981), "Benefits of tree planting in S.W. Victoria", *Trees and Victoria's Resources* 23 (4), 2-6.

BRECKWOLDT, R. (1983), *Wildlife in the Home Paddock: Nature Conservation for Australian Farmers*, Angus & Robertson, Sydney.

CROSS, P. (1984), "A farmer who doesn't care for his land is just a miner", *The Land Magazine*, March 22.

DAVIDSON, R. L. (1982), "Local management to reverse tree decline on farms and crown land", pp. 89-103, in N. M. OATES *et al.* (eds.), *Focus on Farm Trees: The Decline of Trees in the Rural Landscape*, Natural Resources Conservation League, Springvale, Victoria.

DUMSDAY, R. G. (1983), "Policy options for salinity management", pp. 69-86, in N. M. TAYLOR *et al.* (eds.), *Salinity in Victoria*, Australasian Institute of Agricultural Science, Melbourne.

DUMSDAY, R. G., ORAM, D. A. and LUMLEY, S. E. (1983), "Economic aspects of the control of dry land salinity", *Proceedings of the Royal Society of Victoria* 95 (3), 139-45.

FILIUS, A. M. (1982), "Economic aspects of Agroforestry," *Agroforestry Systems* 1, 29-39.

FORESTS COMMISSION, VICTORIA (1972), *The Design of Windbreaks*, Government Printer, Melbourne.

GISZ, P. and SAR, N. L. (1980), *Economic Evaluation of an Agroforestry Project*, Miscellaneous Bulletin No. 33, Division of Marketing and Economics, Department of Agriculture, New South Wales.

GISZ, P. (1982), "Agroforestry: A case study economic evaluation, Southern Tablelands of New South Wales", Workshop paper presented at the Third Annual Conference of the Australian Forest Development Institute, Mount Gambier, South Australia, April 19-23.

GREIG, P. J. and DEVONSHIRE, P. G. (1981), "Tree removals and saline seepage in Victorian catchments: Some hydrological and economic results", *Australian Journal of Agricultural Economics* 25, 134-48.

HELLIWELL, D. R. (1969), "The place of trees and woodlands in lowland Britain", *Quarterly Journal of Forestry* 63, 6-20.

HODGE, I. (1982), "Rights to cleared land and the control of dry land seepage salinity", *Australian Journal of Agricultural Economics* 26, 185-201.

HOWES, K. M. W. and RUMMERY, R. A., (eds.) (1978), *Integrating Agriculture and Forestry*, C.S.I.R.O. Division of Land Resources Management, Perth.

HUFSCHMIDT, M. M. *et al.* (1983), *Environment, Natural Systems and Development: An Economic Valuation Guide*, Johns Hopkins University Press, Baltimore.

HUGHES, K. K. (1984), "Trees and salinity", *Queensland Agricultural Journal* 110 (1), 13-4.

JOHNSTON, P. J. M. and WYLIE, F. R. (1984), "*Casuarina* dieback in the Mary River Catchment", *Queensland Journal of Agriculture* 110 (1), 73-4.

LUMLEY, S. (1982), A partial benefit-cost analysis of the effects of reducing stream salinity in McCallum's Creek catchment, North-Central Victoria. Paper presented at Annual A.A.E.S. Conference, Melbourne.

LYNCH, J. J. *et al.* (1980), "Changes in pasture and animal production resulting from the use of windbreaks", *Australian Journal of Agricultural Research* 31, 967.

MOORE, R. (1983), *Agroforestry: The Integration of Trees and Farming*, Western Australian Forests Department.

MULCAHY, M. J. (1983), "Learning to live with salinity", *The Journal of the Australian Institute of Agricultural Science* 49 (1), 11-6.

NULSON, R. A. and BAXTER, I. N. (1982), "The potential of agronomic manipulation for controlling salinity in Western Australia", *The Journal of the Australian Institute of Agricultural Science*, 48, 222-6.

SIEPEN, G. (1983), *Trees for Farms*, National Parks and Wildlife Service, Sydney.

SINDEN, J. A., JONES, A. D. and FLEMING, P. J. (1983), *Relationship Between Eucalypt Dieback and Farm Income, Stocking Rate and Land Value in Southern New England, New South Wales*, University of New England, Armidale.

SINDEN, J. A. and JONES, A. D. (1983), "Facing the eucalypt dieback problem: A New England case study", *Proceedings of Institute of Foresters of Australia 10th Triennial Conference*, 127-9.

STURROCH, J. W. (1981), "Shelter boosts crop yield by 35 per cent: Also prevents lodging", *N.Z. Journal of Agriculture* 143 (3), 18-9.

TISDELL, C. A. (1970), "On the theory of externalities", *The Economic Record* 46 (113), 14-25.

TISDELL, C. A. (1982), *Wild Pigs: Environmental Pest or Economic Resource?* Pergamon Press, Sydney.

TISDELL, C. A. (1983), "Some questions about caring for seven-tenths of Australia", *Habitat Australia* 11 (4), 32-5.

TISDELL, C. A. and DE SILVA, N.T.M.H. (1983), "The economic spacing of trees and other crops", *European Review of Agricultural Economics* 10, 281-93.

TROTMAN, C. H. (ed.) (1974), *The Influence of Land Use on Stream Salinity in the Manjimup Area, Western Australian,* Technical Bulletin No. 27, Western Australia Department of Agriculture.

UNITED STATES DEPARTMENT OF AGRICULTURE (1957), "Shelterbelts and windbreaks", in *Yearbook of Agriculture 1957,* Washington, D.C.

VAN RENSBURG, C. (ed.) (1983), *Agriculture in South Africa,* 3rd edition, Chris van Rensburg Publications, Johannesburg.

VICTORIAN FARMERS AND GRAZIERS ASSOCIATION AND VICTORIAN GARDEN STATE COMMITTEE (1984), *Financial Benefits of Farm Trees: A Brief Economic Study of Trees on Four Victorian Farms,* Melbourne.

WALSH, C. and TISDELL, C. A. (1973), "Non-marginal externalities: As relevant and as not", *The Economic Record* 49 (127), 447-55.

WYLIE, F. R. and JOHNSTON, J. M. (1984), "Rural tree dieback", *Queensland Agricultural Journal* 110 (1), 3-6.

[8]

Integrating economics within the framework for the evaluation of sustainable land management (FESLM)

Clem Tisdell[*]

Background

In 1972, the United Nations Stockholm Conference brought attention to the importance of achieving sustainable development, but it was not until well into the 1980s that the importance of this goal was widely appreciated. Its significance was underlined by the United Nations Conference on Ecologically Sustainable Development held in Rio de Janeiro in 1992.

The main emphasis of recent sustainability concerns has been on combining conservation and economic growth so as to meet present and future needs, including those of future generations - as, for example, the views expressed in *Our Common Future* (WCED, 1987). The presumption is that current socioeconomic mechanisms governing resource-use and techniques for resource use do not, or may not strike the 'right' balance between the satisfaction of current and future needs, and that some type (s) of public intervention in socioeconomic systems are required, even if these interventions only amount to the provision of a greater amount of knowledge. In turn, this view has generated debate about how much intervention is desirable, and the type of intervention that is likely to be appropriate.

In line with this new policy focus, the Food and Agriculture Organization (FAO) and the International Board for Soil Research and Management (IBSRAM) have developed a framework for the evaluation of sustainable land management (FESLM).[1] Let us consider this framework, and try to assess its economic component, the possible guidance from the economic literature for strengthening this component, the possibility of meaningful economic measures of sustainability, the scope for future economic work in relation to the framework, and finally review some case studies from the IBSRAM/ACIAR land management project.

---

[*] Professor of Economics, the University of Queensland, Brisbane Qld 4072, Australia.

[1] Note that several other organizations also sponsored development of this framework, most notably the International Society for Soil Science; Tropical Soil, Biology and Fertility; the International Fertilizer Development Centre; the International Centre for Research in Agroforestry; the Technical Centre for International Agriculture and Rural Cooperation ACP-EEC; the Australian Centre for International Agricultural Research; the Canadian Development of Agriculture; and the United States Department of Agriculture (Soil Conservation Service).

An overview of FESLM and its economic component

The pillars of FESLM

According to Smyth, Dumanski et al. (1993, p. 8) the evaluation of sustainability within the FESLM involves measuring the extent to which sustainable land management (SLM), as defined below, "can be met by a defined land use on a specific area of land over a specific period of time". According to the FESLM Working Party, there are five objectives or 'pillars' of SLM: (I) productivity, (ii) security, (iii) protection, (iv) viability, and (v) acceptability.

According to the working party, "Sustainable land management combines technologies, policies, and activities aimed at integrating socio-economic principles with environmental concerns so as to simultaneously maintain or enhance production/services (productivity), reduce the level of production risk (security), protect the potential of natural resources and prevent degradation of soil and water quality (protection), be economically viable (viability), and be socially acceptable (acceptability)" (cit. Smyth et al., 1993, p. 7).

Pushparajah and Syers (1993, p. 2) point out that indicators for evaluating the sustainability of land management should address these five pillars (productivity, security, protection, viability, and acceptability), of which economic viability is one. Given this point of view, it follows that (desirable) sustainable land management practices are those that maintain or enhance productivity, increase security, protect the potential of natural resources, are economically viable, and are socially acceptable.

While all these pillars have value, it may be difficult to use indicators based on these concepts to compare different types of land management unequivocally. How, for example, is a ranking to be determined when one type of land management is superior to another on the basis of some pillars, but inferior to the others on the basis of other pillars? This would be a problem even if suitable indicators could be obtained from each pillar.

To resolve the problem, a preference, welfare, or ordering function - putting different weights on the different pillars - would be needed (Menz, 1993). For example, if $x_1, ..., x_5$ specify the value of indicators for each of the five pillars, a preference function of the form $W = W(x_1, x_2, x_3, x_4, x_5)$ would be needed. The main problem, however, is likely to be to get unanimity about the appropriate form of this function, i.e. to what extent should one be prepared to forego protection of the natural environment for economic viability when trade-off is necessary?

Economic viability

In virtually all economies, but particularly in market economies, land management techniques that are not economically viable are unlikely to be sustained. In general, if the effort involved in a type of land management exceeds the perceived benefit or reward obtained, it is uneconomical and not viable. This would seem to apply both to subsistence economies and market economies. In a market economy, revenue obtained from managing land needs to equal or exceed the costs of management, if the form of land management adopted is to be economically viable. Thus the economic break-even point forms a threshold for economic viability. However, the situation is much more complex than might appear at first sight.

The economic viability of a form of land management (the fourth pillar) will be influenced by the first three pillars in the list - plus (usually) market prices for products and services produced by the land and the prices paid for inputs (Table 1). Other things being equal, the lower the productivity of the land, the more difficult it will be to sustain land management, and the degree of yield instability and its insecurity will be reflected in the rapid depletion of the natural potential of the land. For each of the natural or nonmarket components (covered by the first three pillars) relating to sustainable land management, there is also a market or price-related analogue. These analogues are (i) the level of prices received for the produce or services and price levels of any of the purchased inputs (corresponds to productivity); (ii) instability and insecurity of market prices (corresponds to uncertainty of yield); and (iii) long-term trends in market prices (corresponds to sustainability of yield). For example, if input prices tend to rise relative to the prices received for agricultural output, this will increase the economic difficulty of maintaining an existing form of land management using the inputs and producing the output involved.

Table 1. Summary of influences on the economic viability of a land management system.

Natural or physical factors†	Market-related factors
1 (a) Level of productivity	1 (a) Level of price of outputs (and inputs)
2 (a) Uncertainty (and instability) of production	2 (b) Price uncertainty and instability
3 (a) Protection of long-term productivity	3 (a) Trends in prices of commodities

† The three factors in these columns correspond to the first three pillars of FESLM. Social influences, including environmental spillovers, have not been incorporated in this summary.

It is possible for physical indicators of sustainability to go down and for economic viability to rise - for instance if the price of the produce rises sufficiently to offset any decline in natural productivity. However, prices may not rise, and in that case the economic viability of the land management practice is likely to be threatened.

The economic literature and sustainability

Most common economic concept of sustainability

The main emphasis in the economics of sustainability has been on sustaining the welfare (or incomes) of future generations. This has been expressed by saying that "future generations should be left no worse off than current generations" (Tietenberg, 1988, p. 33). The World Commission on Environment and Development (WCED, 1987, p. 43) stated more generally that "sustainable development is development that meets the needs of the present without compromising the ability of future generations to meet their own needs". The time-perspective of this point of view is a relatively long-term one, and would appear to correspond to the "sustainable in the long term" category (25 years +) of FESLM (Smyth et al., 1993, p. 11). This is not to say that economists are disinterested in sustainability over shorter periods.

Nevertheless, this general sustainability perspective has led some economists (e.g. Pearce et al., 1989) to the point of view that it is important to conserve the natural resource base (the productive potential of natural resources) if long-term sustainability is to be achieved. Otherwise, the economic welfare of future generations will be threatened. (However, not all economists are of this view. Some believe that technological and scientific progress will more than counterbalance natural resource depletion). It is also possible that such conservation may be needed to safeguard the future welfare of current generations. Because the perspective provided by this sustainability approach is relatively broad, a considerable amount of additional work is required to make it operational. This concept of sustainability is discussed in some detail in Tisdell (1993), especially in ch. 8-11, and see also ACIL Economics and Policy Pty. Ltd. et al., 1994, ch. 3, pt. I, for a discussion of the measurement of sustainable income.

Conway's evaluation of agricultural techniques and their sustainability

Lynam and Herdt (1989, pp. 382-383) suggest that broad-based approaches of the above type do not provide an adequate operational guide for agricultural researchers - partly because of the ambiguities involved, although they accept that sustainability involves the need to be concerned with future development paths. They consider that Conway's approach in which he states "sustainability is the ability of a system to maintain productivity in spite of a major disturbance, such as caused by intensive stress or a large perturbation" (Conway, 1985) is more promising. More generally, within this type of framework, the sustainability of a system has to do with its ability to maintain desired or desirable properties over a period of time. This is a similar perspective to that adopted in FESLM.

Conway's scheme for evaluating agricultural techniques is actually based on four factors: (i) the level of yield or economic return, (ii) the variability of yield or economic return, (iii) the sustainability of yields - which may depend on the protection of natural resources, and (iv) the income-distribution consequences of a technique.

According to Conway, techniques (or systems of land management) which give higher yields or economic returns, with lower variability of yield or return, greater sustainability of yields, and less inequality of incomes are to be preferred. Very often, however, no technique is superior in all these respects, and Conway provides no guide to how trade-off should be made between his 'pillars'. Furthermore, no simple universal indicators of variability and sustainability are provided. The dimensions of both elements are complex. So at the present time, there are some limits to the application of Conway's approach to evaluating agricultural systems. Note that sustainability is just one of his criteria. Nevertheless it has been used as a framework by Alauddin and Tisdell (1991) to evaluate 'green revolution' technologies.

The operational approach of Lynam and Herdt based on trends in 'output'

Lynam and Herdt, while supportive of Conway's approach, concentrate on the single criterion of sustainability, and attempt to provide it with operational content. They define sustainability as "the capacity of a system to maintain output at a level approximately equal to or greater than its historical average, with the approximation determined by the historical level of variability" (Lynam and Herdt, 1987, p. 394). They go on to say that, "a sustainable system is one with a nonnegative trend in measured output". They even suggest that any technology which adds to a positive trend adds to sustainability, and that this approach is useful for distinguishing sustainability from instability, which refers to variations around the trend line. They state that "measuring the slopes of the trend line of a system's output measure involves specifying the system, the measure of

100   Economics and Ecology in Agriculture and Marine Production

output, the time period of concern and observing the measure over the specified time period" (p. 394).

They propose that "the appropriate measure of output by which to determine sustainability at the crop, cropping system, or farming system level is total-factor productivity, defined as the total value of output produced by the system during one cycle divided by the total value of all inputs used by the system during one cycle of the system; a sustainable system has a nonnegative trend in total productivity over the period of concern" (p. 395). Thus their measure of output is the economic value of output divided by the value of inputs, and will depend not only on physical productivities but also on prices.

Their measure of productivity is given by value of output/value of input. But why not consider value of output minus value of input, that is profit, or profit/value of input as relevant measures? Either of these measures may give a superior indication of economic viability.

Given that there is agreement about how to measure output or productivity, Lynam and Herdt's approach would result in System 1 (Figure 1) being ranked as more sustainable than System 2. The trend line for System 1 is indicated by the line marked AB, and that for System 2 by the line marked CD. Output with System 2 does not fluctuate, but that for System 1 varies in the way indicated by the fluctuating line around AB. In this case sustainability is determined over a 20-year planning period. It is unclear whether the trends can be extrapolated. Note that in this case the least economical form of land management is the most sustainable.

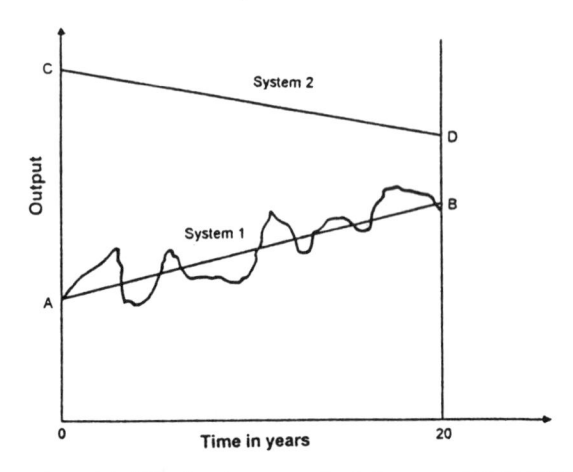

Figure 1. Differences in sustainability of systems according to Lynam and Herdt (1984).

Even if one accepts the operational approach of Lynam and Herdt, it clearly has a number of limitations. (These would also apply even if the alternative measures of output mentioned above are used). First, it cannot be used by itself to determine the desirability of a system. Unlike FESLM, it only takes account of one factor in evaluation, and then in a particular way. Secondly, because of its partial nature, it does not indicate whether a particular system will be adopted and for how long. Thirdly, there are always difficulties in determining trends and the extent to which they can be extrapolated - and this means that problems of best fit of trend lines and of induction arise. For some systems, especially new systems, insufficient observations may be available to be able to determine trends and to extrapolate these with confidence. Furthermore, observations in one area or region may not be readily transferable to another location. These are, of course, generally well-known problems, but it is wise to keep them in mind.

Apart from the above problems, one wonders whether Lynam and Herdt's measure fully accounts for sustainability in the sense considered by Conway. Even over a long period of observations, the system may not have been subjected to some particular type of environmental stress which could occur, and which could permanently reduce the productivity of the system. The occurrence may, for example, be one that is expected to arise with a 2% probability, i.e. on average once in fifty years. Engineers often take such probabilities into account, but it may or may not be possible to test the impact of such a shock experimentally. In many cases, therefore, one must remain uncertain about the extent to which a system is sustainable, and one must be cautious about extrapolating past sustainability.

Integration of sustainability into meaningful economic measures

Already the article by Lynam and Herdt (1989) indicates some ways in which sustainability might take into account meaningful economic measures, even though limitations of such measures have been noted. However, in order to integrate sustainability appropriately into meaningful economic measures we must step back and ask - sustainability of what, and in what sense is it sustainable? Considerable confusion often arises from not considering this matter.

The sustainability of a characteristic of a system and a land management system itself

There appear to be at least two ways in which the term sustainability is used in relation to productive systems, including land management systems. It may be used (i) to describe a property or characteristic of the 'output' of the system; or (ii) to refer to whether the use or adoption of a system will be sustained or maintained. It is used in the former sense by Conway (1985, 1987) and by Lynam and Herdt (1989), whereas it appears to be used in the second (wider) sense in FESLM.

Even when sustainability is used in the second sense, it is in the first sense that it is still likely to be relevant in describing a land management system, because it may influence whether a system continues to be used or adopted. The third 'pillar' of FESLM - protection of the potential of natural resources - is confined to describing a property or characteristic of the 'output' of the system, as suggested in the first definition given above.

Sustainability - an ideal vs. the actual situation

It is also very important when evaluating systems of land management and their sustainability to decide whether the focus is to be on actual choices or decisions, or on ideal choices or decisions.

Decisions of the former type involve a primary focus on the actual situation (a 'positive' focus), and decisions of the latter type involve more considerations of what ought to be (a 'normative' focus). In economics, these are corresponding fields of study sometimes differentiated by terms of 'positive' economics and 'welfare' economics.

From a decision-making point of view, a positive study of land management systems involves studying the factors which lead to a particular system being chosen and being continued to be used. Particularly in modern economies, economic factors will play a large role in such decisions.

The land management systems adopted may not be ideal from a sustainability viewpoint or from some general welfare point of view. To discuss this matter systematically we need to define what the ideal is, and then to compare the actual situation with the ideal situation. Some important questions which arise in this connection are: (i) the degree of divergence between the ideal and the actual practice in land management, and the reasons for this divergence; (ii) the possibility of correcting divergence from the current state of knowledge with regard to the social potential for amelioration; and (iii) the extent to which land management would be less than ideal, even if current knowledge were to be efficiently used - a situation which would require long-term research and development, leading to the extension of the knowledge base.

It must also be recognized that some ideals may not be achievable. Consequently, perspectives in land management must be subject to practical considerations.

It is not absolutely certain whether FESLM is supposed to be a normative or a positive framework or both. If it is a normative framework, it implies that land management systems conforming to the pillars of the framework ought to be adopted and continue to be used. If it is a positive concept, it implies that land management systems satisfying its pillars will (most likely) be adopted and will continue in use. If it is both, then both consequences would follow. On the other hand, if it is a mixture of normative and positive positions, it is necessary to state which pillars are to be interpreted in which way. Thus further clarification appears to be required.

As mentioned earlier, there are some difficulties to be overcome if FESLM is interpreted normatively. The 'welfare' weights to put on the different pillars are likely to be a matter of controversy. Furthermore, simple and widely acceptable measures for all the characteristics of the pillars are not available.

Economic approaches to determining production

In assessing what ought to be done as far as systems of production are concerned, economists usually consider three approaches: (i) cost-benefit analysis (CBA), (ii) social or extended cost-benefit analysis, and (iii) safe minimum standards.

The most widely used approach is CBA, including social CBA. In this approach, no special weight is given to sustainability as such.

Using a variety of economic techniques and taking into account non-economic data (as essential ingredients), costs and benefits are expressed in money terms for the planning interval as a stream, or are estimated in value for each subperiod in the planning period, e.g. for each year, if CBA is used. Sustainability is not completely ignored in such an approach, but it is not a deciding factor. The aim is to adopt the system of land management which gives the highest (economic) benefit to cost ratio, and not to adopt any system for which benefits are less than cost.

CBA

Future costs and benefits in the CBA evaluation are not included at face value, but are usually reduced by a fraction, that is in economic terms (discounted to some extent). Economists usually discount these values by a larger fraction when the prevailing rates of interest are higher and the future values are more distant. The values estimated by economists in

104   Economics and Ecology in Agriculture and Marine Production

this way are called net present values. The lower the rate of interest allowed in these calculations, the greater will be the weight placed on future net benefits compared to current benefits, and consequently there is more emphasis on the long-term sustainability of benefits. Some economists have suggested that it might be appropriate to use a low interest rate or discount rate if sustainability is to be given a greater weight than in normal economic practice.

While traditional CBA does not put special weight on sustainability, it does not ignore the time path of net benefits from processes, and is likely to result in the choice of systems which are relatively sustainable compared to short-sighted profit maximizing. This is illustrated in Figure 2.

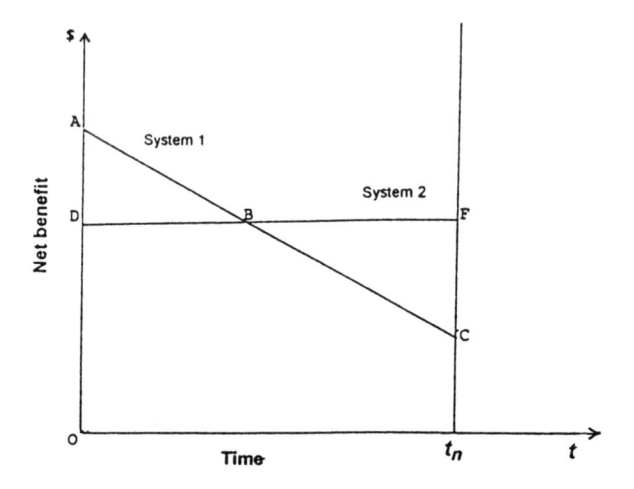

Figure 2. Streams of net benefit from alternative systems of land management.

In Figure 2, the relationship ABC represents the net benefit of adopting a particular land management system, i.e. System 1, on an area of land, and DBF represents the net benefits from an alternative system, i.e. System 2, on the same land. The planning period is supposed to extend from the present to t_n (e.g. 15 years into the future). Suppose that having adopted System 1, it is impossible to switch to System 2. A short-sighted profit-maximizing farmer will adopt System 1, the less sustainable system in terms of net benefits. However, CBA is unlikely to identify this as the optimal economic policy. Even though future net benefits are likely to be discounted by CBA, they will not be ignored, and so the value of System 2 using this approach is likely to exceed that of System 1. The precise conditions for one system rather than the other being recommended by CBA can be determined, but System 1 is only likely to be recommended in preference to System 2, if the interest rate is very high and the discount of future net benefits is therefore high. However, examples can easily be constructed in which economic discounting seems to militate against sustainability, and some are given by ACIL Economics and Policy Pty. Ltd. et al. (1994, p. 14).

Social CBA

The above discussion refers to private CBA, that is to the benefits obtained by the landholder. This may, however, fail to take account of the full social costs imposed, or of the full social benefits obtained from a system of land management. Social CBA and private CBA will only give the same results if there are no significant environmental spillovers or externalities from private decisions. If, for example, a system of land management in the headwaters of a river valley imposes costs on landholders downstream, this should be taken into account in social CBA. For example, clearing of vegetation and more frequent cultivation of land in the headwaters of a river valley may increase the frequency and severity of flooding downstream, and may result in the increased deposition of sand and gravel. This would involve a negative spillover or externality on downstream farming as a result of the form of land management adopted in the upland section of the river valley, and would be factored into social CBA as a cost in excess of the costs borne by landholders in the upland region. Unfavourable externalities or spillovers can help to reduce the sustainability of land management systems as a whole. In FESLM, it is possible that its framers had such factors in mind when they explained (in relation to productivity) that "the return from SLM may extend beyond material yields from agricultural and non-agricultural uses to include benefits from protective and aesthetic aims of land use" (Smyth et al., 1993, p. 8).

Safe minimum standards

The aim of all economic CBA is to express all costs and benefits in monetary units. However, the extent to which this can be done is limited by a number of factors, including uncertainty. This latter factor has led some economists to argue that the use of CBA should be subject to constraint or constraints. These usually take the form of safe minimum standards. It is argued that uncertainty is so great that decision-makers should err on the side of safety in natural resource conservation, especially because natural resource depletion is either not reversible or only reversible at great cost. In the case of soil, this might translate into constraints that particular characteristics of the soil should not be allowed to fall below specified threshold levels, so as to ensure safe minimum standards. However, this leaves open the question of how these thresholds are best determined. The determination of thresholds for the IBSRAM/ACIAR project, especially from an economic viewpoint, would be worthy of future attention.

Sustained use of land management systems and break-even points

FESLM may concentrate on whether in fact a particular system of land management is likely to be adopted and will be sustained in use. A system of land management is unlikely to be adopted and used continuously if it is uneconomic from the point of view of the landholder. Thus the economic break-even point for any land management system might be considered as a critical economic threshold (Coughlan, 1993). The break-even point for a system is the level at which gross receipts or benefits equal total costs. In terms of CBA, it is the situation in which the sum of discounted costs equals the sum of discounted benefits. It is certainly true that land management systems which fail to break even from an economic point of view will not be sustained.

While the above is true, a system which breaks even or yields a net benefit may not be sustained either. It is unlikely to be sustained, for instance, if a more profitable form of land management is discovered. The more profitable form of land management is likely to be adopted. This may take the form of a change from, say, grazing to the growing of bananas in the steeply sloping coastal areas of Queensland if the latter becomes more profitable, or from wheat to cotton on the Darling Downs for a similar reason. Variations in the prices of commodities (inputs or outputs) may also result in an alteration in the type of land management system used. This can create complications for economic thresholds, as observed in the Lethbridge study, Canada (Garreda et al., 1994). There may be economic thresholds at which it is profitable to change from one land management system to another. Consequently, more than one economic threshold can be relevant to the choice of land management system and changes in this choice.

Physical sustainability and economic viability

In the literature, there is some discussion of the extent to which the sustainability of land management systems depends on the sustainability of lower-order elements, which is to say the productivity of physical inputs (including natural resources) in the system. The physical productivity of a system is, for example, lower in the economic hierarchy than its net economic value. Does lack of sustainability of productivity imply that economic viability or economic sustainability is unlikely to be achieved? This is indeed so - unless (i) prices received for the output of this system rise sufficiently, or (ii) the prices of inputs fall enough to compensate for the fall in productivity, or (iii) the preexisting profit margin is already great enough to allow this fall in productivity to be absorbed without going into the red. While lack of sustainability in a lower order of a system does not always spell economic disaster, in many cases this is the ultimate consequence.

Possible future economic work in relation to the framework

Economics has an important position in the framework. In today's predominantly commercial world, it determines to a large extent the type of land management system adopted. Furthermore, even in subsistence or semisubsistence situations, benefits and effort involved in securing subsistence are likely to be balanced against one another, and so economy of effort in relation to benefit remains an issue.

Economics and normative aspects of FESLM

The way in which future economic work could contribute to the development and application of the framework will depend in part on how the framework is interpreted, e.g. the extent to which it is interpreted as a normative framework or as a positive (predictive) apparatus. Economic work can be of value for both interpretations, but one must always distinguish carefully between the normative and the positive point of view.

In the normative sphere, economics would be concerned with what ought to be in relation to land management. As such, it would draw on the field of welfare economics and new developments in environmental economics. The statements arising from such work are of a conditional nature because they are based on particular sets of ethical or value assumptions. Acceptance of the consequences drawn will depend upon acceptance of the underlying value judgments.

Economists usually try to adopt ethical or value judgments which have wide acceptance, but it is sometimes necessary to modify the framework developed, say, in Western countries when it is applied to other countries with different value systems, resulting in differences in social acceptability. The use of social CBA for example is based upon a number of ethical or value judgments. It provides a normative framework that has been widely used for evaluating economic projects and choices about production.

Some economists have argued in recent years that this technique needs to be supplemented by the use of safe minimum standards. The scope for integrating social CBA into the framework and supplementing it by safe minimum standards seems worthy of further investigation. In general, it would be useful to investigate what economics can specifically add to the normative dimension of FESLM.

Economics and predictive aspects of FESLM

The positive side of the issue concerns what choices are made about systems of land management and why they are made. To what extent are they made on a profitability or private CBA basis? What constraints prevent the most economic (i.e. profitable) form of land management being adopted? For instance, financial liquidity or the availability of credit may be a problem, or lack of knowledge may be a problem. The form of land ownership may also be an influence. How does economic uncertainty and instability affect the choice of a land management system? In the light of such economic considerations, what are the chances of a sustainable or relatively sustainable land management system being adopted? Taking into account the influence of economic factors on the choice of land management systems, what types of policies can governments adopt to foster the use of land management systems exhibiting greater sustainability? Is it desirable from a social welfare point of view (normative perspective) to foster such systems? All these questions should be given more attention.

Economics and thresholds or critical points in FESLM

It is also apparent that the notions of economic thresholds and economic viability in relation to land management need further attention. As for economic viability, one must consider it in relation to alternative forms of land use and in relation to particular time periods, e.g. is one concerned about economic viability in the short term or the long term. If the availability of finance is limited and savings are low, as is usually so in subsistence economics, short-term rather than long-term economic considerations are likely to predominate. Often economic viability is associated with whether or not a land management system enables an economic unit to at least break even. However, the break-even point is likely to be different in a short-term situation and in a long-term situation. If a land management system has already embarked on a short-term plan, much of the fixed investment in it can be a sunk or unrecoverable cost, and a landholder may continue with the system even if he/she does not recover capital costs.

From an economic point of view, it is desirable that a land management system should at least break even in the long term. The economic persistence of a land management system may differ in the short term and the long term. Thus the question of economic viability is really quite complicated, and different concepts of break-even in the economics of operations are possible. We also need to decide how opportunity costs (or the value of the best alternative foregone by choosing one alternative out of the set of available choices) is best factored into the economic viability pillar. Considerable scope exists for exploring the above aspects further and considering the specific indicators of economic viability, as suggested in the Lethbridge study in Alberta, Canada, (Garreda et al., 1994; Gomez, 1993b).

Overall scope for contributions from economics

In short, there is considerable scope for articulating the economic factors influencing the choice of land management systems and their sustainability, for linking the pillars of FESLM together in an economic framework (of the type summarized in Table 1) and for exploring the normative dimension of FESLM in terms of welfare economies and environmental economics. To some extent, these objectives may be best achieved by drawing on experience from case studies. If systems identified by FESLM as having greater sustainability are not being widely adopted, it may be useful to identify, through the case studies, the economic impediments to the adoption of these systems. In turn, this may enable appropriate policy action to be identified and taken to deal with the economic constraints.

Economic evaluation in the IBSRAM/ACIAR case studies

The potential value of FESLM is currently being tested in four case studies being partially financed by IBSRAM and ACIAR. There are two case studies for land management on Vertisols, and two for land management on sloping lands. The Vertisols are located in Queensland and Zimbabwe, and the sloping lands are in Queensland and the Philippines. Each study, then, involves a more-developed and a less-developed country, with corresponding respective commercial and subsistence or semi-subsistence approaches to land use. For each type of land, similarities exist in terms of rainfall and climate. The selection of the case studies was deliberate, and by covering a wide range of different situations have added to other case studies throughout the world - hence providing a substantial test for FESLM.

Impacts on economic viability

All the above factors affect the profitability (net present value) of a land management system and consequently they *all* influence economic viability. But economic viability does not depend on profitability alone; it also depends on the liquidity situation of the landholder and his/her savings and access to finance. Therefore, even though economic viability is influenced by each of the first three policies of FESLM, it depends on other factors as well, and notably prices, the availability of finance, liquidity, and savings.

Let us consider each case study in turn and the scope for socio-economic analysis in the light of FESLM.

Vertisols in Zimbabwe (Chisumbanje area)

Nyamudeza (1993) points out that most agriculture in the Chisumbanje area is of a subsistence or semi-subsistence type, with farmers only having a surplus to sell in one year in five on average. Farming is mainly of a dryland (rainfed) type, except on the Chisumbanje Estate (2500 ha), which grows cotton and wheat under irrigation. This also appears to be a commercial property which employs villagers from Chisumbanje on wages.

In relation to subsistence farming, rainfall is low and erratic. Low rainfall often leads to crop failures, and the possibility of crop failure is compounded by the increasing tendency to grow maize, which is less drought-resistant than sorghum. The choice of crops may partly be influenced by economic considerations (which could be further investigated). A social impediment exists to the growing of pearl millet, which is forbidden by tradition (Nyamudeza, 1993).

Livestock already play an important role in local communities in the Chisumbanje area. According to Nyamudeza (1993), cattle are kept for draught power, cash, and milk, and goats are kept for cash, and meat. Cattle provide virtually all the draught power for subsistence cropping, and the availability of such draught power provides a limit to the extent of cropping. Because of human population increases in the area, there has been a need to raise crop production. In turn, this has required an increase in cattle for the provision of draught power - otherwise there may also have been a strong urge to increase livestock numbers. The consequence has been that the amount of grazing land has been reduced, and stocking rates have risen, adding seriously to degradation of grazing lands. Furthermore, because of these increased pressures, the practice of feeding crop residues to livestock has commenced and appears to be spreading. In the long term, this is likely to reduce the fertility of cropped lands (e.g. via soil erosion, loss of plant nutrients, and reduced organic matter). Given the increased pressures to which the land is being subjected as a result of population increase, its natural productivity is in decline, and may very well be declining at a faster than optimal rate.

In this case, human population growth appears to be the basic problem, but it is not a problem that can easily be overcome in the short and medium term. Possibly the land manager has to work on the basis that population in the area will continue to rise for some time yet, and the aim should therefore be to develop land management systems which take this as a given when paying particular attention to sustaining yields. It may even be that the preference for maize as opposed to sorghum is irreversible. If so, this might indicate the need to concentrate research on methods of maize production which go some way towards satisfying FESLM - rather than to promote sorghum research. Sorghum seems to be an inherently more sustainable crop in this area than maize.

The problem of sustainability of subsistence land use in countries like Zimbabwe is a complex communal one, of the sort which is common in many developing countries. The draught power constraints could in principal be overcome by mechanization, and this would reduce some pressures on the land from cattle. However, poverty does not make this feasible for most subsistence farmers. Once again, deforestation is a problem in the area due to overgrazing and firewood needs. The provision of electricity and other forms of 'industrial' power could in principle overcome the last problem, but the income and wealth base of the area may be inadequate to make this possible. An additional problem is the form of land ownership. Economists believe that property rights have important consequences for conservation of natural resources, and that under a system of common (open-access) property there is little or no incentive for conservation. This applies to land as well as to other natural resources. Nyamudeza (1993, p.3) points out: "Small-scale farmers in the Chisumbanje area commonly own the land (own the land in common). This is the case with the majority of people in Zimbabwe. Ownership of land strongly determines the sustainability of land-use systems."

In the case of the area under irrigation, other problems for sustainability are becoming evident, such as salinity and waterlogging. While wheat yields on the Chisumbanje Estate appear to have been maintained over a 25-year period, the trend in cotton yields over a 26-year period has been downward. The reasons for this, and its economic consequences need more investigation. It seems that additional land will be brought under irrigation with the completion of the Osborne Dam, which will displace some small-scale farmers growing rainfed crops. This will enable sugarcane to be grown, and the area involved in cotton/wheat rotation to be extended. Consequently, the sustainability of land management in areas under irrigation will be a growing issue, and can be subjected to economic analysis.

Nyamudeza (1993) lists a number of practices that could add to the sustainability of production in Chisumbanje. The economic constraints to the adoption of these practices (the economic prospects for their adoption) should be further investigated. The general economic viability of adopting recommended practices needs to be considered.

Vertisols in Queensland, Australia (Darling Downs)

Practically all farming in Australia is of a cash/commercial type. Furthermore, while there is virtually no use of commercial off-farm inputs in subsistence farming in Zimbabwe, this is not the case for commercial farming in Australia. In the Darling Downs area, which contains the major city of Toowoomba, medium to high levels of off-farm inputs are used. This can clearly add to business risks if prices received for production or yields vary unexpectedly.

Many farms on the Darling Downs specialize in commercial crop production (some entirely) growing crops such as wheat, sorghum, barley, maize, oilseeds, grain legumes and cotton (Dalal, 1993). However, livestock production (e.g. cattle-based livestock production) is important in some cases.

Dalal (1993) points out that cultivation on Vertisols in the Darling Downs area is dependent mainly on natural rainfall. Factors associated with long-term decline in the fertility of Vertisols under continuous cultivation and cereal cropping in the semi- and subtropical areas of Queensland have been identified (e.g. declining organic matter and the availability of nitrogen). He suggests that: "There is an urgent need, therefore, to adopt fertility restorative practices to maintain sustainable land use. The management options for sustainable land use being examined in the long-term experiment at Warra (Warra Experiment) include alternate grass-legume ley with wheat every four years; alternating various pasture legumes or a grain legume with wheat annually; zero-tillage; and nitrogen fertilizer application" (Dalal, 1993, p.3).

The economic viability of each of these practices would be worthy of investigation. There may also be other land-conservation practices (not being researched at Warra, near Dalby) worthy of investigation from an economic viewpoint for the Darling Downs as a whole, e.g. contour cultivation, and contour banks on sloping land. Dalal (1993) suggests that possibly the clearing of native tree cover has added to problems of land sustainability on the Darling Downs, and that possibly defensive measures might be considered (e.g. contour banks with windbreaks).

Several new developments have occurred on the Darling Downs. Cotton production appears to have expanded following a number of relatively dry years and favourable cotton prices. The implications of this for land sustainability may also be usefully investigated from the point of view of economics. For commercial enterprises of the size of some of the farms in the Darling Downs, the availability of finance and the maintenance of liquidity are major influences affecting economic viability. In Australia generally, farm size is an important influence on economic viability.

Sloping lands in the Philippines (General Santos, Mindanao)

The importance of sloping lands in the Philippines for agriculture (especially uplands) has been documented by Gomez (1993a). Much of this area is under cultivation and deforested, which means that it is subject to considerable soil erosion during wet seasons.

Most of the farmers on sloping lands in the Philippines are engaged in subsistence farming with very little off-farm input. As in Zimbabwe, increasing human population is placing mounting pressure on land resources. Gomez (1993a, pp.4-5) summarizes the problem as follows: "... the need to support an increasing population makes the problem much more difficult. Sloping land farmers usually have poor access to market. Distance to market is usually far and farm-to-market roads are in poor condition. The natural tendency, therefore, is to grow annual cereals and vegetable crops that can be used directly as food. Thus, perennial trees are cut and annual crops are cultivated, resulting in a land management that increases loss of soil and water, reduces productivity, and erodes sustainability. Thus the vicious cycle continues. Trees are cut, annual crops are planted, soil and water resources are lost, productivity is reduced, then more trees need to be cut. The result: an ever-increasing level of unsustainable land management in the sloping land."

Gomez lists four factors that might lead to greater sustainability of sloping land management in the Philippines. In summary these are: (i) the installation of soil- and water-conserving barriers along the contours of slopes planted to annual crops; (ii) a shift to high value perennial crops; (iii) a combination of production and processing at the farm level; and (iv) the improvement of market access.

Gomez (1993a) gives detailed reasons why such developments could lead to greater sustainability of land management. The economic constraints on such developments are worthy of consideration, and the economic prospects of their adoption could be explored further.

Sloping lands, coastal Queensland, Australian (Gympie)

Dwyer and Deuter (1993) identify six major systems of land use in the steeplands of coastal Queensland. They are: (i) annual crops (vegetables, sugarcane), (ii) permanent nonmulched crops (papaws, bananas, pineapples), (iii) permanent mulched crops (avocados, macadamia nuts), (iv) grazing (beef, dairying), (v) forestry (native hardwood, plantation softwood), and (vi) rural residential (low to medium density housing). The levels of degradation in these land-use systems increase with the frequency and intensity of cultivation. The three cropping systems (i-iii), have the highest levels of degradation with annual or frequently cultivated crops most at risk. Soil losses of between 85 and 300 t ha^{-1} yr^{-1} have been measured in newly planted pineapples and sugarcane, the higher soil losses occurring in the wetter northern tropics (Dwyer and Deuter, 1993, p. 3).

As far as horticulture on steepland in Queensland is concerned, annual crops and nonmulched permanent crops dominate as a percentage of steepland allocated to horticulture; so actually the crops which cause the greatest rate of soil degradation predominate. Possibly this is so for economic reasons, and could be a subject for further investigation.

Incidentally, most Australian coastal landholders have relatively easy access to markets. At least in the Australian case, access to markets (unlike the situation in the Philippines) does not appear to have made for greater sustainability. Since banana production appears to be expanding in Queensland, the sustainability issue on steeplands could magnify.

One matter worth investigating is how landholders plan to cope with land degradation. Is it their plan, for example, when the land becomes too degraded to produce profitably annual crops or nonmulched permanent crops, to switch profitably to mulched permanent crops, grazing, or forestry? What is their long-term economic plan?

Furthermore, it may be worthwhile to try to obtain more accurate measures of the off-site economic costs of such land degradation, and notably the economic spillovers or externalities associated with it, especially water-quality issues. Subsequently, appropriate government policies to control this land degradation could be considered.

Concluding comments

FESLM provides a valuable framework for the evaluation of land management systems because it emphasizes the holistic nature of decisions about land management. All of the pillars mentioned in FESLM, either directly or indirectly, have an impact on decisions about land management - and all can be important for the sustainability of a land management system. As far as the framework is concerned, it has been suggested that the extent to which it is to be interpreted as a normative or a predictive structure needs more clarification. Economics and economic considerations can be of value in developing either interpretation.

It has been observed that the term 'sustainability' has been used in a variety of ways in economics and in describing agricultural systems. In FESLM, it refers to the maintenance of a land management system. In some other interpretations, it refers to the maintenance of a property of the system, e.g. the productivity of the system or the natural potential of the land to be productive. Both concepts have value, but need to be carefully distinguished. The latter concept is probably most closely mirrored by the third pillar of FESLM - protection of the potential of natural resources.

In my view, an important question to consider is the economic prospects for adoption of available measures to protect the natural potential of land. Alternatively, what are the economic impediments to the adoption of such protective land measures? Case studies to date show that there are important socioeconomic impediments to their adoption, and that the maintenance of some land management systems would be more sustainable from a natural productivity point of view than those already in use. In many cases, the investigation of the sustainability of a land management system must extend beyond the farm to the local community, and even further afield.

Economics, like FESLM, does not give cut-and-dried answers. It merely provides a set of tools to be used in analysis. Furthermore, no single operational concept of sustainability is available from economics. While some agricultural economists (Lynam and Herdt, 1989) have proposed a relatively simple measure of sustainability, their measure has some difficulties from an operational viewpoint (as was discussed earlier).

There appears to be more scope for measuring how CBA, social CBA and the use of safe minimum standards can be integrated with FESLM. The concept of economic thresholds also seems worthy of further elaboration in relation to the sustainability of land management systems. Such thresholds are not as simple as they may at first appear. Furthermore, if the analysis is to be broadened beyond the farming unit, economic/environmental spillovers and externalities will need to be integrated into the framework. The case studies in Australia, the Philippines, and Zimbabwe provide a valuable empirical basis for considering the feasibility of such integration.

Acknowledgments

I wish to thank Dr. Kep Coughlan for helpful suggestions and references concerning this topic, and Dr. J. Keith Syers for providing valuable guidance and copies of essential reference material. I would also like to express my appreciation to the International Board for Soil Research and Management (IBSRAM), who commissioned this paper through TEMTAC Pty. Ltd., and to acknowledge the cooperation of TEMTAC in producing this survey.

References

ACIL Economics and Policy Pty. Ltd., Fievez, P., Patterson, R. and Wylie, P. 1994. *A Program of R&D into indicators of sustainable crop production systems.* Canberra, Australia: Land and Water Resources Research and Development Corporation.

Alauddin, M. and Tisdell, C. 1991. *The 'green revolution' and economic development.* London: Macmillan.

Conway, G.R. 1985. Agroecosystem analysis. *Agricultural Administration* 20: 31-35.

Conway, G.R. 1987. The properties of agroecosystems. *Agricultural Systems* 24: 95-117.

Coughlan, K.J. 1993. Component research input to FESLM - an example from soil erosion research. Griffith University, Brisbane, Australia. Mimeo.

Dalal, R. 1993. Towards an evaluation of cultivated Vertisols in Queensland. Department of Primary Industries, Warra Experimental Station, Australia. Mimeo.

Dwyer, G. and Deuter, P. 1993. Towards an evaluation of sustainability of steeplands in coastal Queensland - the Lethbridge workshop approach. Department of Primary Industries, Gympie, Australia. Mimeo.

Garreda, S. and Dumanski, J. 1994. Framework for evaluation of sustainable land management: Case studies of two rainfed cereal-livestock land use systems in Canada. In: *Transactions of the XVth World Congress of Soil Science,* vol. 6a, 410-421. Acapulco, Mexico: ISSS.

Gomez, A.A. 1993a. The sloping lands of the Philippines. Southeast Asian Regional Center for Graduate Study and Research in Agriculture. Los Baños, Philippines. Mimeo.

Gomez, A.A. 1993b. The Lethbridge approach to evaluating the sustainability of land use in the sloping lands of the Philippines. Southeast Asian Regional Center for Graduate Study and Research in Agriculture, Los Baños, Philippines. Mimeo.

Lynam, J.F. and Herdt, R.W. 1989. Sense and sustainability: sustainability as a objective in international agricultural research. *Agricultural Economics* 3: 381-398.

Menz, K. 1993. On economics and the framework for evaluation for sustainable land management (FESLM). Australian Centre for International Agricultural Research, Canberra, Australia. Mimeo.

Nyamudeza, P. 1993. Evaluation of Vertisols in Zimbabwe. Save Valley Experiment Station, Chipinge, Zimbabwe. Mimeo.

Pearce, D., Markandya, A. and Barbier, E.B. 1989. *Blueprint for a green economy.* London: Earthscan Publications.

Pushparjah, E. and Syers, J.K. 1993. Background paper on indicators of sustainability. International Board for Soil Research and Management, Bangkok, Thailand. Mimeo.

Smyth, A.J., Dumanski, J.K., Spendjian, G., Swift, M.J. and Thorton, P.K. 1993. *FESLM: An international framework for evaluating sustainable land management.* World Soil Resources Reports no. 73. Land and Water Development Division, Food and Agriculture Organization of the United Nations. Rome: FAO.

Tietenberg, T. 1988. *Environmental and natural resource economics*, 2d ed. Glenview, IL: Scott, Foresman & Co.

Tisdell, C.A. 1993. *Environmental economics: policies for environmental management and sustainable development.* Aldershot, UK: Edward Elgar.

WCED (World Commission on Environment and Development). 1987. *Our common future.* Oxford, UK: Oxford University Press.

[9]
Agricultural Sustainability and Conservation of Biodiversity: Competing Policies and Paradigms

Clem Tisdell

INTRODUCTION

Sustainability issues and matters involving the conservation of biodiversity are no longer new subjects for environmental policy. Nevertheless, they are of continuing interest, there are always new dimensions to consider and unresolved questions remain. In fact, it may only be now that we are starting to have a satisfactory overview of these subjects which have been intensively considered for around two decades. This paper provides an overview of these subjects, paying particular attention to agriculture.

In it the following are considered:

1. Different broad approaches to policy making and implementation applied to environmental policies.
2. Different concepts of and views about sustainability and biodiversity and their dissimilar policy implications.
3. Agricultural sustainability as a concept and as a goal and policies to achieve agricultural sustainability.
4. Important relationships between agriculture and biodiversity.
5. Reasons for sustaining biodiversity, possible methods for doing so and their implications for agriculture.

Let us consider each of these matters in turn.

APPROACHES TO POLICY FORMATION AND IMPLEMENTATION AND THEIR APPLICATION TO ENVIRONMENTAL POLICIES

To a considerable extent mainstream approaches to economic policy, including development policies, have tended to be technocratic. To some extent, this is a natural consequence of econometric model building and the use of mathematical economic models of a relatively deterministic nature. Such models make no allowance for the unexpected and they implicitly suppose a high degree of knowledge on the part of the model builders. Such models can easily become the handmaiden of top down policies and generate mechanistic approaches to economic growth, economic development and environmental policy with unfortunate consequences. The economic literature is not lacking in examples (see Tisdell 1990, Ch. 3).

With growing interest in economics of sustainable development, in ecological economics and in evolutionary economics and with the progressive acceptance that individuals (organizations and groups) are bounded in their rationality (Tisdell 1996a), the limitations of the technocratic approach to policy making and implementation of policies have become more obvious. These considerations have highlighted irreversibility and hysteresis, uncertainty and learning, institutional arrangements, the degree of motivation of actors and several other factors as having an important bearing on successful policy formulation and implementation. Such factors, often overlooked in mainstream economic theory, frequently play an integral part in the success or failure of environmental policy and suggest the relevance of models of an organic rather than a mechanical type.

Those who produce and try to implement policy recommendations in a mechanical manner face serious shortcomings, especially in relation to environmental policy and development. This is clear from recent demands for greater participation of local people in devising environmental policies affecting them, especially if they are required to implement these policies in their local area. Demands for the increased political empowerment of local communities and groups affected by policy making have become commonplace in recent times and academics have become increasingly interested in communitarianism (Etzioni 1991).

Appropriate links between the local community and central policy makers may be important for several reasons. For example, such links may be required to improve the environmental knowledge set of both parties, be needed to motivate the carrying out of policies as planned and may be required to provide appropriate feedback of knowledge between the groups involved and to supply effective governance. These links between policy makers and those affected are important from a motivational and a networking point of view. At the same time, it must be realized that costs are involved in networking and in participatory approaches to policy creation and implementation. Consequently, from a restricted economic viewpoint, participatory policy approaches should only be carried to the point where the additional benefits from these equals the additional cost of such institutional arrangements (Baumol and Quandt 1964 and Coase 1937). From a slightly wider perspective, participation might however, be carried further, for example, when account is taken of factors such as self esteem, sense of belonging. There is still much to be learnt about this area of policy making.

In the development studies literature reference is sometimes made to 'top down' and to 'bottom up' approaches to policy with the latter being preferred by those desiring to empower local communities or other social groups. Another possibility is a 'side by side' approach which involves a joint effort by (central) government and local communities in the policy arena (Tisdell 1995b). These three types are, however, gross simplifications. Nevertheless, even if we keep to the possibility of only 'top down' or 'bottom up' governance, a number of different situations can be imagined once it is realized that a dichotomy is possible between policy formulation and its implementation. For example, policy may be formulated centrally but may be required to be implemented locally. Four possibilities are shown in Table 5.1.

Table 5.1. Central and local responsibility for policies affecting local communities: An initial classification of possibilities

Combination	Party responsible for policy formation	Party responsible for policy implementation
1	Central	Central
2	Central	Local
3	Local	Local
4	Local	Central

Possibility one shown in Table 5.1 is the most centralized one and may show little or no regard to the wishes of the local community. However, Table 5.1 should be extended by taking into account a side by side approach as an additional possibility. If this is done, the additional cases set out in Table 5.2 could arise. In this table, case 5 involves the greatest degree of joint participation by groups.

Table 5.2. Combinations to be added to those in Table 5.1 to allow for side by side approaches to policy

Combination	Responsibility for policy formation	Responsibility for policy implementation
1	Side by side	Central
2	Side by side	Local
3	Central	Side by side
4	Local	Side by side
5	Side by side	Side by side

It should be noted that this classificatory system glosses over a whole range of complexities, many of which should be taken into account in refined analysis of the issues involved. Nevertheless, it makes it clear that even at a relatively superficial level, we need to go beyond the 'top down' and 'bottom up' classification in order to assess the desirability of alternative systems of government and policy creation.

Both for sustainable development, sustainability of land use and conservation of biodiversity, involvement in and empowerment of local communities in policy matters is seen as very important by conservation groups. They believe that such involvement will promote sustainability and be a positive force in conserving biodiversity. In reality, however, the empirical evidence is mixed. For example, with the devolution of control over protected areas from central to local authorities with the demise of the Soviet Union, economic exploitation of these areas has been commenced by some local authorities. Some 'protected' areas are being used for the grazing of livestock and timber is being extracted from others for example. Decentralization of government has proceeded quickly in the Philippines. There is fear that some local politicians will use their enhanced political power to exploit (to their advantage) local natural resources unsustainably. On the other hand, the CAMPFIRE programme has been a success in some parts of Zimbabwe as far as the conservation of elephants are concerned. This programme involves controlled devolution of power and provision of economic rewards to local communities for nature conservation. Loss of local political power has undermined conservation in some countries. Mishra (1982) for example, reports that the replacement of village control over forests by central control exercised by the Forestry Department in Katmandu undermined forest conservation in Nepal. Therefore, it is clear from the conflicting empirical evidence that centralization versus decentralization is only one factor influencing the likelihood of policies being adopted which favour sustainable development and the conservation of nature. Yet it is an important consideration from a conservation point of view. Consider another environmental example.

There is growing interest in social forestry in a number of developing countries and the possibility of harmoniously combining forestry, agriculture and even aquaculture in an integrated system designed to enhance sustainability of land use and counteract unfavourable externalities from agriculture, such as soil erosion. This is especially important on sloping lands. It is of considerable policy significance in many parts of Asia, for example, Northern India and Southwest China.

In China, as in many other countries, afforestation has often been a centralist initiative involving top down decisions and implementation. Plans to afforest hilly areas above the planned Three Gorges Dam on the Yangtze

River are of this nature. Such schemes may have little support at the local village level because they are seen as imposed and they may not be designed with local benefits in mind or may be drawn up without adequate assessment of methods to maximize these benefits. In the longer term, social forestry and agriculture schemes are likely to be more sustainable and effective in achieving the conservation goals being sought than state-imposed forestry. This will be especially so if local people have economic incentives to sustain social forestry and agro-forestry projects.

China appears to be moving towards a less centralist (top down) position in designing and implementing environmental conservation projects. For example, with World Bank support, China has undertaken rapid rural appraisal in Xishuangbanna Prefecture, Yunnan, to identify development projects that may be valued by villagers living near Xishuangbanna State Nature Reserve (Xiang 1995). It is proposed to use this appraisal to identify projects likely to be welcomed by villagers and to offer government aid for these. It is hoped that in return for such aid that villagers will agree to refrain from illegal exploitation of the Nature Reserve. Furthermore, assuming that the projects are an economic success, they will increase village incomes and thus reduce the economic need of villagers to use the Reserve illegally.

One cannot be certain yet whether this new policy will be a success. However, the discussion has raised holistic dimensions of environmental policy which are normally not given much attention in conventional economics. In the final analysis, all policy proposals need to be assessed on a holistic basis.

SUSTAINABILITY AND BIODIVERSITY: ALTERNATIVE CONCEPTS AND THEIR POLICY IMPLICATIONS

Many concepts of sustainability relevant to economic theory and environmental policy exist in the literature and the majority require a holistic approach to policy making. Furthermore, even the concept of biodiversity is not cut and dried. Depending on the measures chosen and on the dimensions of biodiversity stressed, different policy consequences may follow. The same is true of sustainability.

For some, the wide range of concepts of sustainability present in the literature have become a source of confusion and have led to doubts about the value of such concepts. Indeed, some use the term 'sustainability' in a value laden way and propagate the idea that things which are sustainable are desirable. This is clearly unsatisfactory because there are evil or unsavoury things such as poverty and disease which few would believe it desirable to sustain. On the other hand, there are characteristics the sustainability of which might be welcomed such as sustainability of levels of income or of biodiversity. The concept of sustainability gains meaning when it is related to an object such as income, level of returns, biodiversity or community.

Sustainability relates to the ability of a characteristic to maintain itself, that is not to decline with the passage of time. An unsustainable characteristic or variable may decline in varied ways or for varied reasons and the difference between these can be policy relevant.

The decline might for example be due to endogenous factors and might be regular, for example, a regular decline in crop yield due to falling soil fertility because of nutrient mining as a result of the type of cropping engaged in. Or, it might only occur after an exogenous shock. For example as suggested by Conway (1985 and 1987) as a result of ecological stress or an environmental shock, the yield from an agricultural system may be depressed and fail to recover fully once the shock has passed. Such systems are said to lack *resilience*. However, resilience is clearly not the only factor to be taken into account when assessing the sustainability of a system, such as yield or returns from an agricultural system.

An equally important characteristic may be the *robustness* of the system, that is the ability of a system to withstand a shock without being deflected from its path or being significantly deflected from its path (Tisdell 1994a). Some systems require larger minimum sizes of shocks to be deflected than others, deflect differently and so on. It is possible for a system to be robust and not to be resilient or to be resilient but not to be robust. If one has to choose between such systems, how will one choose from a sustainability point of view? The point is that in comparing the desirability of techniques or systems sustainability characteristics additional to resilience should be taken into account. For emphasis, Table 5.3 sets out the four possibilities as far as the resilience and robustness of systems are concerned. However, additional complications can emerge, some of which are mentioned in Tisdell (1994a).

Table 5.3. Resilience and robustness of a characteristic of a system

Possibility	Resilient	Robust
1	Yes	Yes
2	Yes	No
3	No	Yes
4	No	No

Knowledge of the sustainability or otherwise of welfare, or value significant variables can be of practical policy importance from several viewpoints. If, for example, sustainability of such a variable is of positive value and if the variable is predicted to decline, this information may allow;

1. defensive measures to be searched for and taken to avert the decline, or
2. if no such measures can be found and lack of sustainability is inevitable, the information may allow planned adjustment to the decline.

In some cases, both responses may be activated by the knowledge of sustainability problems. Methods may be sought which moderate the decline in a target variable and forward planning may occur to adjust to unavoidable decline.

Again, biodiversity, like sustainability of a variable, is not a straightforward concept and is multidimensional. It may refer to genetic diversity, species diversity, or ecosystem variation and there is far from complete agreement about how biodiversity is best measured (Pearce and Moran 1994, Ch. 1). Again the concepts have varying policy consequences. For example, if a tradeoff is required between ecosystem and species diversity, which should be preferred? Depending upon the type of measurement of biodiversity selected, the focus of policy for biodiversity conservation is likely to be different. If, for instance, genetic variability (within species) is stressed more than that variety of species, species may be more likely to be sacrificed (if necessary) to conserve genetic variability within surviving species. Methods of measurement by directing the focus of attention of researchers to particular characteristics often coincidently bias their policy prescriptions.

The concept of sustainable development and the related concept of ecologically sustainable development continues to be an important backdrop to environmental policy making. For some individuals, conservation of biodiversity is a prerequisite for sustainable development. However, even if we take the most common definition of sustainable development used in economics (namely that it is development ensuring that the incomes, or more generally living standards, of future generations are no less than those of current generations), the concept needs to be fully comprehended if it is to result in appropriate policies. This is so even leaving aside some philosophical difficulties involved in the concept and inadequate attention to the population variable by its exponents (Tisdell 1993, Ch. 10).

Taking the economic concept of sustainable development as given, much of the policy debate has become centred on determining the types of capital stock that might allow sustainable development to be achieved. Man-made capital stock can include physical capital, knowledge, human capital and institutional capital. All of these involve an investment which may be 'funded' by reducing the natural capital stock. In addition, physical capital normally embodies a part of the natural environmental stock. The following question is important: To what extent can man-made capital be substituted for the natural environmental stock and sustainable development still be achieved?

As is well known, a spectrum of views exist about this matter. On the one hand, some believe that the process of substitution of man-made capital for the natural environmental stock can continue without any significant threat to sustainable development. Supporters of this view place relatively *weak conditions* on conservation of the environment but often include the prescription that environmental externalities or spillovers should be taken into account. Indeed, it is possible that some on this side of the spectrum see continuing conversion of natural environmental resource stocks to man-made capital as essential for sustainable development.

On the other side of the spectrum are those who see continuing conversion of the natural environmental resource stock to man-made capital as a serious threat to sustainable development. They point out that the life of man-made capital is relatively short, for example, physical capital, compared to the natural environmental resource stock.[1] Furthermore, the natural resource stock itself provides productive services and in many cases consumptive ones. The destruction of natural capital will inevitably lead to loss of these services and undermine incomes both;

1. directly because fewer environmental services are available for direct consumption and
2. indirectly because the productivity of man-made capital is likely to be reduced once the relative size of the natural environmental resource stock is significantly reduced. In other words, increasing imbalance between factor proportions (an increase in man-made capital *relative* to natural environmental capital) will eventually result in reduced productivity and income.

Some supporters of the above view claim that already the substitution process has reached this critical stage and believe that *strong sustainability* conditions should be imposed to conserve the remaining natural environmental stock. They argue that it is not sufficient to make sure that environmental externalities are fully taken into account. They would favour externalities being taken into account in project evaluation but, in addition, usually want environmental *offset* policies to be implemented so as to keep the environmental resource stock constant.

Pearce (1993) describes those who favour weak sustainability conditions as growth optimists and those who favour strong sustainability conditions as 'dark greens'. However, this does not effectively distinguish between different types of 'dark greens'.

---

[1] In addition, note that the natural environmental resource stock is to a large extent self-reproducing or sustainable but this is not true of man-made capital in the same way. It lacks the degree of autonomy of natural capital in perpetuating itself.

Strong sustainability conditions may be supported for one or both of the following reasons:

1. They are believed to be necessary to ensure that the incomes of future generations do not fall below those of the present. This is a *positive* basis given that the intergenerational equity objective is accepted and that only human beings are to account.
2. Strong sustainability conditions may be supported for a *normative* reason. Some individuals believe that mankind has an obligation to help conserve God's natural creation, especially the living environment. Those holding this view usually support strong sustainability conditions. They are increasingly likely to do so as the availability of natural environments is reduced.

There are a number of variants on the capital substitutability theme. In one simplified version, no serious problem occurs until environmental resources are reduced to a core. However, once the core is reached serious problems arise for sustainability. If the theory is correct, the problem is to identify the core. To what extent for example is biodiversity in the core? To what extent can biodiversity be foregone and incomes be sustained? What is the nature of the core? Is it fuzzy or not, changeable or not? Are there regional cores and a global core? Can they be identified?

As will be noticed, the type of problem being raised here is the nature of the functional relationships involved when man-made capital is substituted for the natural resource stock. For instance are the effects on the sustainability of income continuous or discontinuous, positive, positive up to a point and then negative and so on? The nature of the relationship is of considerable importance from the viewpoint of policy. Furthermore, the fact that the relationship itself is uncertain can have policy implications. Consider for example the precautionary principle. If one believes that preservation of an environmental core is necessary for economic sustainability, but cannot determine the core exactly, one may be inclined to adopt the minimax loss strategy of conserving the environmental stock to a greater extent than is strictly necessary so as to make sure that it contains the core. As time goes on, greater knowledge may be obtained about what constitutes the core. This strategy retains flexibility which is likely to be optimal if irreversibility is present and if learning is expected to take place.

Turning to a slightly different matter, much has been made of the total economic valuation concept. In some respects possibly too much, even though it is a considerable advance on earlier narrower economic practice in relation to valuation. The main problem, as I see it, is the naive belief that it is imbued with superior moral standing. In reality, however, it is limited in its moral dimensions. It is essentially man based and the measuring rod of money used is subject to distortions. It does not satisfy those 'dark greens' who believe that mankind has a moral obligation, at least to some extent, to conserve nature independently of man centred wishes. This group also presumably believes that economic sustainability is not a supreme virtue and in fact, if necessary, would be willing to sacrifice economic sustainability to some extent for conservation of nature. In other words, restrictions on satisfaction of man centred economic welfare are favoured on moral grounds to preserve nature if a tradeoff is required, as some believe is necessary.

The total economic valuation concept encounters difficulties also when individuals possess dual or multiple utility or preference functions (Margolis 1982, Kohn 1993). For example, it is conceivable for an individual to have a self centred preference function and another incorporating a wider moral dimension(s) (Etzioni 1991). In relation to valuation, which of these should be afforded primacy? They could have very different policy implications. An interesting side issue is which of these multiple utility functions will politicians try to satisfy? The different utility functions of individuals change with time and circumstances. What factors determine the formation of such functions and their changing importance over time?[2] What is the consequence of different political arrangements and institutional setups for the influence on policy of these moral dimensions? If it is accepted that multiple utility or preference functions exist for the same individual based on different moralities, then it is clear that it is necessary to go beyond the total economic valuation concept.

AGRICULTURAL SUSTAINABILITY

Considerable discussion of agricultural sustainability and of sustainability of rural communities has occurred in the literature. However, whether sustainability of agricultural activities (or those of any particular industry) is desirable is a moot point. If sustainable development is adopted as the main goal (in the sense that the incomes of future generations should not be less than those of present generations), agricultural activities or other activities will only be considered desirable from a social point of view if they contribute to the main goal. Nevertheless, they could be important from the viewpoint of agricultural communities and could be policy relevant in practice. Politically,

---

[2] It is recognized that institutions may mould preferences to some extent (Kelso 1977). What exactly is their role in this regard? Do they help to establish 'extra' preference functions for individuals or change the degree of dominance or prominence of particular sets of preference functions which an individual may have?

policy makers may be forced to give special attention to the situation of agriculturalists, for example, because of the nature of the voting system.

Sustainability of agricultural characteristics may be of relevance to policy makers from at least two points of view:

1. If lack of sustainability of an agricultural characteristic, such as yields or incomes, is predicted it may be possible to adopt measures to avoid these, for instance commence research to discover ways to avoid the problem.
2. If agricultural sustainability cannot be achieved, then knowledge of this may enable suitable adjustment policies to be devised.

In other words, 'to be forewarned is to be forearmed'. Thus agricultural sustainability, in various forms, is relevant for policy purposes.

Several attempts have been made to specify agricultural sustainability with precision, provide criteria for its evaluation and suggest measures for it. The results have been mixed. There has been a tendency for writers to concentrate on the sustainability of different characteristics. Those chosen in many cases seemingly reflect the values of the individuals involved in choosing them. Given that sustainability often involves a variety of characteristics, simple measures of it usually fail to receive widespread support.

Lynam and Herdt (1989), for example, attempt to measure the sustainability of an agricultural system by the trend in the ratio of the value of agricultural output from the system divided by the value of inputs used by it. If the trend is non-declining, the system is said to be sustainable. This indicator will be non-declining if the value of output minus the value of input is not declining, that is if net income from the system is not declining.

This measure has a number of limitations (Tisdell 1996b). Firstly, past trends cannot necessarily be extrapolated. Second, economic viability depends upon biophysical and market factors. It is possible for yields to be declining and for profitability to be improving for market reasons. For some, this might not be regarded as a sustainable system.[3] In any case, in this circumstance the system would most likely violate strong conditions for sustainable development. Third, it is unclear from the Lynam and Herdt (1989) formula how opportunity cost and opportunity return are taken into account. For example, the net income from use of a technique may be non-declining but it may be relatively unprofitable to continue with its use because an alternative technique gives a higher net income or rate of return. Use of the former technique fails to be sustained for economic reasons.

It is useful to consider the Framework for the Evaluation of Sustainable Land Management (FESLM) suggested by the Food and Agricultural

---

[3] This is true for the Framework for Evaluation of Sustainable Land Management outlined below.

Organization (FAO) and the International Board for Soil Research and Management (IBSRAM).

The FELSM Working Party declared that:

> Sustainable land management combines technologies, policies and activities aimed at integrating socio-economic principles with environmental concerns so as to simultaneously:
> - maintain or enhance production/services (productivity)
> - reduce the level of production risk (security)
> - protect the potential of natural resources and prevent degradation of soil and water quality (protection)
> - be economically viable (viability)
> - and socially acceptable (acceptability).
> (quoted in Smyth *et al.* 1993, p. 7).

As indicated above, it would be possible for Lynam and Herdt's (1989) criterion to be satisfied and for FESLM not to be. It is *possible* for productivity to decline and economic viability not to decrease; or for natural resources to deteriorate and economic viability to be sustained. On the other hand, such deterioration could lead to lack of economic viability of an agricultural system. It all depends!

A common claim is that if an agricultural technique is to be sustainable, it must be simultaneously sustainable from a biophysical, economic and social viewpoint. In reality, however, agricultural techniques may remain economically sustainable for a very long period at the same time as biophysical characteristics are declining. Furthermore, a technique that is not entirely socially acceptable may be economically viable and in some cases, as time passes, may become more acceptable socially (social transformation occurs). This is not to suggest that holistic dimensions involved in agricultural sustainability should be ignored. Not at all. Nevertheless, one should be careful about drawing conclusions about what is needed for sustainability.

There has been considerable debate about the sustainability of different categories of agricultural techniques and their environmental impacts. It is worthwhile considering briefly some of the issues. Categorizations include:

1. Modern versus traditional techniques.
2. Conservation farming versus conventional farming.
3. High external input agriculture (HEIA) versus low external input agriculture (LEIA).
4. Organic agriculture versus non-organic.
5. Extensive versus intensive use of land for agriculture.

These classifications overlap to some extent but not entirely.

In order to place the discussion of these alternative agricultural systems in context, note that agricultural activities may prove to be economically (and in some cases, biophysically) unsustainable because of the type of techniques used or because of the nature of activities engaged in on a property (that is because of *internal* effects) or because of *external* effects. External impacts can include waterlogging and salinization of land from irrigation schemes, reduced water availability to particular properties due to demands by competing users and lack of appropriate methods for allocation of the water and so on. A large number of examples could be catalogued but I shall not do that here. Agricultural practices on particular properties may affect other agriculturalists, those engaged in other industries, or impact on consumers directly as a result of environmental spillovers. And other industries and consumers (individuals) can have adverse environmental consequences for agriculture. Overall economic efficiency in satisfying human wants and long term sustainable development requires that external effects be accounted for in policy making. (However, because of the scale factor, this is not sufficient.) I shall not on this occasion discuss the type of policies which might be adopted to cope with externalities because they have been the subject of a major part of economic research on environmental policy design.

Returning to alternative agricultural methods categorized above, consider the following matters:

1. Conway (1987) argues that modern agricultural systems are less sustainable than traditional ones. While this may be broadly so, there are exceptions. Minimum till and no-till systems for example are a modern type of technique and may be more sustainable than some tillage systems especially those which leave ploughed land to fallow, thereby exposing the soil to the elements. Trickle irrigation systems and irrigation systems using moisture sensors (modern techniques) add to conservation of water and may promote sustainability of yields. Furthermore, it is quite possible that 'modern' techniques will be discovered which greatly reduce non-point pollution from fertilizer. Already slow release pelletized fertilizers help do this. In any case, traditional agricultural systems are sometimes not sustainable or become unsustainable with changing socioeconomic conditions. Schultz (1974) has identified some modern agriculture systems which seem more environmentally sustainable than traditional ones. In reality, the sustainability of a technique does not depend solely on whether it is modern or traditional.

2. Many traditional agricultural systems are typified by a low level of external inputs (LEIA) whereas many modern ones exhibit a high degree of reliance on inputs external to the farm or village (HEIA). At first sight, some may believe that LEIA is very sustainable (Reijntjes *et al.* 1992). However

this is not necessarily the case. With growing population LEIA can intensify and result in an expanding area of land being cultivated. For example, swidden or shifting agricultural systems (called *jhum* in Northeast India) involve a low level of external inputs. However, as population increases and the need for providing greater economic needs makes itself felt, cycles of shifting agriculture become shorter and larger land areas are exposed to the elements as a result of this form of cultivation. The consequence is rapidly declining soil fertility, severe soil erosion and escalating loss of biodiversity. Such systems eventually become uneconomic and are unable to maintain the incomes of a rising population.[4] The dynamics of overall social change cause them to become unsustainable.

3. Agricultural conservation techniques, such as the use of hedgerows with alley cropping on steep slopes, can add sustainability to agricultural yields and reduce adverse externalities. However, they are often not as economic as other methods and frequently involve an initial capital investment which farmers in less developed countries find difficult to make. There are traditional conservation methods using integrated methods and crop rotation but some conservation methods have been or are being developed in modern times. These modern conservation techniques do not necessarily rely on organic methods and need not involve low external inputs. Nevertheless they can be the source of considerable environmental improvement.

4. Interest in organic agriculture has grown, especially in Germany mainly because of its perceived health benefits for humans (Lampkin and Padel 1994). In some quarters, there also appears to be a presumption that organic agriculture is environmentally benign or favourable. However, it is possible for organic agriculture to be intensive and not favourable to nature conservation. Furthermore, if organic agriculture results in reduced yields (which it need not do in the longer term), then for the same output it will require a larger land area to be used for agriculture so the environmental impact of agriculture will be more widely felt.

5. It is true that most modern agriculture is intensive and reliant on a high level of external inputs, several of which are non-renewable. Several advocates of increased nature conservation support a return to more extensive systems

---

[4] Ramakrishnan (1992) has shown that for *jhum* cycles of 20-25 years, slash-and-burn agriculture in Northeast India is both very economic and sustainable. However, because of pressures, mainly as a result of rising population, the length of these cycles have in many cases fallen to 4-5 years. The method is now relatively uneconomic, biophysically unsustainable, a source of very serious adverse externalities and a major source of biodiversity loss.

of agriculture, especially in Europe (Hampicke 1996). In some countries, however, reliance on extensive systems would result in agriculture spreading over a larger land area. Even extensive agriculture can cause serious disruption to natural ecosystems. Extensive grazing by cattle and sheep in many parts of Australia has substantially changed natural ecosystems and has been implicated in the disappearance of at least one native species, a small marsupial or wallaby.[5] When extensive systems of agriculture replace intensive ones in a given area, biodiversity may increase and greater conservation of nature can occur. However, extensive systems often spread onto marginal lands with very adverse environmental consequences. Furthermore, if the choice is between (1) a small area under intensive agriculture with the remaining area not used for agriculture but left in a relatively natural state and (2) extensive agriculture over the whole area, which is best? In Australia for instance, some conservationists have supported the establishment of plantation forests on the grounds that this will reduce harvesting pressure on natural forests. It may do so but the economic benefits from increased economic productivity are not always utilized for greater conservation of nature (cf. Tisdell 1994b). They are often used to raise exploitation of nature even more.

I am sorry if the above makes simple suggestions for environmental improvements in agriculture appear to be problem ridden. However, it seems short sighted to ignore the type of issues raised above. Nevertheless, I accept that much of modern agriculture is over dependent on high external inputs, too intensive, gives less attention to conservation methods than is desirable and could make greater use of organic materials. However, there appears to be a strong argument for some use of artificial fertilizers but on a smaller scale than hitherto and in a way which reduces their leaching from the soil. Nevertheless, it is of concern when less developed countries like China try to indicate their agricultural progress by their rate of use of artificial fertilizer.

China is now using artificial fertilizer at one of the highest rates in the world and appears not to be recycling organic wastes including human excreta to the same extent as in the past. The latter is partly a result of increased urbanisation of China. For this and other reasons, water supplies in China have become nutrient rich; high in nitrates, phosphorous and organic matter. The increasing frequency of 'red tides' in the China Sea is partly blamed on discharge of such water. Many other adverse environmental effects are also being generated by this problem. It is surprising that a country which traditionally emphasized balanced agricultural systems and polyculture should

---

[5] Possibly a more dramatic example of the biodiversity loss due to extensive agriculture is the clearing of the tropical rainforest of the Amazonian Basin for cattle ranching; the so-called hamburger connection.

have allowed this modern trend to proceed so far. In the end (even now) it endangers its very large aquaculture industry (the largest in the world); a significant source of animal protein for its people.

AGRICULTURE AND THE CONSERVATION OF BIODIVERSITY

There are at least three angles from which we might be interested in agriculture and the conservation of biodiversity. These are:

1. The impact of agriculture on biodiversity conservation.
2. The benefits to agriculture from conservation of biodiversity.
3. The constraints placed on agriculture by decisions and policy measures to conserve biodiversity.

Let us consider each.

The intensification and spread of agriculture has been a major source (probably the prime source) of loss of biodiversity. This has mainly occurred because of destruction of wildlife habitat as a result of the conversion of wildlands to agricultural use and the increasing intensity of use of lands already used to some extent for agriculture. Where wild species of animals and plants compete with domesticated ones, they are seen as pests by the farmer and destroyed where possible.

Sometimes biologists see farming as a way of saving endangered species, for example, farming of turtles and of giant clams. If a commercially viable industry can be established, farming is an effective possible means of saving a species. However, profitable farming, depending upon its nature, may result in increasing displacement of the farmed species from the wild. Areas suited to farming the species concerned may also be the habitats favoured by the wild species. These may consequently be appropriated for farming displacing wild members of the species and in some cases thereby endangering their continued existence in the wild. There may also be other mechanisms which result in farming of species endangering wild stocks of the same species (Tisdell 1991). Furthermore, many of those who favour the retention of biodiversity do not consider domesticated stocks of a species to be an adequate substitute for wild stocks.

Just as the development of innovations and techniques used in economic production involve evolutionary aspects (Tisdell 1996a) and introduce inflexibility and hysteresis into systems, so can the development of the farming of species. Those species which have begun to be farmed early in the history of mankind tend to be increasingly advantaged for commercial purposes over others (Swanson 1994). This is because considerable learning and accumulation of knowledge about the capacities of such species in a domesticated situation takes place, which is reinforced today by formal research. Furthermore, the farmed species are selectively bred over a long period of time to become fitter for the tasks which human beings have assigned to them. In addition, their products become well known to consumers who discover an increasing range of ways to use them, such as varied recipes, find their use increasingly to be socially acceptable and develop personal tastes in their favour. All of these factors make it very difficult to develop economically the farming of a species not previously farmed, for example, kangaroo farming, giant clam farming or the growing of a new food crop. At the same time, the fact that one species is used commercially at an earlier stage than another may be to a large extent a matter of chance. The selection of species and techniques for use being in part myopic is often not the optimal from a long term viewpoint. This occurs for many innovations (Tisdell 1996a).

While agriculture has been and in many parts of the world continues to be, a major force destroying biodiversity, it can also be an economic beneficiary from the conservation of biodiversity. It is claimed that genetic diversity within species can provide a valuable bank to be drawn on to conserve the viability of cultivated plants and domesticated animals. Many cultivated crops depend on a narrow genetic base and from time to time, due to occurrence of diseases and other factors, lose their vigour and economic viability. By drawing on a wide gene bank, new varieties of a species can often be developed which at least for a time, show resistance to the problem.

Nature conservation can widen the scope for agriculture, sometimes species and their varieties with no apparent use now turn out to be useful in the future and profitable to cultivate or to husband. In so doing, they extend the range of agricultural possibilities or future options for agriculture.

It is, however, very difficult or impossible to place an exact value or even possibly a reasonable approximation, on the economic value of conserving biodiversity.[6] Nevertheless, we do know that if for genetic reasons, one of our major crops such as wheat or rice should fail to survive, or fail to survive productively from an economic viewpoint, that the economic costs would be

---

[6] Pearce and Moran (1994, Ch. 6) provide some monetary estimates for the value of conservation of diversity of medicinal plants in rainforests. The estimates, however, are very uncertain and problematic. This is especially true of those forecasts involving value of human lives saved or income loss avoided. Value is dominated by the needs of those in high income countries; OECD countries. In effect the lives of those in higher income countries are more valued than those in lower income countries. The effect could be to save the genetic pool which most favours higher income earners. This type of ethics disturbs me. It appears inequitable and unjust. It would for example violate Rawls's principle of justice (Rawls 1971) and would not appear to accord with the Christian principle that all are basically equal in the sight of God.

phenomenal. It is of course possible to give estimates *ex post* of the benefits of certain species having survived. Rubber for instance could be taken as an example. Rubber plays a very important role in modern transportation and although synthetic rubber exists, natural rubber is an essential component of radial tyres and has many other uses. I understand that the rubber plantations of Southeast Asia were only saved from a disastrous disease by drawing on genetic reserves present in the rainforests of Brazil.

Given that there are demands from the public for the conservation of biodiversity and rare species, agriculture activity is being increasingly constrained to help accommodate their demands. For example, in many countries agricultural properties are more frequently subject to preservation orders. The clearing of land or of habitat suitable for particular species is more and more restricted. In Australia, grazing of livestock on protected or relatively natural areas is increasingly banned or severely restricted. Furthermore, land use on many agricultural properties is subject to mounting limitations. Therefore, growing environmental concerns are imposing extra costs on at least some sections of the agricultural community. Agriculturalists are finding it necessary to adjust to this changing social climate.

SOCIOECONOMIC METHODS FOR SUSTAINING BIODIVERSITY AND THEIR AGRICULTURAL IMPLICATIONS

With the general increase of interest in the state of the environment, there has been growing interest in how economic and social mechanisms can be used to sustain biodiversity and attain environmental goals.

Substantial attention is being given to the possibilities for using economic incentives, to extending property rights and to harnessing markets to conserve biodiversity and natural resources. Indeed, in some quarters there is a state of euphoria about the likely effectiveness of these policies.

However, in relation to biodiversity, market related systems are liable to be very selective in the saving of species (Tisdell 1995a). Those species that can favourably be used for economic purposes in the relatively short run are liable to be favoured. Often this is at the expense of other species and as in agriculture, this can result in a reduction in biodiversity.

Maximization of economic benefit can lead to the rational elimination of some species. Discounting of economic gains, for instance favours the elimination of species which increase in value at a slower rate of growth than the interest rate (Clark 1976). Other things equal, these are slower growing or slower reproducing species. Observe however that opportunity rates of return provide another economic rational for extinguishing selected species. Where for example, two species are in competition for the use of the same resources and both are of commercial value, the one with the highest rate of return will be favoured. This will be so even if the internal rate of return from both is well in excess of the rate of interest. Elimination can occur in many ways, for example, by directly destroying the population of the economically less favoured species or by altering habitats to favour the economically more profitable species. In practice, the latter has had the greatest impact in reducing biodiversity.

It might be thought that the above failure of the market mechanism to conserve a species is a consequence of there not being total economic evaluation of a species. However, even if total economic valuation takes place and discounting of the estimated monetary flow occurs *both* of the above types of situation can occur within the extended framework. Economics can support the elimination of species of relatively low total economic value. Such a recommendation is anathema to many conservationists. I personally do not accept the total economic valuation test as a final arbiter of whether a species should survive. Its ethical foundations are too narrow. It does not sit well with strong (or even relatively strong) conditions for sustainability and is being increasingly challenged in terms of community values.

Transaction costs are likely to hamper the creation of property rights and the effective operation of market systems in conserving biodiversity and uncertainty further adds to market failures. Even methods which reward local communities according to economic use of species, such as rewards for use of elephants locally for trophy hunting, can lead to selective conservation of species. Whether such methods truly conserve biodiversity is therefore arguable. My own view is that considering the situation overall, they reduce biodiversity.

There is no guarantee that increased commercial use of wildlife or the greater commercial appropriation of total economic value will foster biodiversity. In my view the opposite is more likely as those species come to be favoured for which the largest total of economic value can be appropriated. However, the situation is complex. Nevertheless, I want to make it quite clear that I dissent from the blanket view that greater economic appropriation of total economic value of species and of their varied forms should be encouraged from a policy point of view in order to conserve biodiversity; it can lead to the opposite result.

A further observation may be in order. Some economists (Hampicke 1996) suggest that species conservation might be a non-economic decision but that economists can nevertheless be involved in terms of cost effectiveness analysis. While this may be so up to a point, the cost of saving some species may be the disappearance of others. So it is not clear that cost minimization can be divorced from evaluation after all.

If biodiversity is at least considered in part to be a type of merit good (or to some extent a type of public good) then there is a role for the state in financing its provision and/or in helping to supply protected areas and services supporting conservation of biodiversity. Furthermore, if biodiversity conservation is the goal, efforts to make *multiple use* of protected areas need to be resisted if the likely consequence of such multiple use would be to make for more uniformity of environments. In general, diversity of environments is needed to support biodiversity.

There is increasing pressure to establish biosphere reserves. These can assist in the conservation of biodiversity provided that they are not used as a means to reduce the size of core protected areas. Biosphere reserves do, however, place increasing restrictions on land use, for example, by agriculturalists. Furthermore, both the presence of protected areas and the use of biosphere reserves can result in increased populations of species regarded by many farmers as pests. This is a serious problem for farmers when these species are protected and farmers are either not compensated for the damage caused or are inadequately compensated, as is often the case.

CONCLUDING COMMENTS

New approaches to environmental policy making are needed which are less mechanical and more organic than some neoclassical approaches appear to be. Concepts of sustainability and of biodiversity are complex but useful. It was observed that sustainable agricultural systems need not promote sustainable development as such. Low external input agricultural systems, as well as traditional ones, are not necessarily sustainable, especially when dynamic exogenous changes, such as rising human population, occur. Extensive agricultural systems are not necessarily favourable for biodiversity conservation although on the same land type, such systems can result in greater preservation of biodiversity than intensive agricultural systems.

While agriculture is in some respects a beneficiary of biodiversity conservation it is also disadvantaged by it in a number of ways pointed out above. Consequently, many equity or income distribution problems are raised by biodiversity conservation.

That use of policies involving market mechanisms, economic incentives and property rights for protection of nature has significant limitations for conservation of biodiversity. This is not to say that advantage should not be taken of such mechanisms but state intervention to conserve biodiversity by direct means is still required given merit good and public good arguments, the presence of fundamental uncertainties and the occurrence of unavoidable market failures.

REFERENCES

Baumol, W. and Quandt, R. 1964. 'Rules of thumb and optimally imperfect decisions', *American Economic Review*, 54, 23-46.

Clark, C.W. 1976. *Mathematical Bioeconomics*. New York: Wiley.

Coase, R.H. 1937. 'The nature of the firm', *Economica*, New Series 4, 386-405.

Conway, G.R. 1985. 'Agroecosystem analysis', *Agricultural Administration*, 20, 31-55.

Conway, G.R. 1987. 'The properties of agroecosystems', *Agricultural Systems*, 24, 95-117.

Etzioni, A. 1991. 'Contemporary liberals, communitarians and individual choices', in Etzioni, A. and Lawrence, P.R. (eds), *Socio-economics: Towards a New Synthesis*. New York: M.E. Sharpe.

Hampicke, U. 1996. 'Opportunity costs of conservation in Germany'. Paper presented at the OECD International Conference on Incentive Measures for Sustainable Use, Cairns, March, 1996.

Kelso, M.M. 1977. 'Natural resource economics: The upsetting discipline', *American Journal of Agricultural Economics*, 59, 814-23.

Kohn, R.E. 1993. 'Measuring the existence value of wildlife: Comment', *Land Economics*, 69, 304-08.

Lampkin, N.H. and Padel, S. 1994. *The Economics of Organic Farming: An International Prospective*. Wallingford, UK: CAB International.

Lynam, J.F. and Herdt, R.W. 1989. 'Sense and sustainability: Sustainability as an objective in international agricultural research', *Agricultural Economics*, 3, 381-98.

Margolis, H. 1982. *Selfishness, Altruism and Rationality*. Cambridge: Cambridge University Press.

Mishra, H.R. 1982. 'Balancing human needs and conservation in Nepal's Royal Chitwan National Park', *Ambio*, 11, 246-51.

Pearce, D. 1993. *Blueprint 3: Measuring Sustainable Development*. London: Earthscan.

Pearce, D. and Moran, D. 1994. *The Economic Value of Biodiversity*. London: Earthscan.

Ramakrishnan, P.S. 1992. *Shifting Agriculture and Sustainable Development: An Interdisciplinary Study for North-Eastern India*. Carnforth, UK: Parthenon and Paris.

Rawls, J. 1971. *A Theory of Justice*. Cambridge, MA: Harvard University Press.

Reijntjes, C., Havekort, B. and Waters-Bayer, A. 1992. *Farming for the Future: An Introduction to Low-External-Input and Sustainable Agriculture*. London: Macmillan.

Schultz, T.W. 1974. 'Is modern agriculture consistent with a stable environment?', International Association of Agricultural Economics, *The Future of Agriculture: Technology, Policies and Adjustment*. Oxford: Agricultural Economics Institute.

Smyth, A.J., Dumanski, J.K., Spendjian, G., Swift, M.J. and Thorton, P.K. 1993. *FESLM: An International Framework for Evaluating Sustainable Land Management*. World Soil Resources Reports No. 73. Rome: Land and Water Development Division, Food and Agriculture Organisation of the United Nations.

Swanson, T.M. 1994. 'The economics of extinction revisited and revised: A generalised framework for the analysis of the problems of endangered species and biodiversity loss', *Oxford Economic Papers*, 46, 800-21.

Tisdell, C.A. 1990. *Natural Resources, Growth and Development*. New York: Praeger.
Tisdell, C.A. 1991. *Economics of Environmental Conservation*. Amsterdam: Elsevier.
Tisdell, C.A. 1993. *Environmental Economics*. Aldershot, UK: Edward Elgar.
Tisdell, C.A. 1994a. 'Biodiversity, sustainability and stability: An economist's discussion of some ecological views', Economics Discussion Paper, No. 139, Department of Economics, The University of Queensland, Brisbane, Australia.
Tisdell, C.A. 1994b. 'Conservation, protected areas and the global economic system: How debt, trade, exchange rates, inflation and macroeconomic policy affect biological diversity', *Biodiversity and Conservation*, 3, 419-36.
Tisdell, C.A. 1995a. 'Does the economic use of wildlife favour conservation and sustainability?' in Grigg, G.C., Hale, P.T. and Lunney, D. (eds), *Conservation Through the Sustainable Use of Wildlife*. Brisbane: Centre for Conservation Biology, The University of Queensland.
Tisdell, C.A. 1995b. 'Issues in biodiversity conservation including the role of local communities', *Environmental Conservation*, 22, 216-27.
Tisdell, C.A. 1996a. *Bounded Rationality and Economic Evolution*. Aldershot, UK: Edward Elgar (in press).
Tisdell, C.A. 1996b. 'Economic indicators to assess the sustainability of conservation farming projects', *Agriculture Ecosystems and Environment*, 57, 117-31.
Xiang, Z. 1995. 'Rapid rural appraisal (RRA), participatory appraisal and their application in the Global Environmental Facility (G-EF-B) Program in China', Biodiversity Conservation Working Paper, No. 19, Department of Economics, The University of Queensland, Brisbane, Australia.

[10]

Genetic selection of livestock and economic development

C.A. Tisdell and C. Wilson[1]

Introduction

According to the FAO, animal production accounts for between 30 and 40 per cent of the total value of global agricultural production (Hammond, 2002). Domesticated animals supply meat, milk and eggs and, in many developing countries, provide animal power for transport, cultivation and the operation of machinery, even though the use of livestock in that regard declines as motorization spreads. Livestock is also a source of hides and leather and its manure is used for fuel and fertilizer. In addition, domesticated animals are often able to consume crop residues and agricultural by-products to economic advantage. Furthermore, domesticated animals supply products used in medicine and food additives, and have additional economic values, including, in some societies, cultural values.

As a result of projected economic growth globally, demand for animal products such as meat, milk and eggs is predicted to rise substantially during this century. Most of this increased demand will come from developing countries as many experience (1) faster rates of population growth than in higher income countries, (2) growing urbanization, and (3) rising levels of purchasing power (Delgardo et al., 1999). These factors can exert a strong upward influence on the demand for meat, milk and eggs in lower income countries. In higher income countries, growth in demand for these products is likely to be much slower, with greater emphasis on quality attributes, such as low-fat meat, rather than quantity.

These changes can be expected to occur against a background of increasing economic globalization. This will result in changes in the utilization of available genetic resources present in domestic animal stocks and, in all probability, will result in continuing loss of livestock diversity due to concentration on fewer breeds in order to maximize economic gains. According to the World Conservation Monitoring Centre (1992, p. 397),

> Pursuit of higher production targets, the commercial success of particular breed promoters, and in developed countries, changes in consumer preferences have led to livestock development activities becoming concentrated in few breeds and breed groups. The corollary of this is that more breeds are declining in importance, many have been lost and the survival of many others is in considerable doubt. Concern for rare breeds has been most marked in northern temperate countries with a history of specialised livestock production, but it is becoming increasingly evident that declining breeds in less developed countries also represent genetic resources of great significance.

According to the UNDP (undated, p. 2) about half of all Europe's breeds of domesticated animals disappeared in the last century, and a third of the remaining livestock breeds in both Europe and North America are believed to be endangered. The majority of the world's surviving breeds are now located in developing countries where there is still heavy reliance on locally adapted breeds (Hammond, 2002). Just as market extension and the increasing ability of mankind to reduce the dependence of agricultural production on local natural environments have been major factors in the loss of livestock breeds in Europe, similar losses are likely to become more prevalent in developing countries.

In this chapter, market changes and scientific advances likely to reduce diversity of livestock breeds are outlined and associated aspects of market failure are discussed. Factors that influence the genetic selection of livestock in developing countries and the survival of breeds are given particular attention, taking Vietnam as an example.

Market changes, scientific and technological advances in animal husbandry and loss of livestock breeds

Many economic factors influence the survival of livestock breeds as economic development occurs. Initially, the overall historical pattern of evolution of livestock breeds appears to have been one of a globally growing number of breeds increasingly adapted to local natural environmental conditions. But with the extension of markets and technical change, this pattern seems to have been replaced by one of a declining number of livestock breeds globally.

The extension of markets has resulted in less dependence of individuals on local natural environments for the bulk of their economic requirements, and greater international (inter-regional) specialization in production. Furthermore, technological advances have made the productivity of agriculture (including animal husbandry) less dependent on local natural environments and local natural resources. This factor, combined with greater inter-regional trade, tends to reduce the economic value of breeds adapted to local natural environments. For example, a breed not well adapted to a local natural environment may become the best economic choice in the local area if the environmental conditions experienced by livestock of this breed can be economically modified to make these animals more suitable in the locality than the local breed. Modern agriculture increasingly decouples agricultural production from surrounding natural or environmental conditions, even though complete decoupling may be difficult. Consequently, exotic breeds of livestock that are intolerant of local environmental conditions (see the discussion of the biological law of tolerance in Chapter 3) may, if they have desired attributes, increasingly replace locally adapted breeds because suitable artificial environments are created for them.

However, it is not only technical advances, such as those involving improved housing and management of livestock, that help decouple animal husbandry from local environmental conditions. Increased trade in agricultural inputs over ever increasing distances also means that agricultural production is less dependent on local resources, or the natural ecological niche locally. Man's intervention in controlling environments, combined with growing trade, results in greater uniformity of (manipulated) environments in which agricultural production takes place. This growing uniformity is a major factor responsible for a reduction in the number of livestock breeds and varieties of individual crops.

At the same time, the extension of markets is associated with growing inter-regional specialization in production and subsequent exchange. Therefore, in some regions, livestock production may be replaced by a different form of production if a region has an international comparative advantage in the latter, as economic globalization proceeds. This can result in the disappearance of livestock breeds locally adapted to the region in which a shift in production patterns occur. For example, the dwarf cattle of Southern Nigeria have almost disappeared because tree crops grown to earn cash income via international sales have replaced pre-existing subsistence agricultural systems.

Inter-regional trade fosters specialization in production according to comparative advantage. Where a region's comparative advantage is not in livestock production or production involving particular livestock, extension of trading possibilities may result in production from livestock being replaced by other forms of economic production. Where the livestock concerned are locally adapted breeds, these breeds are likely to disappear.

In addition, just as expanded possibilities of market exchange increasingly result in humans specializing in economic production according to their comparative advantage, greater specialization of livestock breeds is also encouraged. For example, some breeds of cattle are comparatively more efficient producers of milk or of meat than others, whereas some are quite efficient in producing a combination of milk and meat. The latter are 'all-rounders'. These 'all-round' breeds may be preferred economically in semi-subsistence economies and/or those economies that do not engage in much trade. But in market-exchange economies, short-run economic efficiencies may be maximized by concentrating on special purpose breeds and exchanging their products.

The basic idea can be crudely illustrated by adapting Lancaster's (1966a, b) characteristics approach to it. In Figure 1, three breeds of cattle are considered, A, B, and C. A unit of resources, when allocated to breed A, produces the combination of milk and meat shown by point A, and similarly points B and C show the combination of milk and meat for the same amount of resources allocated to breeds B and C. In the absence of trade, the all-round breed B may be preferred because of a family's preferred combination of milk and meat. But if trade becomes possible and involves no transaction costs, instead of consumption possibilities being represented by the points A, B and C, the equivalent possible set is that for the triangular set ABC, the frontier of which is the line AC involving linear combinations of production only from breeds A and C. Consequently, in a free trade situation the all-round breed B is no longer economic and is likely to disappear. It is no longer economic, since more milk and more meat for the same use of resources can be obtained by relying only on breeds A and C for production rather than utilizing breed B. In other words, points to the northeast of B, such as α, in Figure 1 become possible by reliance on specialized breeds when exchange is easy.

Technological progress also plays a role in the loss of livestock breeds by increasing substitutability. For example, motorization, increasing wage costs and affordable fuel prices have reduced the demand to use livestock for draught, transport and similar purposes. Breeds that were specialized for these purposes are thus disappearing, or

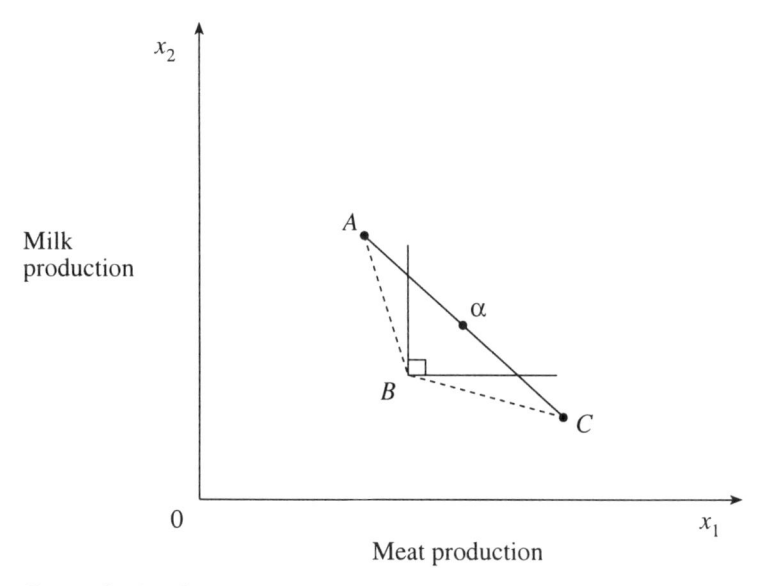

Economics and Ecology in Agriculture and Marine Production 143

[Figure: graph with axes x_2 (Milk production) vertical and x_1 (Meat production) horizontal, showing points A, B, C and α on a frontier]

Figure 1 Market extension may lead to the loss of all-round livestock breeds and their replacement by breeds more specialized in production

rely on preservation societies for their continued existence. Similarly, animal fat has become of less value as economical vegetable oils have been increasingly developed as substitutes. Cheap chemical fertilizers also reduce the reliance on animal manure. Furthermore, changes in cultural values erode dependence on the use of livestock.

Apart from supply-side developments that influence substitution, changing tastes and lifestyles also play a role in the survival of livestock breeds. For example, health considerations have increased the demand for low-fat meats and reduced demand for fatty meats. Therefore, breeds of livestock that tend to produce more fat lose favour, as is, for example, happening even in Vietnam, as discussed in the next section.

Another important influence on the array of livestock breeds that survive and develop is a form of path dependence (see Swanson, 1995). Research and development may become concentrated on breeds of livestock that are already well established in higher income countries and account for considerable sales of livestock products. This can result in potentially superior breeds[2] in less developed countries being neglected in research and development and eventually disappearing as the gap in productivity between breeds widens on a global scale (see Tisdell, forthcoming). Path dependence may also reinforce use of specialized breeds for particular purposes, for example, for milk or meat, as technology is developed and tailored to these different purposes (see Tisdell, forthcoming).

While market systems do have the potential to raise short-term economic gains because they select livestock breeds for survival that are able to provide extra economic gains in the short to medium term, these systems may fail from a long-term point of view. Because of path-dependence, the set of livestock breeds surviving may not be the one able to provide the greatest total economic benefits in the long term. Secondly, if production becomes dependent on very few breeds of livestock, most of which show little biological tolerance and thereby rely heavily on the ability of mankind to manipulate their environments, a high-risk situation can emerge. If it eventually becomes impossible for mankind to control an environmental/ecological factor to which these breeds are susceptible, production may not be sustained. On the other hand, this situation might have been avoided or moderated by conserving some breeds tolerant to the changed environmental circumstances; for example, due to climate changes arising from global warming or a new disease. Genetic erosion reduces mankind's ability to respond to changed environmental circumstances. This suggests that it is optimal to err on the side of conservation of genetic stocks of livestock. Nevertheless, at the same time, it is extremely difficult to decide in advance the optimal set of genetic stocks to conserve because of the long periods of time that have to be considered and the extreme degree of long-term uncertainty involved. However, there is little doubt that the extension of market systems and scientific/technical progress that decouples agricultural systems from surrounding natural environments pose a risk of lack of economic sustainability for mankind.

Breed selection in developing countries and a case study of Vietnam's pig stocks
In developing countries, many breeds of livestock are still kept for multiple economic purposes. For example, some breeds of buffalo are kept for draught purposes, milk, meat and hides, and their manure may be used for fertilizer and fuel. The same is true of cattle. As economic development proceeds, the economic value of the array of breeds being utilized in a country is liable to change. For the types of reasons outlined in the previous section, some breeds may no longer prove to be economical and, in some cases, exotic breeds or crosses of exotic breeds and local ones may prove to be more valuable and eventually replace local breeds.

One might expect this to occur naturally as the market system spreads but the process of breed selection and replacement may also be influenced by government policies (often driven by economic factors) that are, for example, formulated or applied by departments of livestock development (or their equivalent) in developing countries. In turn, the policies and practices of developing countries may be influenced by foreign aid from more developed countries. Such aid may be tied to the transfer of breeds of the donor country and associated technology even though neither the breeds nor technology may be appropriate to the aid-receiving country. This can create biases in breed selection and survival. According to an FAO report (FAO, no date), inappropriate breed selection may be promoted in developing countries because donor countries give preference to high-input, high-output breeds developed for benign environments. Commercial interests in donor countries promote the use of relatively temperate-adapted breeds and create unrealistic expectations in tropical developing countries, according to this report. An example might be the introduction of the Pietrain breed of pig from Belgium to Vietnam as a result of Belgian foreign aid. This breed suffered initially from heat stress in the Vietnamese tropical climate. This was so severe that air-conditioning was installed for boars held for artificial insemination purposes at the Bin Thang Animal Research and Training Institute of the Institute of Agricultural Sciences of South Vietnam, located near Ho Chi Minh City. Nevertheless, the breed and crosses of it have continued to diffuse in Vietnam.

Australia has also been involved in transferring pig breeds to Vietnam. The main transfer from Australia has been that of Australian Yorkshire pigs (also known as Australian Large Whites) from Queensland. These had been bred in tropical Queensland and were well adapted to tropical heat and humidity. They also had the capacity to produce lean meat efficiently when fed on concentrated pig food with appropriate lysine (protein)/energy content.

A study by Tisdell and Wilson (2001) found that the transfer of this genetic stock from Australia and Australian-associated sponsored research into the feeding and breeding of pigs in Vietnam give a high rate of economic return.

As discussed by Tisdell and Wilson (2001), pigs are extremely important in Vietnam. Pork is a basic part of the meat diet of the Vietnamese and is estimated to provide around 70 per cent of their animal protein intake. Pig hides provide extra value for farmers and pig manure is an important and cheap source of fertilizer for crops. Pigs are also a cultural asset to most Vietnamese and often act as an indicator of a family's prosperity. Hence, farmers keep pigs for other than commercial reasons. However, in recent times specialized commercial pig farms have grown rapidly in numbers in Vietnam, and these are mostly located near and in urban centres.

The pig population in Vietnam greatly exceeds that of buffalo and that of cattle which are used for draught purposes as well as having other uses. A comparison of pig numbers between Vietnam and Australia (a major producer of pork in the Asia/Pacific region) also shows that Vietnam has a much larger pig population than the Australian level of stock,[3] but Australia's production of pork relative to its pig population has remained higher than the Vietnamese level in recent times (Tisdell and Wilson, 2001). The introduction of foreign pig genes to Vietnam, however, has brought about increases in productivity and quality of pig meat, although Vietnam still lags far behind major pork producing countries. For instance in 1994, Australia slaughtered 1.87 pigs per year in relation to its stock, but by 2000 this had risen to over two (Table 1). The comparable figures for Vietnam are 0.99 and 1. Hence, the Australian slaughter rate is about twice that in Vietnam. Large differences also exist in the amount of pork production on average in relation to pig stocks as shown in Table 1. As shown in this table, in 1994, Australia produced 121 kg of pork annually per pig in its stock whereas Vietnam produced 61.8 kg, around half that of Australia. By 2000, Australia's annual supply of pork in relation to its stock of pigs was 137.5 kg per member of its pig population; the corresponding figure for Vietnam was 67.3 kg.

Table 1 Number of pigs slaughtered and weight of pork produced per member of the pig stock for Australia and Vietnam, 1994 and 2000

	1994	2000
Pigs slaughtered in relation to the stock (number)		
Australia	1.87	2.03
Vietnam	0.99	1.00
Weight (kg) of pork produced per member of the pig population		
Australia	121.0	137.5
Vietnam	61.8	67.3

Source: Tisdell and Wilson (2001, p. 11).

Vietnam has long had the goal of increasing the productivity of its pig industry, its volume of pork production and the quality of its pork. From 1961 to 1981, its supply of pig meat was relatively static. During this period, a period in which its communist government and Vietnam itself were relatively isolated from the rest of the world, the carcass weight of its slaughtered pigs tended to decline, the number of pigs slaughtered annually was relatively constant and pig stocks showed only small increases in their numbers. Vietnam's per capita pig meat consumption fell from about 7 kg per head in 1960 to around 5 kg per head for most of the 1970s. Thus the already low level of animal protein in the Vietnamese diet in the 1960s was reduced even further by the poor performance of the pig industry in the 1970s. However, beginning in 1981, Vietnam made extra efforts to import pig breeds to improve the pig genotypes available to it. The feeding and housing of pigs were also improved. All these measures increased its production of pig meat. However, this was inadequate to meet its animal protein requirements. Continuing genetic improvements along with further research and dissemination of new techniques were found to be necessary for Vietnam to continue increasing its pork supplies. This was because the Vietnamese were facing a relatively low level of consumption of animal protein and because they have a strong preference for pork as meat. Pork exports were also a possibility. In addition, Vietnam's population was growing at a relatively rapid rate; in 1999, it was approaching 80 million, more than double that in 1965.

Given all these factors, it was inevitable that Vietnam's government would want to increase food supplies, and that its programme would include raising the productivity of its pig industry. As a result further foreign pig genes were introduced to Vietnam in the 1990s from Belgium and Australia by means of foreign aid with almost immediate results on yields. Without the introduction of foreign pig genes, it is almost certain that significantly less growth in pig meat supplies in Vietnam might have been achieved. In addition, introduced foreign pig genes, to a large extent from Queensland, Australia, have improved the quality of Vietnamese pork. The new genes have helped to reduce stress[4] for the animals during transport and excessive fattiness. Traditional pig breeds in Vietnam produce fatty meat. Low-fat pork makes for greater acceptability by consumers, may have associated health benefits and could make it easier in the future for Vietnam to export pork. Vietnam at present is not a significant exporter of pig meat.

Furthermore, in urban areas of Vietnam, lean pig meat sells at a price up to almost 30 per cent higher than that of fatty pork. In addition, Australian introduced breeds, namely Yorkshires and to a lesser extent Duroc breeds, have a more favourable feed-conversion ratio for concentrated pig food than local pig lines in Vietnam and other exotic breeds. They also have higher growth rates. Furthermore, the size of litters of Australian Yorkshire pig genotype sows surviving to weaning are larger than local Vietnamese breeds and other exotic breeds. This is an economic advantage because fewer sows have to be kept to produce the same number of slaughter pigs. Alternatively, the saving in the stock of breeding sows can be used to produce extra pigs for slaughter. Thus pork production can be increased without increasing costs. Again, Australian Yorkshire boars, as selected in Vietnam, also have greater reproductive ability than local and other exotic breeds (see Figure 2). This attribute also has positive economic implications. It allows some reduction in breeding stocks to occur in relation to the quantity of slaughter pigs produced, and the better reproductive performance of Australian Yorkshires compared with local and other exotic breeds means that fewer resources have to be tied up in breeding stock. All these factors further reduce costs of producing pork.

Figure 2   Australian Large White (Australian Yorkshire) boar used in Vietnam for artificial insemination. Artificial insemination in Vietnam has been used to ensure rapid diffusion of this exotic genotype

As a result of these developments a substantial fraction of Vietnam's pig stock is now based on imported genetic material. Moreover, in order to prevent inbreeding, further imports of the exotic boars may be needed in the future. This means that the indigenous breeds will continue to be diluted with exotic, high performing breeds. In addition, it is still not known how these exotic breeds would cope in the event of a disease outbreak where usually the local breeds are more resilient. If the dilution of breeds continues, it is inevitable that pure local Vietnamese breeds will give way to exotic and mixed breeds and eventually disappear. For example, the Mong Cai, a local breed of pig in Vietnam (see Figure 3), may be one such casualty.

Sometimes this process of replacement of local breeds occurs at the village level. Figure 4 provides an example of this for a poor rural village north of Ho Chi Minh City. In this village, the Vietnamese government has assisted villagers to replace local pig breeds by Australian Large Whites. However, introduced exotic livestock breeds often prove to be inferior at the village level. Hence, a separate commercial livestock industry of an intensive nature based on exotic breeds evolves in developing countries relying on similar production techniques to those used in developed countries. This

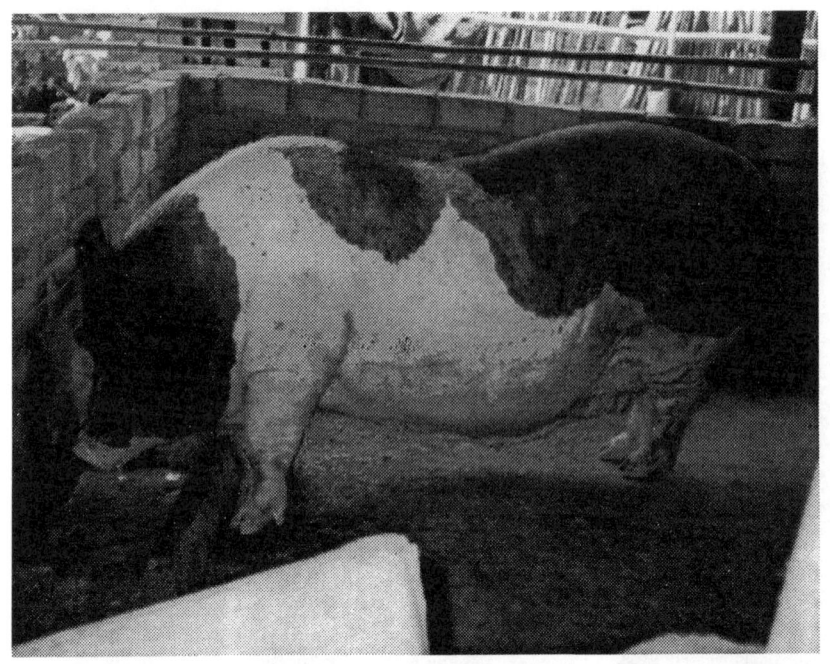

148 *Economics and Ecology in Agriculture and Marine Production*

Figure 3   The Mong Cai, a local breed of pig in Vietnam. This breed may be lost due to the introduction of exotic breeds and economic change

Figure 4   An Australian Large White sow and her litter in a sty in a poor village in Vietnam. Government assistance has fostered the recent introduction of this breed to such villages and thereby accelerated local breed replacement

has happened in the Vietnamese pig industry (see Figure 5). A dualistic pig industry has evolved as economic development has gathered momentum in Vietnam.

Figure 5 Intensive pig production on the outskirts of Ho Chi Minh City based on Australian Large Whites. Many intensive commercial piggeries have grown up around Ho Chi Minh City and are based on exotic breeds, or crosses involving exotic breeds, and similar production techniques to those used in higher income countries

Although the rate of economic return from introduced exotic breeds is very high in Vietnam, observations by one of the authors (Clem Tisdell) indicates that the main beneficiaries from such introductions are pig producers involved in relatively large-scale intensive pig production on the urban periphery of Ho Chi Minh City in particular. Intensive pig production in Vietnam tends to be in the neighbourhood of larger cities that depend on concentrated food and imported lysine (protein derived from soy beans). Lysine is a significant component in concentrated pig feed. These observations accord with those of Delgardo et al. (1999) for developing countries; namely, that in many developing countries, increasing urbanization is a major trend and this is accompanied by growth of intensive livestock industries located in urban peripheries.

Concluding comments

Just as the survival of livestock breeds in higher income countries has been influenced by their process of economic development, similar patterns might be expected to emerge in developing countries as they experience economic growth. This will undoubtedly result in continuing loss of livestock breeds worldwide. Unfortunately, we cannot be confident that market and political systems will leave future generations with the best long-term stock of genetic animal resources. This will no doubt lead to increasingly 'path dependent' production systems. At the same time, heavy reliance on a limited gene pool could expose such production systems to collapse with heavy economic losses. Some countries, especially developing countries, may also be locked into a 'choice limited' production system with heavy reliance on a foreign-based gene pool while indigenous gene stocks continue to be eroded. This is now becoming evident in countries such as Vietnam, where the processes discussed in this chapter are unfolding.

Notes

1. School of Economics, The University of Queensland, Brisbane, 4072. We wish to thank scientists in Australia and Vietnam who assisted us with our study of the breeding and nutrition of pigs in Vietnam.
2. From an economic perspective.
3. Vietnam's pig population in 2000 was eight times that of Australia's.
4. Stress in animals tends to make their meat tough.

References

Delgardo, C., Rosegrant, M., Steinfeld, H., Ehui, S. and Courbois, C. (1999), *Livestock to 2020: The Next Food Revolution*, Food and Agriculture and Environment Discussion Paper 28, Washington DC: International Food Policy Research Institute.

FAO (no date), *Extensive Pastoral Livestock Systems: Issues and Options for the Future*, <http://www.fao-kyokai.or.jp/edocuments/document2.html>

Hammond, K. (2002), 'Overview of domestic animal diversity – international implications', *ANGR Briefs*, Rome: FAO.

Lancaster, K. (1966a), 'Change and innovation in the technology of consumption', *American Economic Review Supplement*, May, 14–23.

Lancaster. K. (1966b), 'A new approach to consumer theory', *Journal of Political Economy*, **74**, 132–57.

Swanson, T.M. (1995), 'Why does biodiversity decline? The analysis of forces for global change', in T.M. Swanson (ed.), *The Economics and Ecology of Biodiversity Decline: The Forces Driving Global Change*, Cambridge: Cambridge University Press.

Tisdell, C.A. (forthcoming), 'Socioeconomic causes of loss of animal genetic diversity: analysis and assessment', *Ecological Economics*.

Tisdell, C. and Wilson, C. (2001), *Breeding and Feeding Pigs in Vietnam*, ACIAR Impact Assessment Series No. 17, Canberra: Australian Centre for International Agricultural Research.

UNDP/CSOPP (no date), 'Conserving indigenous knowledge', < http://www.undp/csopp/CSO/New Files/dociknowledge2.html> Accessed 29 October 2001.

World Conservation Monitoring Centre (1992), *Global Biodiversity: Status of the Earth's Living Resources*, London: Chapman and Hall.

PART III

PEST AND DISEASE CONTROL AND AGRICULTURAL PRODUCTION – BIOECONOMIC ASPECTS

[11]

Crop Loss Elasticity in Relation to Weed Density and Control

C. A. Tisdell

Department of Economics, University of Newcastle, 2308, Australia

B. A. Auld

Agricultural Research and Veterinary Centre,
Forest Road, Orange, 2800, Australia

&

K. M. Menz

Bureau of Agricultural Economics, PO Box 1563, Canberra, 2601, Australia

SUMMARY

Relationships between yield loss in crops and weed density are analysed using an elasticity function $((dL/L)/(dW/W))$. In general, loss in crop production caused per weed is higher in low density weed populations than in higher density weed populations. Therefore low density weed populations which are widespread could cause significant crop loss.

Control of low density weed infestations will often not be economic by chemical and mechanical methods because of fixed control costs per unit area. In contrast, a control method with costs largely independent of area, such as classical biological control, could provide economic control in these situations of widespread low density weed infestations.

INTRODUCTION

Yield loss in crops in relation to weed density could conceivably take a number of forms; three are shown in Fig. 1. (i) In the simplest case (line OA, Fig. 1) increasing weed density results in a proportionate crop yield loss. (ii) As weed density increases there is a proportionately greater yield

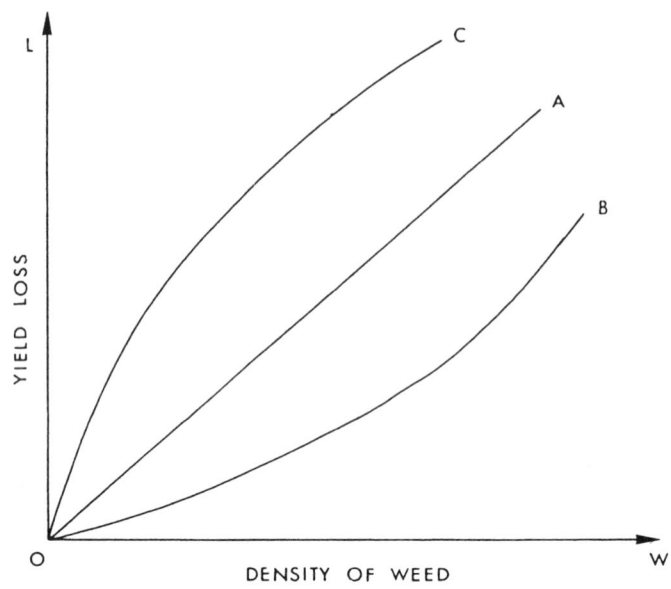

Fig. 1. Putative yield/weed density response curves showing yield loss per unit area (or fixed area).

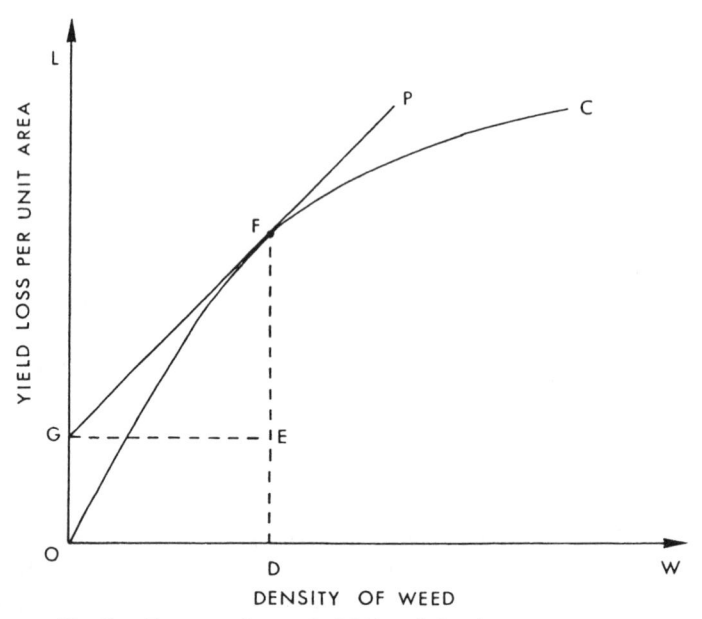

Fig. 2. Common form of yield/weed density response curve.

loss (curve OB, Fig. 1). (iii) Incremental yield loss decreases as weed density increases (curve OC, Fig. 1).

Here we consider, by simple graphical analysis, the elasticity of these functions and the economic implications for weed control. Although our discussion is basically restricted to crops, the argument also applies to pastures (in which a weed species has a net negative effect on pasture yield).

YIELD LOSS RESPONSE

There is a good deal of evidence to show that the most usual form of the relationship of crop yield loss to weed density is curve OC in Fig. 1. It has been found for a number of crop/weed systems in different locations (Berglund & Nalewaja, 1971; Weatherspoon & Schweizer, 1971; Dew, 1972; Reeves, 1976; Chisaka, 1977; Medd et al., 1981). A similar relationship has been shown for weight of weeds and crop yield (Medd et al., 1981).

The elasticity (e) (Brennan, 1970) of yield loss (L) due to increasing weed density (W) can be expressed as:

$$e = \frac{dL}{L} \bigg/ \frac{dW}{W}$$

$$= \frac{dL}{dW} \frac{W}{L} \quad (1)$$

If $e = 1$, increasing weed density on existing infested land by a small proportion, say 1%, results in the same relative yield loss as increasing the infested area by 1% and keeping the weed density constant. If $e < 1$, greater *overall* loss would be incurred from diffusion of weeds rather than their concentration; if $e > 1$, the reverse applies.

The elasticity of the usual response function (OC Fig. 1) is represented at any W value, say D (Fig. 2), as:

$$e = \frac{dL}{dW} \frac{W}{L}$$

$$= \frac{EF}{GE} \frac{GE}{FD}$$

$$= \frac{EF}{FD} < 1 \quad (2)$$

As the tangent (dL/dW) to a strictly concave and increasing curve passing through the origin will always pass through the L axis, the relevant elasticity is less than unity in this most common form of crop loss/weed density interaction. Thus, loss per weed is greater as density of weeds decreases (although, at extremely low densities, the magnitude of the total loss may be undetectable).

The elasticity of curve OB (Fig. 1) is greater than unity. However, this response, and the straight line response (OA Fig. 1; $e = 1$), do not appear empirically relevant.

RELEVANCE TO ECONOMICS OF WEED CONTROL

As we have shown above, greater total loss of crop yield or agricultural production may commonly result when weeds are dispersed at low density over a large area than when the same population is concentrated over a smaller area. (In this discussion we are ignoring the consequences of population build up which may occur as a result of not treating low density infestations.)

The cost of some weed control methods, especially at low density, is proportional to the area to be covered. Application of herbicides or mechanical control in many cropping systems may require the whole area to be treated when an infestation of any density occurs. In weed control in pastures a similar situation often applies. Costs per unit area treated then approximate to a fixed cost and are independent of weed density (line UV, Fig. 3). If we assume that the treatment eliminates the weed completely, the gross value of the weed eradication will be equal to the gained yield times its value (shown as curve RST Fig. 3; RST is OFC from Fig. 2 multiplied by the price of the product).

The break-even point where costs equal revenue gained from treating the weed is S. For densities of less than W_1 treatment is uneconomic, but treatment is economic for greater densities.

The higher the price received for the produce (which would shift RST upward) or the greater the loss in yield caused by weed (which would steepen RST), the lower the density at which treatment is economic (as S moves closer to U).

Weed control is therefore less likely to be economic: (a) the lower the density of the weed; (b) the greater the cost of the treatment; (c) the smaller the revenue productivity of it. For *widespread* weeds of low

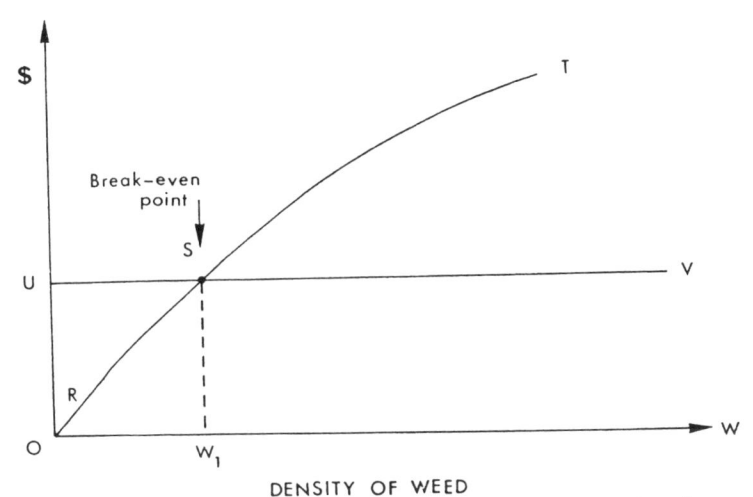

Fig. 3. Generalised fixed cost (USV) and revenue/yield (RST) functions.

density, a control method whose costs are largely independent of area, such as classical biological control, could be economically advantageous compared with other methods. The greater the area of a weed infestation the more likely is classical biological control to be economic because its total costs are largely independent of the area to be treated and therefore control cost per hectare falls with the size of the area benefiting from this treatment. This, together with the other points made in the paper, suggests that weeds which occur at low density and are widespread should not be overlooked as candidates for biological control in extensive agriculture.

CONCLUSION

Motivation for initiating weed control programmes frequently stems from observations of conspicuous high density infestations. However, weeds which occur at low density and are widespread can cause significant economic loss: they should not be overlooked, especially as potential targets in biological control in extensive agriculture.

REFERENCES

Berglund, D. R. & Nalewaja, J. D. (1971). Wild mustard competition in soybeans. *Proc. North Central Weed Control Conference (U.S.A.)* p. 83.

Cited by Zimdahl, R. L., *Weed crop competition—A review*, IPPC, Cornvallis, 195 pp.

Brennan, M. J. (1970). *Theory of economic statistics*. (2nd Edn.) Prentice-Hall, Englewood Cliffs, N.J., x + 437 pp.

Chisaka, H. (1977). Weed damage to crops: Yield loss due to weed competition. In: *Integrated control of weeds*. (Fayer, J. D. & Matsunaka, S. (Eds)), University of Tokyo Press, Tokyo, pp. 1–16.

Dew, D. A. (1972). An index of competition for estimating crop loss due to weeds. *Canadian Journal of Plant Sciences*, **52**, 921–7.

Medd, R. W., Auld, B. A. & Kemp, D. R. (1981). Competitive interactions between wheat and ryegrass. *Proc. Sixth Aust. Weeds Conference, Broadbeach*, Vol. 1, 39–43.

Reeves, T. G. (1976). Effect of annual ryegrass (*Lolium rigidum* Gaud.) on yield of wheat, *Weed Research*, **16**, 57–63.

Weatherspoon, D. M. & Schweizer, E. E. (1971). Competition between sugarbeets and five densities of kochia. *Weed Science*, **19**, 125–8.

Economic threshold/critical density models in weed control[1]

B.A. Auld[2] and C.A. Tisdell

Summary
A basic economic threshold model is described. The influence of reinvasion, time intervals, spillovers and uncertainties is discussed. The impact of yield improvement, carryover effects, control costs and price effects, including quality factors, are discussed in relation to a specific example, wild oats (*Avena fatua* L.) in wheat.

Economic threshold density model
Figure 1 indicates the economic threshold at which weed treatment becomes economic in a basic case. Curve OBD represents the value of additional yield from weed elimination and line CBE represents the costs of weed elimination, it being assumed that the weed treatment ensures virtual elimination of the weed in the field. W_1 is the threshold weed density. At weed densities higher than this, gains exceed the cost of treatment and at lower weed densities gains from weed treatment fail to cover costs. It is possible that the value of the productivity (yield) increase curve does not cross the cost curve. In that case, the economic benefits of treatment are less than costs for all densities of the weed.

Reinvasion and carryovers
The greater the expected intensity of reinvasion/reinfestation and the speedier it is, the smaller is the anticipated gain in revenue productivity from treating weeds in a field unless the control affects future weed populations. This would apply particularly to new cohorts of a weed species germinating after control treatment of earlier cohorts. In essence, treatment of the weed can be considered to be less effective if reinfestation occurs in a crop. When the revenue productivity curve is adjusted for the reinfestation effect, it is lower than otherwise and lower the greater is the intensity (degree) of reinvasion. Hence, the threshold density at which it becomes economic to treat a weed is higher the greater is the intensity and speed of weed reinvasion.

However, control of a weed species in one crop in one season can affect the ability of that weed to reinvade that crop in following seasons from buried seeds. Hence, there is a direct spillover of weed control in one year to the next and future years. Thus the weed density at which control is economic is less the greater this carryover effect. The importance of carryover populations in wild oats control in wheat has been highlighted by several workers (e.g. Wilson et al., 1984).

In principle, it is not difficult to take account of the economic influences of time-dependent events if the net present value criterion is adopted. The net present value of

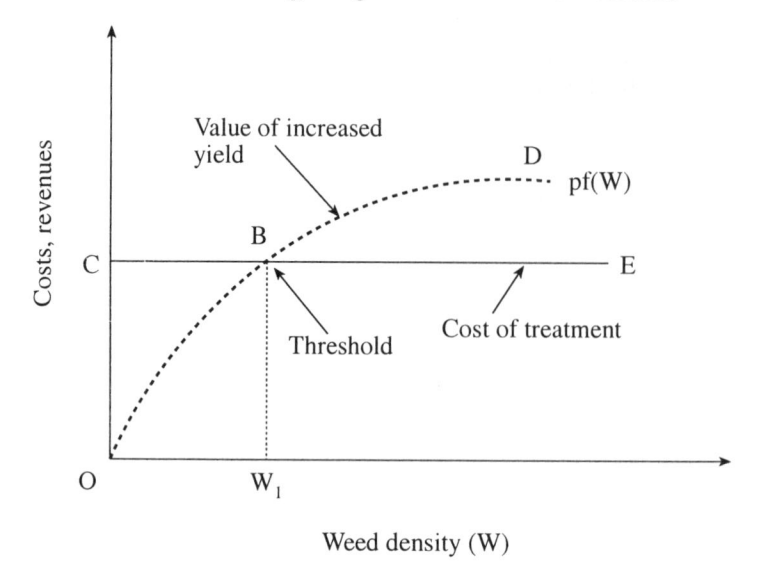

160   Economics and Ecology in Agriculture and Marine Production

Figure 1   The economic threshold for treatment of a weed assuming that treatment is fully effective. In the case shown, it is economic to treat the weed if its density exceeds W_1; treatment is not economic at lower weed densities

a weed control strategy is equal to the sum of the stream of discounted future net benefits or net gains from it. The discount rate is usually based on a market rate of interest and is larger the higher is the rate of interest.

The time element is significant when there are substantial delays between outlays on weed control and the reaping of economic benefits from them. It is especially important where carryover benefits are significant, for instance in establishing perennial crops. Other things being equal, the greater are carryover net benefits from weed control the lower is the weed density at which weed control is economic.

Uncertainties

Many uncertainties can arise in weed control including effectiveness of treatment, possible phytotoxic effects on crops, prices, productivity gains, reinvasion and the magnitude of carryover effects. Marra and Carlson (1983) suggest that carryover effects from weed control are sometimes so uncertain that it is appropriate to ignore them. Sometimes increased uncertainty about the future is taken into account by applying larger discounts to future costs and benefits thereby putting a reduced weight on these in decision-making.

Particularly where pre-emergence herbicides or other prior treatments are adopted, a farmer may be uncertain about the level of weed infestation to be expected in the absence of such treatment. Weed densities and therefore loss in yield to be anticipated from weed may be uncertain. This uncertainty can influence a farmer's decision to control weeds.

For example, suppose that a farmer wishes to maximize the mathematical expectation of his net profit from weed treatment, after allowing for the cost of treatment (C). His net gain function (J) can be written as:

$J = R - C$ (where R is net profit other than for weed control costs).

It is assumed that the farmer wishes to maximize the expected value, E(J).

If J as a function of W increases at a decreasing rate, then the greater the uncertainty about the level of weed densities, the more likely is weed treatment to be uneconomic (i.e. the more likely expected net gains from treatment are negative). In the simplest case:

$J = pf(W) - C$ (where p represents the price per unit of yield of the crop).

If yield loss, f(W), increases at a decreasing rate, $J'(W) > 0$ and $J''(W) < 0$ (Figure 2). It implies that the net gain function is an increasing strictly concave function of the weed densities. In these circumstances, greater uncertainty about weed densities, with the mean anticipated density held constant, increases the likelihood that E(J) is negative and therefore the probability that weed control is uneconomic. For instance in Figure 2, if weed density is known to be W_2 then $E(J) = 0$, gains from treatment of weeds would just cover costs. However, if there is a 0.5 probability that weed density is W_3 and a 0.5 probability of it being W_1 (where $W_2 = 0.5W_1 + 0.5W_3$) the expected value of J is J_1 and is negative (see Arrow, 1971). This uncertainty reduces expected gain from weed treatment and renders weed control uneconomic on average in this case.

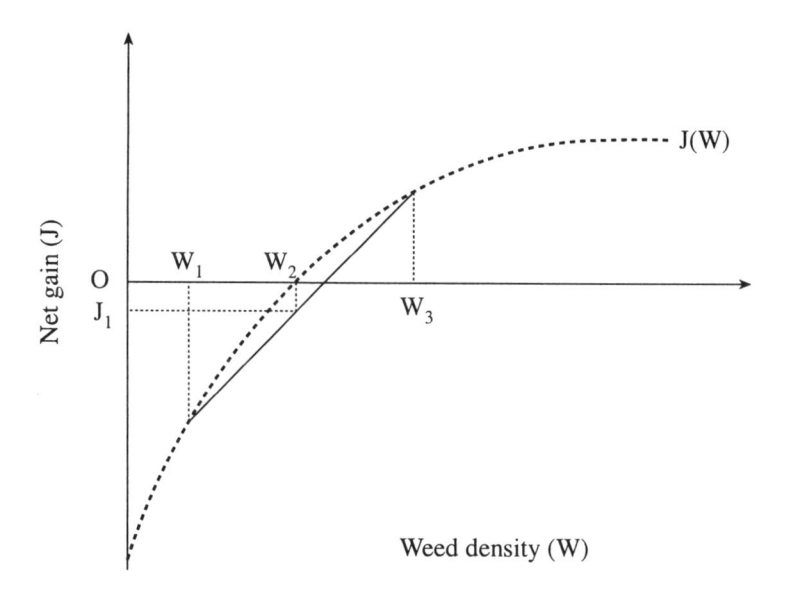

Figure 2 Illustration of a case in which greater uncertainty about a weed density reduces expected net gain from weed control

It is also possible for the yield loss function itself to be subject to uncertainty (Chisaka, 1977). Using Chisaka's approach, we can identify a maximum yield loss function and a minimum yield loss function in relation to weed densities. Let W_L represent the threshold weed density at which weed treatment is just economic if the maximum loss function prevails and let W_U represent that if the minimum loss function prevails (Figure 3). Given that functions for the value of the increased yield from weed treatment are positively sloped throughout (as in Figure 1), it follows that if the (initial) weed density is less than W_L it is not economic to treat the weed but it is always economic to treat it if the (initial) density exceeds W_U. In cases where $W_L < W < W_U$, the attitude of the farmer towards uncertainty will influence the decision about whether it is optimal or not to treat the weed.

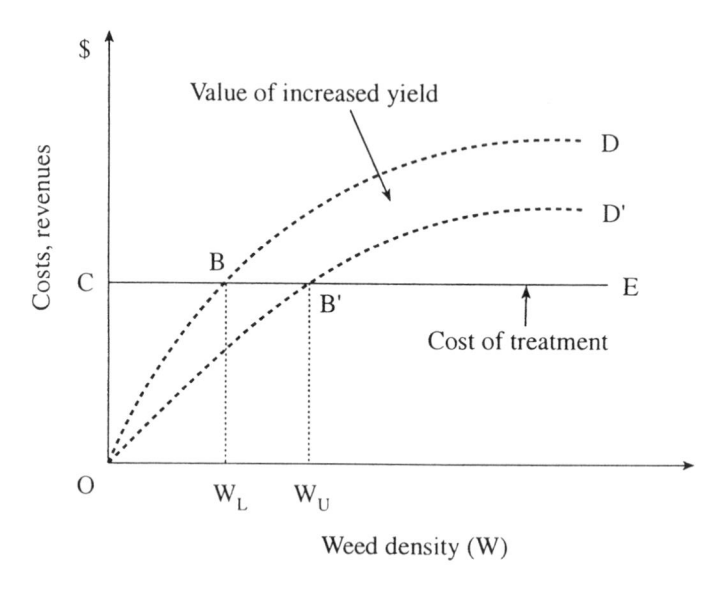

Figure 3 The economic threshold is represented by the range W_U to W_L where there is uncertainty about the loss function (depicted by shading)

The consequences of applying the maximization of expected net gain criterion to the decision of whether or not to engage in weed control (Figure 2) seems to go against the conventional wisdom concerning the effect of uncertainty. On the usual assumption that production loss from weeds increases at a decreasing rate with their density, this criterion implies that with greater uncertainty about densities of weeds, control is less likely to be an optimal strategy. That this is so can be seen from a different representation of the problem to that in Figure 2, a representation that is also useful for the purpose of introducing other criteria for choice under uncertainty.

In Figure 4, DFG represents profit per unit area if weeds are not treated. Given the usual assumptions, it declines at a decreasing rate with increased weed density. The shaded area represents the loss from the presence of weeds. If the weeds are treated and treatment is fully effective, profit may be as is indicated by line AFB. The threshold level of weed density for economic treatment is W_1. DA represents the cost of treatment. It is clear that if uncertainty about weed density increases, the expected value of profit in the absence of treatment rises (due to the strict convexity of curve DFG), whereas the expected gain from treatment remains constant. Hence, the earlier point is supported by this representation of the problem. For instance, if weed density is W_1 a farmer would be indifferent between treating weeds or leaving them untreated if maximizing expected profit is his objective. But suppose the farmer is uncertain about the weed density and believes it 0.5 probable that weed density is W_0 and 0.5 probable that it is W_2 where $0.5W_0 + 0.5W_2 = W_1$. Then the expected profit from non-treatment of weeds, OL, exceeds the expected profit with treatment, OA.

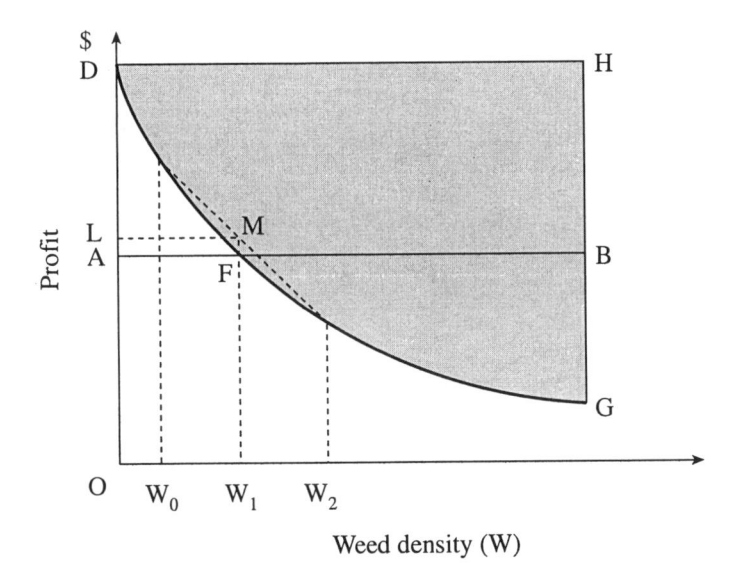

Figure 4   Profit is a function of weed density in relation to the alternative strategies of treatment and non-treatment of weeds

It is conceivable that a farmer may wish to maximize expected utility (Luce and Raiffa, 1957, Ch. 2) from profit rather than expected profit. Utility may increase with profit but at a diminishing rate. Consequently, since profit in the absence of treatment of weeds tends to decline with increased weed density, utility can also be expected to decline with greater weed density. If this happens, the farmer's utility curve as a function of weed density becomes strictly concave. In these circumstances, greater uncertainty about weed density tends to favour treatment of weeds as weed densities become more uncertain.

Although it has been argued that the expected utility maximization criterion is likely to be the most appropriate criterion for choice under uncertainty (Luce and Raiffa, 1957, Ch. 2) (this criterion reduces the expected profit maximization criterion if the utility function is linear), other criteria are defensible.

Riskiness in agricultural production is generally regarded as greater than with other commercial operations, because it is more subject to the vagaries of weather, and the prices received for agricultural produce are frequently unstable. In this risky environment, farmers may adopt management plans which are conservative in the sense that some expected long-run average profit is sacrificed, in order to avoid the possibility of short-term economic disasters. Reichelderfer (1980) notes that risk perception and aversion may be a stronger motivation than profit maximization.

If a farmer is extremely risk-averse (has a great desire for security), his or her choice may be characterized by the minimax gain criterion (Luce and Raiffa, 1957, pp. 278–80). Given the possible states of nature, a farmer employing this criterion should adopt the strategy that maximizes the minimum possible gain associated with each strategy. In this case, there are two alternative possible strategies – treat the weed or do nothing to control it.

The optimal minimax gain strategy can be seen from Figure 4. If there is no possibility of the weed density exceeding W_1, the optimal minimax strategy is not to treat the weed. If there is a possibility, no matter how remote, of the weed density exceeding W_1, the optimal minimax strategy is to treat the weed. As the degree of uncertainty about the density of the weed increases, the chance of a weed density greater than W_1 rises. Therefore, the likelihood that treatment is optimal increases if the minimax rule is adopted. This appears to accord more closely with what seems to be the conventional wisdom. Clearly also, sellers of herbicides would have an incentive to promote the adoption of minimax gain strategies by farmers.

Economic threshold – an example

The following notional example of the effect of wild oats (*Avena fatua* L.) on wheat yield illustrates some of the principles already outlined.

Assumptions

Potential wheat yield 5 t/ha; wheat grain weight 0.9 kg/L. The effect of wild oat density on yield is as described by Dew (1972) (no account of wheat sowing rate or time is taken; no wild oats germinate after herbicide treatment). Control by post-emergent herbicide is 90 per cent effective with no effect on wheat. Costs including herbicide, labour and machinery are fixed ($22/ha). Untreated populations of wild oats increase by 1.35 in the second year, under straw burning and tine cultivation (Wilson et al., 1984); a discount rate of 10 per cent is used to calculate the effect on yield in year 2 (in year 1 dollars). A deduction of $15/t for wheat grain contaminated with > 100 wild oat seeds/L (Australian Grain Handling Authority standard) (wild oat plants produce 15 seeds/plant and 90 per cent fall before wheat harvest). Two prices for wheat, $100/t and $150/t, are considered.

Results

The simple case illustrated in Figure 1 is shown here (Figure 5) in curves (1) and (2). The economic threshold is at 7 wild oat plants/m^2 if the value of wheat is $100/t. However if the price of wheat rises, the economic threshold decreases, illustrated here by an increase to $150/t for wheat producing an economic threshold of between 3 and 4 wild oat plants/m^2 (curve 3).

If the effects of (not treating) wild oats are carried over into a second year of wheat cropping, the effect of a given wild oat density measured in year 1 increases so that the threshold falls, in this case to between 1 and 2 wild oat plants/m^2 (curve 4).

The effect of the weed on quality is to produce a discontinuity in the value of increase yield curve, in this case jumping from curve (2) to curve (5) between 7 and 8 wild oat plants/m^2. However, it is (already) economic to control the weeds in the field, only taking account of their competitive effects in year 1, at densities greater

```
Costs of control, revenues from control ($/ha)
100 ─┤                  (5) Value of increased yield (yr 1)
 90 ─┤                      (wheat $100/t) plus value of
                             increased grain quality
 60 ─┤                  (4) Value of increased yield (yr 1 + 2)
 50 ─┤                                     (wheat $100/t)
 40 ─┤                  (3) Value of increased yield (yr 1)
                                             (wheat $100/t)
 30 ─┤                  (2) Value of increased yield (yr 1)
                                             (wheat $100/t)
 20 ─┤                  (1) Control costs
 10 ─┤
  0 ─┼──────┬──────┬
  0       5     10
  Wild oat density (plants/m²)
```

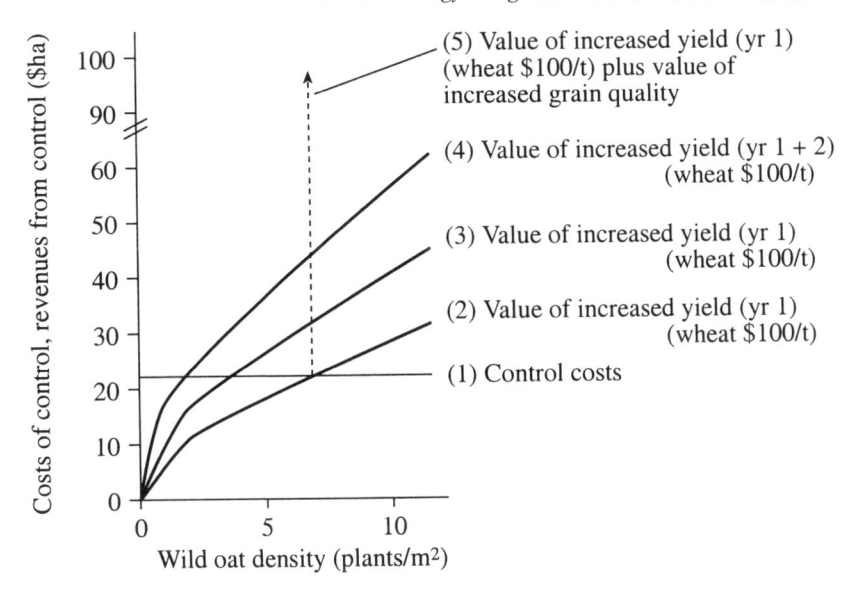

Figure 5 An example of a practical application of the economic threshold model: wild oats, Avena fatua, in wheat

than 7 plants/m². If a farmer is adopting the minimax strategy, in the simplest case, if there is any chance of more than 7 wild oat plants/m², the optimal minimax strategy is to treat the weeds.

With some other weed species, in Australia at least, the official contamination tolerance is 1/50 of that for wild oats and there is nil tolerance for some other weed seeds. The threshold density for these species, taking only competition in year 1 into account, may therefore be considerably higher than the threshold when quality effects are taken into account. In these cases, it may be more economical to clean the contaminated grain after harvest than attempt to control contamination by treating weeds in the field. This would particularly apply in Australia in areas where low wheat yields (e.g. 2 to 3 t) were expected and grain-cleaning costs are $8.50 to $12/t.

Conclusion

A range of factors influences the economic/critical density model. There are several other complications including taking into account the whole-farm context and the occurrence of mixed weed species populations that can have a major bearing on decisions to control weeds. These and other broader issues of weeds in regional contexts as well as social factors in weed control are discussed in the book, *Weed Control Economics* (Auld et al., 1986).

Notes

1. This article originally appeared in B.A. Auld and C.A. Tisdell (1986), 'Economic threshold/critical density models in weed control', *Economic Weed Control*, Proc. EWRS Symposium, pp. 261–8. The content of this article is the same as in the original (except for minor changes) but its presentation has been improved. Consequently, pagination is no longer the same as in the original.
2. Agricultural Research and Veterinary Centre, Orange, New South Wales, Australia.

References

Arrow, K.J. (1971), *Essays in the Theory of Risk-Bearing*, Amsterdam: North-Holland.

Auld, B.A., Menz, K.M. and Tisdell, C.A. (1986), *Weed Control Economics* (Applied Botany and Crop Science Series), New York: Academic Press.

Chisaka, H. (1977), 'Weed damage to crops: yield loss due to weed competition', in J.D. Fryer and S. Matsunaka (eds), *Integrated Control of Weeds*, Tokyo: Japan Scientific Societies Press, pp. 1–16.

Dew, D.A. (1972), 'An index of competition for estimating crop loss due to weeds', *Canadian Journal of Plant Science*, **52**, 921–7.

Luce, R.D. and Raiffa, H. (1957), *Games and Decisions: Introduction and Critical Survey*, New York: Wiley.

Marra, M.C. and Carlson, G.A. (1983), 'An economic threshold model for weeds in soybeans (*Glycine max*)', *Weed Science*, **1**, 604–39.

Reichelderfer, K.H. (1980), 'Economics of integrated pest management: discussion', *American Journal of Agricultural Economics*, **62**, 1012–13.

Tisdell, C.A. (1983), 'The optimal choice of a variety of a species for variable environmental conditions', *Journal of Agricultural Economics*, **34**, 175–85.

Wilson, B.J., Cousens, R. and Cussans, G.W. (1984), 'Exercise in modelling populations of *Avena fatua* L. to aid strategic planning for the long term control of this weed in cereals', *Proc. 7th E.W.R.S. Int. Symp. on Weed Biology, Ecology and Taxonomy*, Paris, pp. 287–94.

[13]

Influence of spatial distribution of weeds on crop yield loss[1]

B.A. Auld[2] and C.A. Tisdell

Summary
Crop yield/weed density functions generally show a decreasing rate of yield loss with increasing weed density (i.e. they are strictly convex functions). We demonstrate the influence of spatial distribution of weeds on crop yield for generalized convex function and give a numerical example for barnyard grass (*Echinochloa crus-galli* var. *oryzicola*) in rice. Plants generally display some degree of contagious or clumped distribution. Where this occurs in crop weeds, estimates of yield loss, based on mean densities over large areas, will usually be over-estimates.

Introduction
Crop yield/weed density models generally indicate a decreasing rate of yield loss with increasing weed density (Cousens, 1985), i.e. strictly convex functions (see Figure 1). Estimates of crop loss are typically obtained by estimating the mean weed density for a large area, e.g. 1 ha or a field of many hectares. However, mean loss values for similar weed densities often vary widely and thus there is difficulty in establishing economic thresholds with precision (Auld et al., 1987). This may be partly explained by the influence of uneven spatial distribution of weeds, as plants frequently have clumped or contagious distributions (Kershaw, 1973).

General derivation
Suppose that an area under study is divided into 1 m² quadrats and that weed density is uniform within each quadrat. Let $f(w)$ represent the yield in a quadrat as a function of its weed density. Furthermore, let n be the number of different weed densities observed in relation to all the quadrats making up the field.

If p_i represents the proportion of a field of size A m² with a weed density of w_i, total yield is

$$Y = \sum_{i=1}^{n} p_i A f(w_i) \qquad (1)$$

and

$$\sum p_i = 1 \qquad (2)$$

Average yield m⁻² is:

$$y = Y/A = \sum_{i=1}^{n} p_i f(w_i) \qquad (3)$$

If all the w_i are not equal (i.e. if weed density per m² is not uniform) and since $f(w)$ is strictly convex, it follows that

$$\sum_{i=1}^{n} p_i f(w_i) > f\left\{ \sum_{i=1}^{n} pw_i \right\} \qquad (4)$$

(Hardy et al., 1934; Karlin, 1959) i.e. non-uniformity of weed density compared to uniformity results in a greater estimate of yield on average m^{-2}.

This is illustrated in Figure 1. If weed density is uniformly distributed at W per square metre, yield on every m² (and therefore on average m^{-2}) is Y_2. Now suppose that weed density in half the m² quadrats is W_1 and in the other half is W_2. In those square metres with a density of W_1, yield is Y_4 and in those with a density of W_2 yield is Y_1. Hence yield on average m^{-2} is

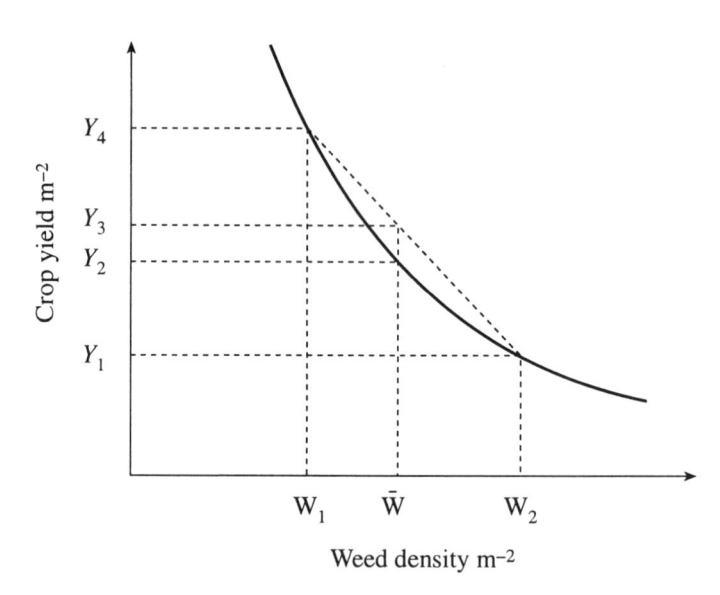

$$Y_3 = 0.5Y_4 + 0.5Y_1 \qquad (5)$$

Figure 1   Generalized crop yield/weed density convex function

Because of the strict convexity of $f(w)$ and as can be observed from Figure 1, $Y_3 > Y_2$, the yield obtained when weeds are uniformly distributed. Hence because the yield estimate is greater for clumped distributions, the yield loss estimate is lower. Thus the current assumption in crop loss models of an even distribution of weeds over-estimates yield loss.

A numerical example

Some appreciation of the impact on physical yield loss can be gained from a specific example. Chisaka (1977) presented a typical yield/weed density relationship that was specific for barnyard grass (*Echinochloa crus-galli* var. *oryzicola*) in rice (although he gave no data on distribution of the barnyard grass we will assume it to have been uniform). Using his function, let us consider two possible situations where the mean weed density over an area of 1 ha was one barnyard grass plant m^{-2}, i.e. the total weed population in each instance was 10 000 plants. In one case the weeds are distributed evenly so that one plant occurs in each m^2. In another case (ignoring edge effects) weeds are distributed in 100 discrete 1 m × 1 m patches each with 100 weeds. The predicted yield loss in the first case is 51 kg ha^{-1} and in the second it is 36 kg ha^{-1}.

Conclusion

Plants frequently display some degree of clumped or contagious distribution. Where this occurs with crop weeds, current estimates of crop yield loss based on mean densities per acre or hectare will usually be over-estimates; the more uneven the spatial distribution of weeds the greater the discrepancy. Consequently, the ability to determine crop loss functions with more precision, and therefore economic thresholds, would require data on weed distribution. In the absence of greater precision, uncertainty about weed distributions may influence decision-making in weed control (Auld and Tisdell, 1987). There is clearly a need to establish the magnitude of this distribution effect in field studies.

Notes

1. This article originally appeared in B.A. Auld and C.A. Tisdell (1988), 'Influence of spatial distribution of weeds on crop yield loss', *Plant Protection Quarterly*, 3(2), p. 81. The content of this article is the same as in the original (except for minor changes) but its presentation has been improved. Consequently, pagination is no longer the same as in the original.
2. Agricultural Research and Veterinary Centre, Forest Road, Orange, New South Wales, 2800, Australia.

References

Auld, B.A. and Tisdell, C.A. (1987), 'Economic thresholds and response to uncertainty in weed control', *Agricultural Systems*, **25**, 219–27.

Auld, B.A., Menz, K.M. and Tisdell, C.A. (1987), *Weed Control Economics*, London: Academic Press.

Chisaka, H. (1977), 'Weed damage to crops: yield loss due to weed competition', in J.D. Fryer and S. Matsunaka (eds), *Integrated Control of Weeds*, Tokyo: Japan Scientific Societies Press, pp. 1–16.

Cousens, R. (1985), 'A simple model relating yield loss to weed density', *Annals of Applied Biology*, **107**, 239–52.

Hardy, G., Littlewood, J.E. and Polya, G. (1934), *Inequalities*, Cambridge: Cambridge University Press.

Karlin, S. (1959), *Mathematical Methods and Theory in Games, Programming and Economics*, Vol. 1, Reading, MA: Addison-Wesley, pp. 404–5.

Kershaw, K.A. (1973), *Quantitative and Dynamic Plant Ecology*, 2nd edn, London: Edward Arnold.

INTERDEPENDENT PESTS:
THE ECONOMICS OF THEIR CONTROL

CLEM TISDELL
University of Newcastle
New South Wales, Australia

ABSTRACT

Population levels of two or more pest species are frequently interdependent and this has consequences for the optimal control of any single one targeted for control. When the controlled pest species is a predator on another pest or is in competitive relationship with it, the optimal (most economic) control of the target species is smaller than in the absence of the interdependence. On the other hand, if the controlled pest species is the prey of predator that is also a pest, or is in symbiosis with another pest, greater control of the target species is required (from an economic standpoint) than in the absence of this interdependence. Conditions for the most economic control of a targeted pest species are outlined and it is observed that governments sometimes fail to take account of the interdependence of pest populations in their pest control policies.

The populations of some species that are pests to man (such as wolves and wild pigs or dingoes and wild pigs) appear to be interrelated. Any scheme to control the population of one of these pest species needs to take account of the impact of this control upon the population of the other pest species and the economic consequences of this interdependence.

The purpose of this short paper is to outline some of the economic principles that need to be considered in controlling species of pests when their populations are interdependent. Although some of the argument will be cast in terms of dingoes and wild pigs to make it more concrete, the argument is a general one. Incidentally, in my study of wild pigs I found that several foresters argued that wild pig populations in Australia depend on the population level of dingoes and vice-versa [1]. The dingo is regarded by many as an important predator of the wild pig. This has induced me to look at theory of this matter.

© 1982, Baywood Publishing Co., Inc.

MODELLING THE INTERDEPENDENCE OF PEST SPECIES

Predator-Prey Model

Typical relationship between the population of two pest species, one of which is a predator and the other its prey, might look like those shown in Figure 1. The population of dingoes (y), the predator species, is shown as a rising function of the population of its prey — wild pigs, x. Graphically it is illustrated by curve AB. Mathematically,

$$y = f(x) \text{ and } f' > 0. \tag{1}$$

The population of wild pigs is shown as a declining function of the number of dingoes. In explicit form

$$x = \phi(y) \text{ where } \phi' < 0, \text{ and} \tag{2}$$

in implicit form

$$y = g(x) \text{ where } g' < 0. \tag{3}$$

which graphically is illustrated by curve CD in Figure 1. The solution of

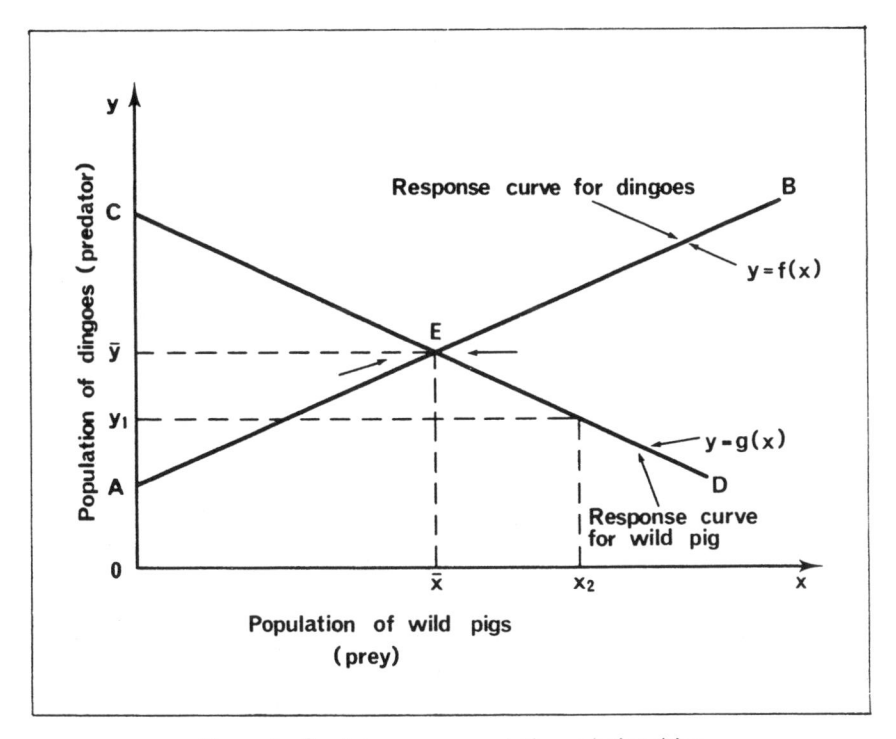

Figure 1. Predator-prey population relationships.

equations (1) and (3) give the equilibrium population of the pests in the absence of human intervention. It corresponds to the intersection of curves AB and CD and in the case shown corresponds to a population of x̄ of pigs and ȳ of dingoes.

Note that in the chosen example the population of dingoes does not disappear if wild pigs are eliminated but remains at level OA. The dingo (the predator) does not depend exclusively on the wild pig for its diet or for survival. In practice dingoes do eat a variety of animals, including kangaroos.

Consider some implications of this model. If the system is in equilibrium and the number of dingoes is reduced, for example by a control campaign, the population of wild pigs rises. For example, if the population of dingoes is reduced by y_1 the population of wild pigs increases to x_2. Given the abundance of prey, there is likely to be a *tendency* for the predator species to increase its population rapidly and this may make it costly to hold the predator population at y_1.[1] However, one of the costs of reducing the dingo population is an increase in the population of another pest, the wild pig.

Shifts in the response curves also alter the equilibrium populations of the pest species. A shift upwards in the response curve for wild pigs (for example because environmental conditions and available food become more favorable for them) increases the equilibrium population both of wild pigs and dingoes. A shift downwards in the dingo response curve because of more human control of their populations, leads to an increase in the population of wild pigs. Both of these cases are illustrated in Figure 2. In the former case the equilibrium shifts from E_1 to E_2 and in the latter case from E_1 to E_3.

However, one must be careful in generalizing from the above model. A predator population may be almost independent of *a* prey species [y = f(x) may be almost vertical] and yet the availability of the prey may be important from the point of view of pest control. For example if pigs become more readily available they may be substituted for sheep more frequently in the diet of wild pigs. Even if the wild dog population remains stationary this is of significance to graziers.

Competitive Pest Species

It is possible for populations of different pest species to be in competition for food and/or habitat. For example, if kangaroos are regarded as pests, red and grey kangaroos could be in competition in some regions. The relationship between two competitive pest species may be like that in Figure 3. The curve CD indicates the population of species, Y, for example red kangaroos,

---

[1] The rate of change of the production function might be of the form
$$\frac{dy}{dt} = k\,[g(x) - f(x)]$$
where k is a positive constant.

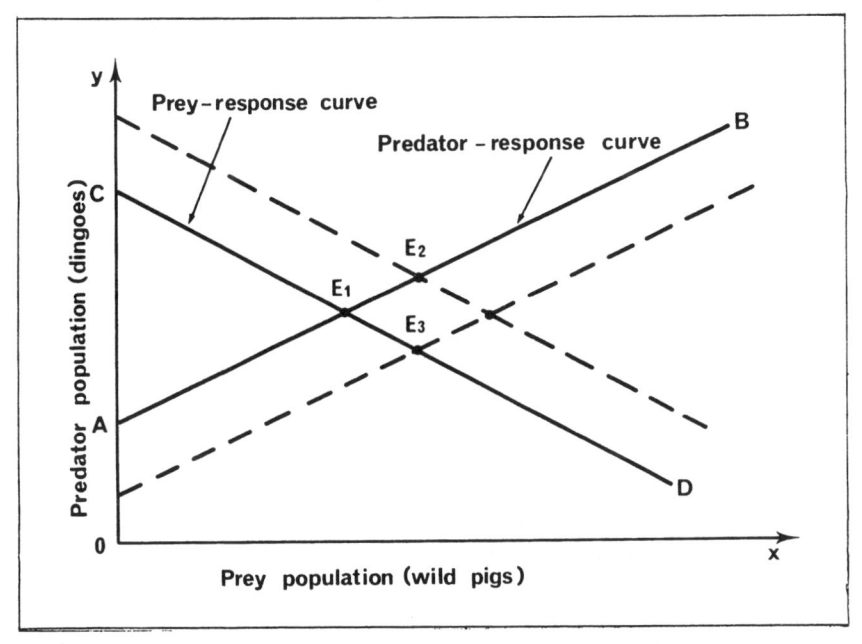

Figure 2. Shifts in predator-prey population equilibriums due to shifts in the population response curves.

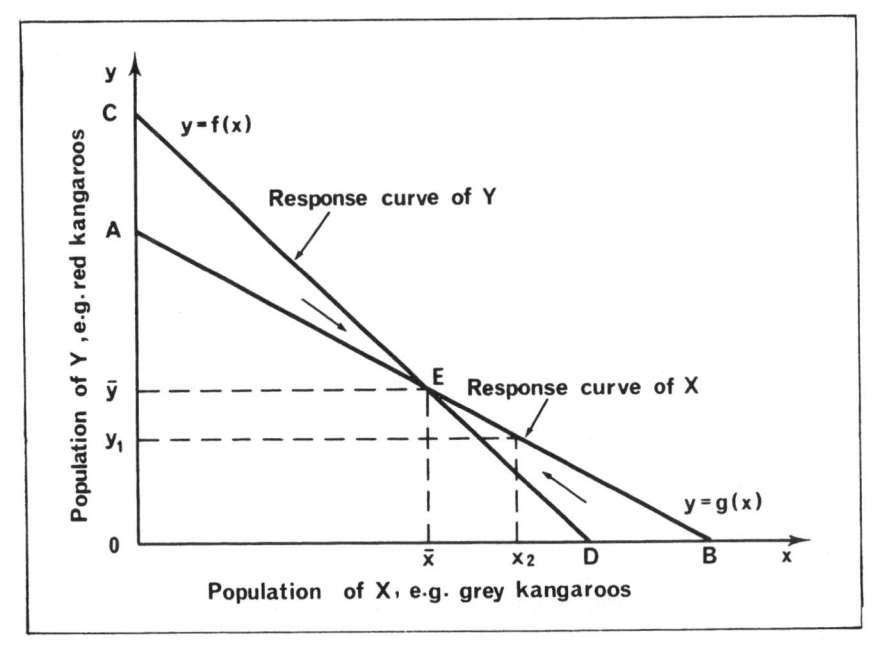

Figure 3. Population relationships for competitive pest species.

as a function of the population of another species, X, say grey kangaroos. The curve AB represents the population of X as a function of that of Y. The two populations are in equilibrium at E, that is when $y = \bar{y}$ and $x = \bar{x}$.

One implication of this model is that if the population of one of the pests is reduced by control measures (and the other is not controlled) the population of the other pest rises. Thus in Figure 3, if the population of Y is reduced from $\bar{y}$ to y_1 that of X rises from $\bar{x}$ to x_2. A reduction in the population of one of the pests is compensated for to a certain extent by an increase in the population of the other. This influences, as discussed below, the economics of controlling the pests.

Symbiotic Pests

Some pests are in a symbiotic relationship, for example ants and aphids or scale insects. A relationship of this kind is illustrated in Figure 4. The line AB represents the response of species X to the population of Y and curve CD the response of species Y to the population of species X. The equilibrium levels of the populations occur at E, that is for a population of $\bar{y}$ for species Y and $\bar{x}$ for species X.

In circumstances of this type by reducing the population of one of the pests one also lowers the population of the other pest. From the point of view of pest control, control of one of the pests yields a bonus because of the control it exerts on the other pest.

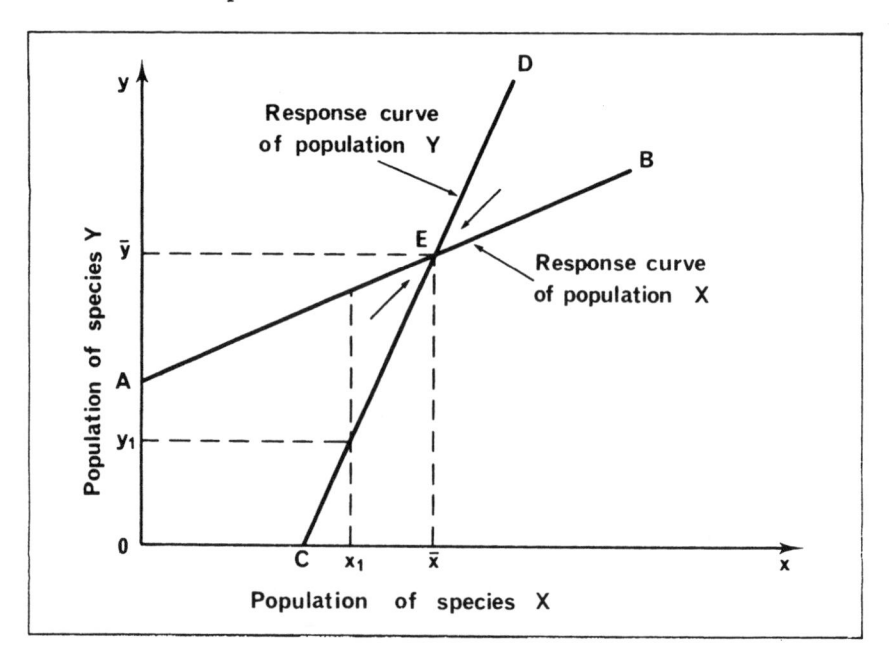

Figure 4. Population relationships for symbiotic pests.

Note that although population Y is shown in Figure 4 as being dependent upon a minimal quantity of x, in fact OC, for its continued existence, it is possible for the response curve of population Y to intersect the Y-axis. If it does so, population Y can exist in the absence of X. However, stability of equilibrium requires CD to intersect AB from below. In the linear case, this requires that the intercept of CD with the Y-axis is below point A.

Economic Consequences of the Interdependence

If one is only controlling the population of one pest species in a group of interdependent pest species, the optimality of its control requires that account be taken of its interdependence. While profit maximization requires that the population of a pest be reduced until the marginal cost of its reduction equals the marginal gain from it [2-4], in the case of interdependence, species account must be specifically taken of this interdependence. If one is controlling a predator species or a competitive species, the marginal cost of its reduction must be compared with 1) the direct marginal gain from the reduction in its numbers *less* 2) the marginal loss from an increase in the other pest species. Profit maximization requires that the reduction in the target species proceeds until the marginal cost of its reduction plus the marginal loss from increase in the non-target species equals the direction marginal gain from the reduction in the target species. This is illustrated in Figure 5. Curve KL represents the marginal cost of reducing the target species, curve MN the combined marginal cost of reducing the species (taking account of the non-target species) and ST is the marginal gain from the reduction. Net gain is maximized for a reduction of r_1 in the population of the target species.

However, if account is not taken of species interdependence, a larger reduction than r_1 of the target species appears to be optimal, namely a reduction of r_2. The greater the marginal loss from an increase in the non-target species, the smaller the justified reduction in the target species. This will tend to be so if the non-target species increases considerably as the target species is reduced and/or creates considerable economic damage. In these circumstances, curve MN is higher in Figure 5.

If the target species is either the prey of another pest or in symbiotic relationship with another pest, greater control is likely to be optimal than would appear from ignoring this interdependence. In this case, the cost of reducing the target species needs to be compared with 1) the direct marginal gain from a reduction in the population of the target species *plus* 2) the marginal gain from a decrease in the population of the dependent pest-species. Profit maximization requires that the population of the target species be reduced until the marginal direct cost of that reduction equals the marginal direct gain from this reduction plus the marginal gain from the decrease in the population of the dependent species. This is illustrated in Figure 6.

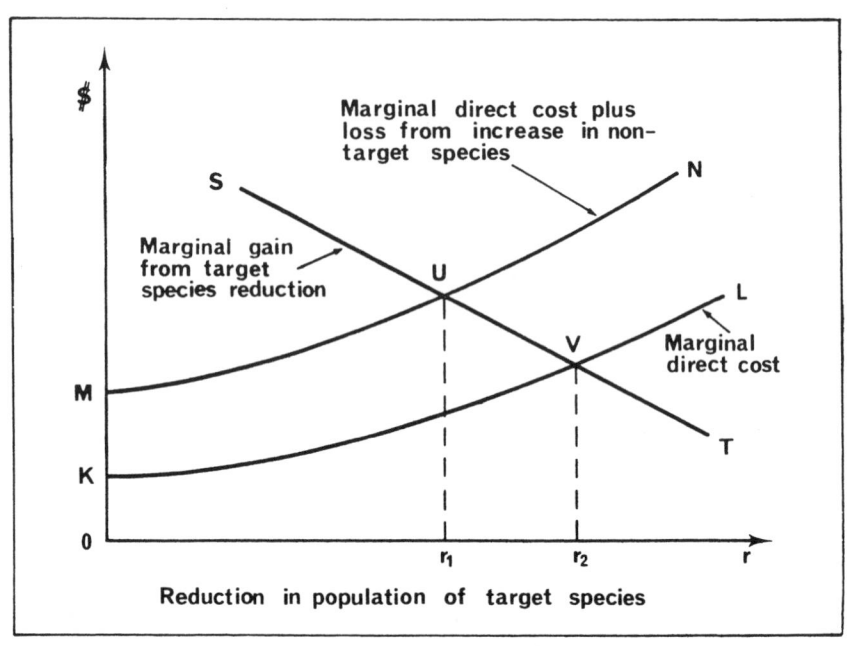

176 Economics and Ecology in Agriculture and Marine Production

Figure 5. Profit-maximizing reduction in the population of a predator-pest or a competitive pest.

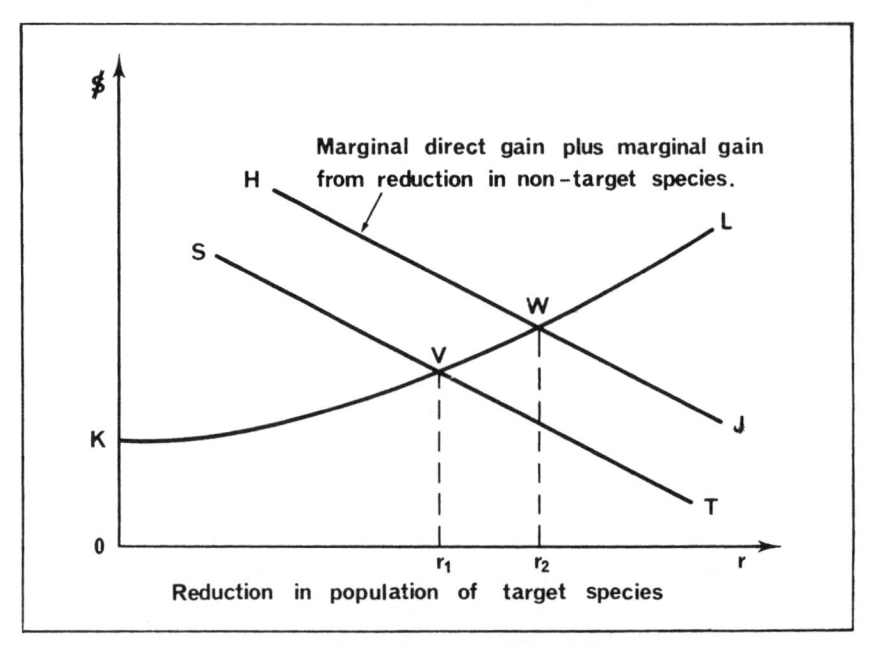

Figure 6. Profit-maximizing reduction in the population of a prey-pest or a symbiotic pest.

In Figure 6, curve KL represents the marginal cost of reducing the target pest, curve ST is the direct marginal gain from this and HJ equals this marginal gain plus the marginal gain from a reduction in the dependent non-target species. Profit from pest control in this case is maximized when the population of the target species is reduced by r_2. A greater reduction in the target species is justified than when its interdependence with another pest is ignored, the most profitable reduction in the target species amounts to r_1.

CONCLUSION

It can be seen that where a predator-pest or a competitive-pest is being controlled that the mere consideration of the gains from control of that target species (itself) is likely to overstate the benefits from its control if its prey or its competitive species is a pest. Control of the target species is liable to be on a scale greater than the most profitable scale. For instance, this could be so for the dingo or wild dog if its main prey is also a pest. Conversely in the case of a prey species (the prey of a pest) or one in symbiotic relationship with another pest, the benefits of its control are liable to be understated if account is only taken of the direct benefits of a reduction in its population. The most profitable level of reduction in its population can be expected to be greater than suggested by the narrow approach because a reduction in its population also lowers the population of the pest dependent on it.

In the ecological world as in the economic world interdependence between components of systems is important. But as yet little regard appears to have been paid to this in policies for vertebrate pest control in Australia. The benefits of controlling many species appear to be judged in isolation from the web of interrelationships of species. This may reflect the difficulties of modeling the interrelationships. However, serious consideration needs to be given to these questions in the major pest control policies, such as dingo destruction, undertaken by government agencies in various Australian states.

REFERENCES

1. C. A. Tisdell, *Wild Pigs: Environmental Pest or Economic Resource*, Pergamon Press, Sydney, in press.
2. J. C. Headley, Defining the Economic Threshold, *Pest Control Strategies for the Future*, National Academy of Sciences, Washington, D.C., pp. 100-108, 1972.
3. C. Tisdell, Simple Economic Models of Pest Control, *Research Report or Occasional Paper 33,* Department of Economics, University of Newcastle, Australia, May 1977.

4. _____ , Observations on the Wild Pig Problem in N.S.W., *Research Report or Occasional Paper 41*, Department of Economics, University of Newcastle, Australia, March 1978.

Direct reprint requests to:

Clem Tisdell
Department of Economics
University of Newcastle
New South Wales 2308 Australia

BIOLOGICAL WEED CONTROL — EQUILIBRIA MODELS

B.A. AULD

Agricultural Research and Veterinary Centre, Forest Road, Orange, N.S.W. 2800 (Australia)

C.A. TISDELL

Department of Economics, University of Newcastle, Newcastle, N.S.W. 2308 (Australia)

(Accepted for publication 7 November 1984)

ABSTRACT

Auld, B.A. and Tisdell, C.A., 1985. Biological weed control — equilibria models. *Agric. Ecosystems Environ.*, 13: 1—8.

Interactions between weeds and biological control agents are examined by simple predator/prey biomass analysis. The possibility of discontinuously stable systems is demonstrated. The shape of consumption functions for individual predators and populations is shown to be important in their success. The potential for augmentative biological control, where discontinuous stability exists, is shown and the relevance of an ungrazable fraction in weed population persistence is also described.

INTRODUCTION

Analyses of success or failure in biological weed control are comparatively rare. Occasionally parasitism or environmental constraints are suggested as factors responsible for failure or limited success. Moreover there have been few thorough studies in the field or laboratory or speculative analyses (Caughley, 1976) on the dynamics of interacting species. Little progress has been made in providing a rational basis for selecting the most effective biological control agents (Winder and van Emden, 1980).

Populations of weeds (prey) and biological control agents (predators) can be modelled by predator—prey systems, although numbers of weeds and agents will be affected by factors other than their respective densities. For the agents these include: aggregative response to host patches, dispersal powers, search efficiency (Beddington et al., 1978) and size of plants being attacked. For the weeds they include: climatic and edaphic factors and interference from other plants. Yet no matter what mechanisms are involved it is clear that there must be adequate numbers and biomass of biological control agents in relation to weeds (Huffaker et al., 1976) for successful biological control programmes. Moreover, the fact that there can be direct interdependence between weed biomass and biological control agent's populations and consumption functions often appears to be overlooked. However, Meyers

(1980) has recognised this in the cinnabar moth—ragwort system in which spatial distribution is also important and White (1980) has observed this interdependence in *Cactoblastis cactorum* Berg. and *Opuntia* spp.

It is characteristic of effective arthropod biological control projects that predator—prey equilibria appear to be stable (Hassell, 1978). The same appears to be ultimately true in the biological control of weeds (e.g., Nakamura and Ohgushi, 1981). In this paper, possible discontinuity of stability is investigated using simple graphical models based on the predator—prey theory. The interaction between grazing intensity and ability of plants to respond in regrowth, given that they cannot influence the rate of renewal of principal resources (water and sunlight), is explored in this paper. Noy—Meir (1975) adopted the classical predator—prey analysis of Rosenzweig and MacArthur (1963) to study grazing systems. Noy-Meir's (1975) approach is followed to investigate a particular case of grazing systems, the biological control of a weed (prey) by a host-specific predator (grazer).

ASSUMPTIONS

We shall assume that the growth in biomass of the weed species concerned (i.e., combined growth in biomass per plant and in population) is of a logistic form

$$\frac{dW}{dt} = rW(1 - W/K) \qquad (1)$$

where W = weed biomass, r = net proportional growth rate of weed biomass and K = environmental carrying capacity or saturation level. The general form of plant/sward growth is well established (Brougham, 1955; Donald, 1961) and there is evidence that plant population growth is also of that form, although a constant maximum rate may be maintained for long periods in perennials (Harper, 1977). Thus the relationship of weed growth to weed biomass-density has a single maximum (unbroken line in Fig. 1) which may form a plateau. Its amplitude and maximum value would be affected by seasonal conditions as well as, for instance, the addition of fertilizer.

Various forms of the relationship between predators and prey intensity have been postulated (Huffaker et al., 1976) and found experimentally for predators/herbivores and plant biomass (Noy-Meir, 1975). These are reducible to three main forms: linear, Michaelis and sigmoid. Zimmerman and Malan (1980) have shown a Michaelis type response for *C. cactorum* feeding on *Opuntia ficus-indica* L. We shall, initially, consider the Michaelis form

$$C = k \frac{W}{W + K_s} \qquad (2)$$

where C = level of consumption of the weed biomass, k = maximum rate of the consumption process and K_s = half saturation constant. At higher predator population increases, Michaelis curves with higher maxima and steeper

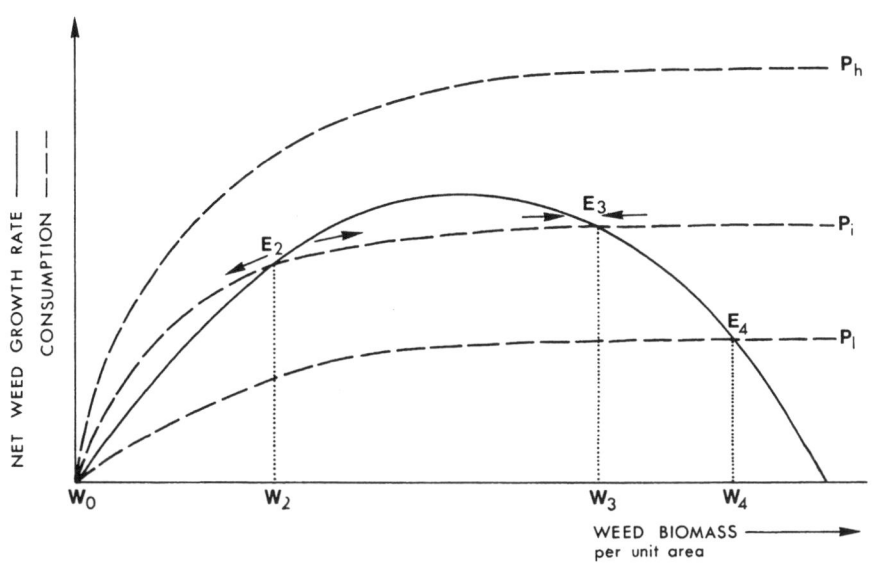

Fig. 1. Interaction between weed growth rate (solid line) and three population densities of predators (broken lines) with Michaelis consumption functions. High predator populations (P_h) lead to weed eradication (W_0); low predator populations (P_1) lead to stable equilibrium (E_4) at high weed biomass (W_4); intermediate populations (P_i) may lead to stable equilibrium (E_3) at W_3 or unstable equilibrium (E_2) at W_2 which could revert to W_0 or E_3.

slopes near the origin are produced (Fig. 1), i.e., the parameters k and K_s increase.

The higher consumption curves (P_h in Fig. 1) correspond to higher levels of predator population, but there may be a limit to the extent to which consumption curves can be pushed up by increasing predator release. Each curve represents a particular level of predator release and in the time considered mortality of the predator can occur. However, no significant change in its population size is envisaged, even though the weed biomass may be altered. Thus, the model is especially relevant to short-term ecological situations (e.g., one-seasonal) such as may be encountered in initial, inundative or augmentative releases (Tisdell et al., 1984). For simplicity, changes in production of weed biomass and in consumption as a function of weed biomass are assumed to be instantaneous.

ANALYSIS

Four equilibria situations are possible (Fig. 1) (tangential cases are omitted for simplicity). At high predator populations (P_h) the weed is eradicated (W_0); for low predator populations (P_1), and equilibrium position E_4 is reached at a high weed biomass W_4. (Similar equilibria situations could occur for linear consumption functions.) For an intermediate population (with a Michaelis consumption function) (P_i), a stable equilibrium E_3 is possible at a relatively high weed biomass W_3, but at W_2 there is an unstable equilibrium E_2 which may lead to eradication of the weed; as soon as $W < W_2$ the population is too low to maintain itself above the consumption rate, and extinction to W_0 follows.

A sigmoid consumption function can produce two steady states separated by an unstable equilibrium point (Fig. 2). Thus a given population of preda-

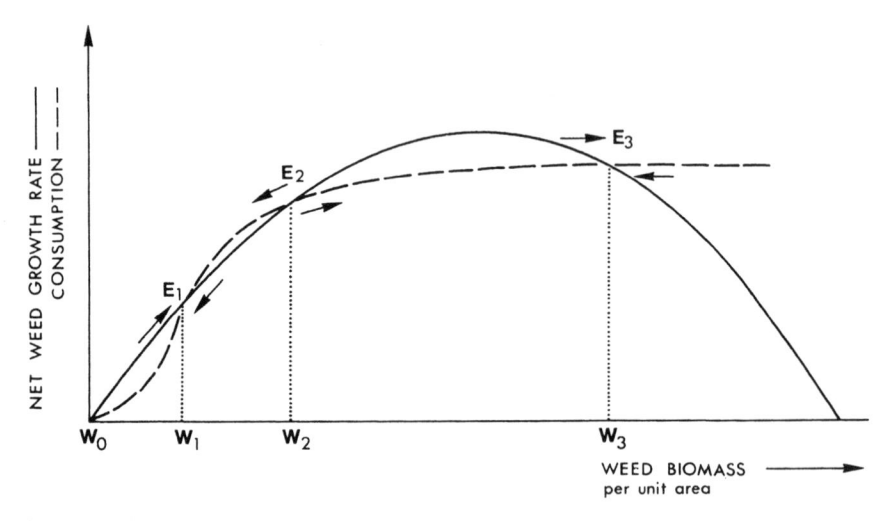

Fig. 2. Interactions between weed growth (solid line) and a predator population (broken line) with a sigmoid consumption function. Two possible stable equilibria biomasses W_1 and W_3 are separated by an unstable equilibrium point E_2 at weed biomass W_2.

tors could maintain two quite different weed biomass, one high (W_3) and one low (W_1); from the unstable equilibrium position, E_2, consumption could move in either direction, depending on whether a slight shift was given to either function. A similar multiple equilibria possibility has been observed in relation to fisheries management (Peterman, 1977; Holling, 1978).

The above graphical examples can be presented in a more general form, where N represents the weed biomass after consumption,

$$\frac{dN}{dt} = \frac{dW - C(t)}{dt} \tag{3}$$

and in equilibrium

$$\frac{dN}{dt} = 0 \tag{4}$$

It is assumed that,

$$\frac{dW}{dt} = F(W) \tag{5}$$

and

$$C = H(\alpha, W) \tag{6}$$

where W is the biomass of the weed at time, t, and α is a parameter or set of parameters varying with the release level of the predator. For a particular release level, it might be that $\alpha = \hat{\alpha}$ and

$$C = H(\hat{\alpha}, W) = G(W) \tag{7}$$

Equilibria points are found by solving (5) and (6) simultaneously for different values of α.

Given instantaneous adjustments, an equilibrium value is stable for a small displacement in a particular direction if forces return the system to its original equilibrium. An upward displacement in weed biomass from an equilibrium is unstable if the production of weed biomass exceeds predator consumption of the biomass for the displacement, and stable if the opposite is so. A downward displacement of weed biomass is unstable if the displacement consumption by the predator exceeds biomass production of the weed and is stable if the opposite is the case (Figs. 1 and 2). The introduction of lagged adjustments complicates the analysis of stability of the equilibrium — an equilibrium that is stable for an instantaneous adjustment may be unstable for a lagged one (Tisdell, 1972; May, 1973, 1975).

A further modification of these main possibilities is possible when ungrazable portions of the weed (W_p) remain (Fig. 3), (e.g., rhizome, seed or individuals inaccessible to predators in space or time).

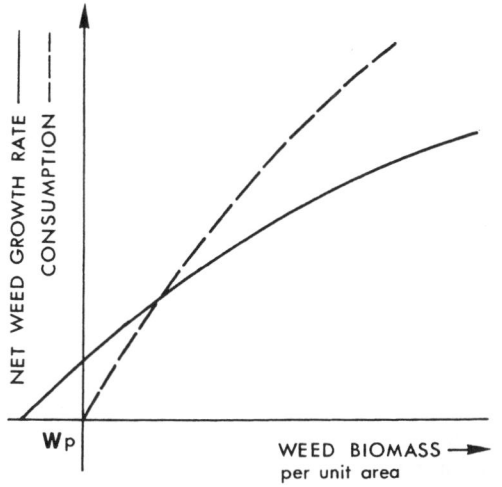

Fig. 3. Interaction between predators (broken line) and weeds (solid line) with an ungrazable fraction (W_p).

DISCUSSION

Equilibria at high weed biomass (W_3, W_4 in Figs. 1 and 2) in biological control of weeds is common (e.g., *Rhinocyllus conicus* Froelich on Italian thistle; Goeden and Ricker, 1978). Harris (1980) argues that these low levels of damage still put stress on the target and are therefore useful, especially when combined with additional stress. Other forms of stress may be environmental, reducing weed biomass production, thus ultimately producing an equilibrium at a much lower biomass. Such a situation could apply at the edges of a spreading population. Auld (1969) has shown how relatively low infestations of predators play a part in reducing spread of *Ageratina* (= *Eupatorium*) *adenophora* King and Robinson, although higher infestations of predators in areas favourable for the weed do not reduce local population growth rates. Hosking and Deighton (1980) have demonstrated that the effect of *Dactylopius austrinus* De Lotto on *Opuntia aurantiaca* Lindley is enhanced by decreasing moisture availability to the plant, which would reduce biomass production rate.

The steeper the initial part of the consumption function (i.e., the greater the searching efficiency of predators; Fig. 1) the greater the likelihood of reaching an equilibrium at low weed biomass or eradicating the weed. This indicates that given the assumptions of the model, r-type biological control agents should be preferred to K-type agents, particularly for scattered weed populations, but spatial aspects of the weed population are not taken into account in this model. It also suggests that selecting for higher per capita consumption may be worthwhile.

For all forms of consumption function, increasing predator density will reduce weed biomass. However, if the consumption function has a small, constant, initial gradient, the predator population may have to be increased substantially to reach low equilibrium weed biomass.

When an unstable equilibrium is reached small increases in predator population will cause a dramatic reduction in weed biomass (cf. Peterman, 1977; Holling, 1978).

Berryman (1982) has recognised the possibility of more than one equilibrium and unstable points in analysing pine beetle dynamics. This phenomenon has also been demonstrated in experimental grazing studies with domestic animals, where small differences in plant productivity have caused large differences in animal production (Morley, 1966).

The possible existence of intersection points at E_2 (Figs. 1 and 2) means that supplementation of predator populations could have dramatic effects. Using an augmentative approach in the biological control of weeds, supplementing predator populations has recently proved useful in controlling *Opuntia aurantiaca* in areas where classical biocontrol has been limited in Australia (Hosking and Deighton, 1979) and in *Cyperus rotundus* L. control in the U.S.A. (Frick and Chandler, 1978).

Additionally, some reduction in weed biomass at these points by other control methods could have a profound synergistic effect. In this latter case, however, clumped areas are more likely to be treated and this may affect the predator population (Zimmerman, 1979). Conversely, if weed growth were increased (i.e., upward shift in biomass function) at these unstable points, for instance by favourable climatic conditions, a stable equilibrium at a high biomass would be reached (Zimmerman and Malan, 1980).

Ungrazable fractions (Fig. 3) may exist in the form of buried weed seeds or some inaccessible perenniating tissue as occurs in *C. rotundus* control (Frick and Chandler, 1978). The ability of predators to survive periods when no grazable fraction is present is crucial in successful classical biological control. In some cases it could be worthwhile maintaining some of the weed population in a grazable state to ensure predator survival, rather than incur the costs of reintroductions of the predator.

ACKNOWLEDGEMENT

The authors thank Dr. K.M. Menz for his comments on the draft manuscript.

REFERENCES

Auld, B.A., 1969. Incidence of damage caused by organisms which attack crofton weed in the Richmond-Tweed region of New South Wales. Austr. J. Sci., 32: 163.
Beddington, J.R., Free, C.A. and Lawton, J.H., 1978. Characteristics of successful natural enemies in models of biological control of insect pests. Nature, 273: 513—519.
Berryman, A.A., 1982. Biological control, thresholds and pest outbreaks. Environ. Entomol., 11: 544—549.
Brougham, R.W., 1955. A study in rate of pasture growth. Aust. J. Agric. Res., 6: 804—812.
Caughley, G., 1976. Plant—herbivore systems. In: R.M. May (Editor), Theoretical Ecology. Blackwell Scientific, Oxford, pp. 94—113.
Donald, C.M., 1961. Competition for light in crops and pastures. In: F.L. Milthorpe (Editor), Mechanisms in Biological Competition. Symp. Soc. Exp. Biol., 15: 282—313.
Frick, K.E. and Chandler, J.M., 1978. Augmenting the moth (*Bactra verutana*) in field plots for early-season suppression of purple nutsedge (*Cyperus rotundus*). Weed Sci., 26: 703—710.
Goeden, R.D. and Ricker, D.W., 1978. Establishment of *Rhinocyllus conicus* (Col.: Curculionidae) on Italian thistle in southern California. Environ. Entomol., 7: 787—789.
Harper, J.L., 1977. Population Biology of Plants. Academic Press, London, 892 pp.
Harris, P., 1980. Evaluating biocontrol of weeds projects. Proceedings of the 5th International Symposium on Biological Control of Weeds, Brisbane, pp. 345—353.
Hassell, M.P., 1978. The dynamics of Arthropod Predator—Prey Systems. Princeton University Press, Princeton, NJ, 237 pp.
Holling, C.S., 1978. Adaptive Environmental Assessment and Management. Wiley, New York, 363 pp.
Hosking, J,R. and Deighton, P.J., 1979. The distribution and control of *Opuntia aurantiaca* in New South Wales. Proceedings of the 7th Asian—Pacific Weed Science Society Conference, Sydney, pp. 195—200.

Hosking, J.R. and Deighton, P.J., 1980. Biological control of moisture stressed *Opuntia aurantiaca* using *Dactylopius austrinus*. Proceedings of the 5th International Symposium on Biological Control of Weeds, Brisbane, pp. 483—487.

Huffaker, C.B., Simmonds, F.J. and Laing, J.E., 1976. The theoretical and empirical basis of biological control. In: C.B. Huffaker and P.S. Messenger (Editors), Theory and Practice of Biological Control. Academic Press, New York, pp. 42—78.

May, R.M., 1973. Stability and Complexity in Model Ecosystems, Princeton University Press, Princeton, NJ.

May, R.M., 1975. Biological populations obeying difference equations: stable points, stable cycles and chaos. Theoret. Biol., 51: 511—524.

Morley, F.H.W., 1966. Stability and productivity of pastures. Proc. N.Z. Soc. Anim. Prod., 26: 8—21.

Meyers, J.H., 1980. Is the insect or the plant the driving force in the cinnabar moth—tansy ragwort system? Oecologia, 47: 16—21.

Nakamura, K. and Ohgushi, T., 1981. Studies on the population dynamics of a thistle-feeding lady beetle *Henosepilachna pustulosa* (Kôno) in a cool temperate climate forest. II. Life tables, key-factor analysis, and detection of regulatory mechanisms. Res. Popul. Ecol. (Kyoto), 23: 210—231.

Noy-Meir, I., 1975. Stability of grazing systems: an application of predator—prey graphs. J. Ecol., 63: 459—481.

Peterman, R.M., 1977. A simple mechanism that causes collapsing stability in salmonoid populations. J. Fish. Res. Board Can., 34: 1130—1142.

Rosenzweig, M.L. and MacArthur, R.H., 1963. Graphic representation and stability conditions of predator—prey interactions. Am. Nat., 97: 209—223.

Tisdell, C.A., 1972. Microeconomics: The Theory of Economic Allocation. Wiley, Brisbane.

Tisdell, C.A., Auld, B.A. and Menz, K.M., 1984. On assessing the value of biological control of weeds. Prot. Ecol., 6: 169—179.

White, G.G., 1980. Current status of prickly pear control by *Cactoblastis cactorum* in Queensland. Proceedings of the 5th International Symposium on Biological Control of Weeds, Brisbane, pp. 609—616.

Winder, J.A. and van Emden, H.F., 1980. Selection of effective biological control agents from artificial defoliation/insect cage experiments. Proceedings of the 5th International Symposium on Biological Control of Weeds, Brisbane, pp. 415—439.

Zimmerman, H.G., 1979. Herbicidal control in relation to distribution of *Opuntia aurantiaca* Lindley and effects on cochineal populations. Weed Res., 19: 89—94.

Zimmerman, H.G. and Malan, D.E., 1980. The role of imported natural enemies in suppressing re-growth of prickly pear, *Opuntia ficus-indica*, in South Africa. Proceedings of the 5th International Symposium on Biological Control of Weeds, Brisbane, pp. 375—381.

ON ASSESSING THE VALUE OF BIOLOGICAL CONTROL OF WEEDS

C.A. TISDELL[1], B.A. AULD[2] and K.M. MENZ[3]

[1] *Department of Economics, University of Newcastle, N.S.W. 2308 (Australia)*
[2] *Agricultural Research and Veterinary Centre, Department of Agriculture, Forest Road, Orange, N.S.W. 2800 (Australia)*
[3] *Bureau of Agricultural Economics, P.O. Box 1563, Canberra, A.C.T. 2601 (Australia)*

(Accepted 18 May 1983)

ABSTRACT

Tisdell, C.A., Auld, B.A. and Menz, K.M., 1984. On assessing the value of biological control of weeds. *Prot. Ecol.*, 6: 169—179.

A range of factors is outlined that need to be taken into account in assessing the economic and social value of biological weed control. These factors, which are important for any social cost-benefit analysis of biological weed control, vary in significance depending upon whether the biological control is of the classical, augmentative, or inundative type. The market-institutional system is 'biased' in favour of chemical and mechanical controls rather than biological ones, and favours inundative biological controls rather than classical ones. The community as a whole may gain by governments actively tipping the balance towards biological controls, for example through its research efforts and its dissemination of information.

INTRODUCTION

In this paper we consider economic and social factors in the biological control of weeds, especially in comparison with alternative methods of control, such as the use of herbicides. Although a number of these considerations are mentioned in the literature, it is important to bring them together systematically. We also discuss the reasons why biological control is often not adopted even when it is socially and economically the most advantageous form of control.

Three basic types of biological control can be distinguished: classical, augmentative and inundative.

(1) Classical biological control involves the release of a natural enemy (predator, parasite) of a weed or pest. The biocontrol agent is imported and once established, the population of the agent is self-sustaining and reaches a long-term equilibrium with that of its (weed) host.

(2) Augmentative control refers to a situation in which the biocontrol agent is unable to sustain itself in the long-term in all areas of pest infestation without augmentation of its population. Periodic or seasonal supplementation of the control agent may be required at least in some regions.

Additional populations of the control agent or carriers of it may need to be released. For example, European rabbit fleas *(Spilopsyllus cuniculi* (Dale) carrying myxomatosis virus need to be periodically released in certain areas of Australia (where there are few mosquitos as carriers) to maintain control of rabbit populations (Davis et al., 1976; Ecos, 1978).

(3) Inundative biological control refers to relatively short-term control achieved by releasing a biocontrol agent in 'heavy' concentrations such as suspensions of fungal pathogens applied in the same way as herbicides (mycoherbicides). The population of the control agent is not self-perpetuating, or only self-perpetuating at a level that gives insignificant control of the pest. The control effects of inundative biocontrol agents tend to be short-term in much the same way as chemical controls. While they are often thought of as being most suited to control of pests of annual crops (Templeton et al., 1979), their range of application is wider.

CLASSICAL BIOLOGICAL CONTROL

Spillovers or externalities are a very important economic and social consideration in classical biological control. The control agent is introduced to a limited number of places and spreads to other areas, 'searching out' the pest and controlling it there; it does so of its own accord. From a cost point of view, this ability of the control agent to search for or to come independently into contact with the pest can be regarded as an advantage compared with chemical and mechanical pest control where human effort and additional expense and energy use are required to link the agent with the pest. Furthermore, a biological control agent may be able to spread to otherwise inaccessible areas.

However, these spillover advantages from 'searching' or autonomous movement of the biocontrol agent mean that the private profit obtained by an individual farmer were he to instigate classical biological control on his farm is less than its social value. (In certain cases but not all, this social value is the increase in profit accruing to all farmers.) This impedes the private exploitation of such a technique. It reduces the demand for the biocontrol agent from any manufacturer or supplier and ultimately results in the classical biological control technique being used to a lesser extent than is socially optimal from an economic viewpoint. To correct for this problem, the government may need to subsidise private use of the biological control or, more usually, the state itself may supply and apply the biological control.

One difficulty in evaluating the classical biological control of weeds is that the weeds may not necessarily be regarded as a pest by all landholders or members of the community or by the same individual at all times. Given the spillover of control results from the mobility of the control agent, this conflict involves a difficult social evaluation problem, particularly where those suffering losses from biological control are not the ones deriving benefits (Andres, 1980). For instance, in Australia a government institution, the Commonwealth Scientific and Industrial Research Organisation (CSIRO) planned to introduce several insects to control the weed, "Paterson's curse", *Echium plantagineum* L. However, in some regions, for example in the Flinders Ranges of South Australia, graziers regard Paterson's curse as a useful drought fodder reserve for sheep (despite its cumulative adverse impact on the liver of sheep fed on it) and in these regions the plant is called "Salvation Jane". Honey producers also gain from the presence of the weed. A court injunction against CSIRO, restraining it from releasing the control agents, was obtained by some of those who derive benefits from the weed and the matter is still in dispute. Andres (1980) cites a similar conflict of interest with *Tamarix ramosissima* Ledeb. control in the U.S.A.

Although the benefits of a weed control should be easier to determine if the supply of *marketable* products is affected by such control (Andres, 1980), this does not appear to have reduced the complexity of the Paterson's curse conflict. While it seems that the monetary gains of the group of farmers who would benefit from a reduction in the amount of Paterson's curse would far outweigh the monetary losses of the losers, the distribution of income would be altered by the introduction of the control agent. On the basis of the Kaldor-Hicks principle (Winch, 1971) that a social change (in this case the introduction of the control agent) is desirable if the gainers could compensate the losers, biological control appears to be socially desirable. But this ignores the income distribution question: Is the distribution of income adversely affected and, if so, is the net overall gain sufficient to compensate from a social point of view for a worsening of the distribution of income? There is no 'scientific' economic answer to these questions which, in practice, are resolved by the political process.

In addition to 'searching out' and attacking target weeds, biological control agents may also 'search out' and attack non-target species. This could be regarded as an environmental spillover, somewhat akin to environmental spillovers associated with chemical or mechanical weed control. The consequences of the environmental spillover from biological control may be less reversible than with mechanical control, (e.g., resulting in soil erosion) or with chemical control, (e.g., affecting non-target species) because the biological control agent is self-perpetuating. A fungal pathogen, *Puccinia xanthii* Schw. accidentally introduced to Australia from the U.S.A. aroused interest because of its apparent specificity to *Xanthium* and *Ambrosia* species weeds in the U.S.A. and its potential for biological control of *Xanthium* species in Australia (Alcorn, 1976). However since 1976 it has been recorded in the field on two sunflower *(Helianthus annuus)* cultivars (c.v. Hysun 30, c.v. Sunfola 68-2) even though it has never been recorded on *Helianthus* in the U.S.A. (J. Walker, personal communication, 1982). The bulk of successful introductions of biological control agents have involved highly host-specific 'enemy' species (Huffaker et al., 1976), so even where an agent has proven itself in a country overseas, it still needs to be specificity-tested for its suitability in any other country contemplating its introduction (Charudattan, 1982).

Another important factor to be considered in evaluating a control agent is whether the target-weed or pest will develop resistance to the control agent and at what rate (Tisdell, 1982b). It is sometimes believed that resistance of the pest is likely to build up more slowly when biological controls are used rather than chemical ones. There is already some evidence for growing resistance to herbicides in some weed species (Muir, 1978).

However, even if the host species develops resistance or an unfavourable environment is experienced by the biocontrol agent, the possibility exists for the control agent to evolve, becoming more virulent. The opportunity for evolution by the biological control agent is probably greatest in the case of classical biological control, where *Chrysolina quadrigemina* has apparently adapted to the climatic conditions of British Columbia and is now providing improved control of *Hypericum* (Andres et al., 1976). On the other hand, myxomatosis in Australia has apparently not kept pace in developing virulence commensurate with the growing resistance of rabbits to the disease, even though genetic resistance and virulence are difficult to measure (Davis et al., 1976).

To assess the net economic benefit of the control, account ought to be taken of the possibility that a weed may not have reached its ultimate geographic range, or density. A comparison must be made of the costs in the absence of control, including likely spread patterns (Auld et al., 1979; Menz et al., 1980), with the costs in the presence of control. Other things equal, the faster the spread of a weed, the greater the economic benefits from its control.

Risk considerations play a role in the acceptability and evaluation of biological control. In general from the point of view of landholders, the value of controlling a pest, whose incidence varies in an uncertain way, is greater than the average loss caused to them by the pest. Risk-averse farmers are willing to pay a premium or insurance charge to reduce the risk of uncertainty they face (Tisdell, 1982a). If biological controls are not as certain or reliable in their effectiveness as, say, chemicals, this reduces their relative value in the eyes of landholders. Thus farmers will tend to favour the use of herbicides and may even use more herbicide than is needed. Thus society is likely to be exposed to more serious risks than otherwise (Oelhaf, 1978).

From an economic point of view, classical biological control is often seen as advantageous because it involves a once and for all initial outlay whereas inundative and augmentative biological controls (as well as chemical and mechanical controls of weeds) require recurring outlays. In the latter cases, regular spreading of the control agent is needed and this involves recurrent costs and continuing energy use. Forms of control other than classical biological control also require continuing inputs from advisory staff in government and industry, and have a longer lag period between a useful research finding and its practical implementation.

AUGMENTATIVE BIOLOGICAL CONTROL

Classical biological control agents have the advantage that they are self-perpetuating and self-spreading but it may be socially profitable to aid the spread of a newly introduced control agent by creating a number of foci for its spread, an/or modifying the environment to favour the control agent and augmenting the agent's population.

Augmentative biological control embraces features and raises economic questions which are also involved in both classical and inundative biological control. Augmentative control may involve supplementing a classical biological control programme on a regular basis so as to maintain a higher (more effective and/or economic) degree of control. In that case it is underpinned by a classical process, but involves recurrent costs. In some cases, augmentation takes place on a regular basis, thereby approaching the inundative method. *Opuntia aurantiaca* Lindley, tiger pear, control by *Dactylopius austrinus* De Lotto, cochineal, in central western New South Wales, is an example. Low winter temperatures reduce insect populations to a level at which no weed control is achieved (N.S.W. Gov. Printer, 1967); supplementation of the insect population in spring is carried out by government-employed field workers.

The type of spillovers and issues discussed in classical control continue to be relevant. However, the need for recurrent supply and application of the control agent means that many of the economic considerations involved in inundative control apply. For example, the ease with which the agent can be 'artificially' increased in supply or stored is relevant. In the case of tiger pear control, air conditioned sheds are required at a number of locations for raising cochineal.

POTENTIAL FOR INUNDATIVE BIOLOGICAL CONTROL

There are several similarities between inundative biological control and chemical control but biological control is usually more specific. Although claimed as an advantage of biological control, specificity can be undesirable from a commercial standpoint if the micro-organism is specific to only one pest and consequently has a small market potential (Bowers, 1982). However, research is now extending the range of options available in microbial biological control agents. It is possible to combine a variety of host-specific fungal pathogens into one mycoherbicide where broad-spectrum weed control is required (Quimby and Walker, 1982).

The actual costs involved in applying mycoherbicides appear to be similar to those incurred in using herbicides. However, biological control agents generally have the advantage of no mammalian toxicity and less impact on the environment and this may help to reduce handling and application costs. Biological agents may also be usable closer to harvest date than many chemicals. But success of inundative biological control may be more dependent on favourable ambient or environmental conditions than chemical control methods. This adds to the risk (of ineffective control) involved in using biological control compared with a chemical control as far as the individual farmer is concerned - thus reducing the probability of his choosing biological control in preference to chemical control.

Where inundative biological controls are used in regularly disturbed environments (for instance, on cropland that is periodically ploughed) the control agent is unlikely to evolve naturally to counter any increasing resistance in the target species. Thus the maintenance of genetic stability and virulence of the control agent may be a greater problem than in the case of classical biological control. Efforts may be required to develop more virulent strains of the control agent as resistance builds up in the host. The problem may therefore be not unlike that of growing resistance of weeds or pests to chemicals.

Inundative biological control agents are unlikely to search for or autonomously reach, and attack, target weeds in non-target areas as do classical biological control agents. This will be an advantage where the inundative control agent has potentially pathogenic effects on useful plants outside the area of application. It also means that each user is likely to appropriate all, or nearly all, of the benefits of their use for himself.

In the absence of government supply of biological control, the success of a biological control method depends upon its commercial viability. In deciding whether to supply a biocontrol agent, private enterprise can be expected to take into account: (a) the volume of demand for the agent; (b) the regularity fo demand for it; (c) the shelf-life of the agent (how easily it can be stored and transported); (d) the costs of producing the biological control agent and the ease or costs involved in expanding production to cope with any peak demand for the agent. Bowers (1982) mentions market potential, stability and shelf-life and the potential for mass production as being important considerations for a manufacturer in deciding whether or not to commercialize a biological control agent. Microbial control agents (fungi, bacteria, viruses and protozoa) have particular potential for commercialisation because they can usually be produced in stable form and stored (as spores, scletoria or resting cells). Currently the U.S. Environmental Protection Agency (EPA) has ten microbial pest control agents registered. These include five bacteria, three viruses, one protozoa and one fungus.

Despite the economic difficulties involved, particularly in commercialisation of biocontrol agents, biological control agents will probably assume an increasingly prominent role in pest management (Bowers, 1982). In the United States a number of microbial biological control agents have already been successfully commercialised and the range offered can be expected to expand.

GENERAL ECONOMIC, SOCIAL AND INSTITUTIONAL BARRIERS TO THE USE OF BIOLOGICAL CONTROL

The fact that a biological control may be the least-cost method of control from a social standpoint does not guarantee its use. Institutional, social and economic arrangements provide constraints.

Considerable costs are involved in discovering biological control agents and testing their suitability for a new environment when they are to be introduced from abroad. However where a biocontrol organism has already been discovered in one country, its specificity tested and it has been successfully used in another country, the cost of introducing it to a third country is considerably reduced.

Where a biological control agent which may be suitable for commercialisation has only been discovered, much further work in developing production techniques, distribution systems and marketing the product is needed for its economic success. Because of the considerable risks involved, companies may be reluctant to invest in such developments even though from the community's point of view the risks may be worthwhile taking. Although the proportion of synthetic chemicals screened to those registered as pesticides is of the order of 10 000 to 1 (Oelhaf, 1978), according to Bowers (1982) the road to the discovery of chemical pesticides is more straightforward and more attractive to private firms than that for finding microbial pesticides.

The risks involved for private industry in investing in the development and marketing of new pesticides can be reduced by opportunities for patenting discoveries. The degree of patent protection required would vary widely according to the particular project and the risk involved (Bowers, 1982). Biological control organisms have not been patentable in the past. However, in the United States, as a result of a Supreme Court decision, the way is now open for patenting novel forms of living organisms. Many countries have not as yet extended their patent system to cover organisms and this may retard the development of biological control by private industry. Nevertheless, it would appear to be possible to patent a process for producing an organism or a method for its use in, say, pest control.

The effective life of the patent can also influence the willingness of companies to take risks in bringing new products to the market. R.B. Brett (1982) Executive Director, Agricultural and Veterinary Chemicals Association of Australia (AVCA) claims that in Australia "under current regulations, half the patent life of a product has expired before marketing commences and a return on investment is received. Continuation of the situation is likely to affect adversely the availability of new products and technologies to Australia's rural industries." The AVCA suggests that patent life be extended from the present 16 years from date of lodgement to 20 years. While this could well stimulate private inventions and innovation to some extent, the impact of the patent system is uncertain and complex (Mandeville et al., 1982; Tisdell, 1983). Nevertheless, absence of patentability (for example for some biological discoveries) is seen as a disadvantage by most pesticide companies, and can be expected to influence their dicisions.

In the pest control industry, regulation and registration barriers to new pesticides are significant. The cost involved in these procedures and the time taken to complete environmental and toxicological testing of pesticides prior to release can significantly erode patent time and reduce the returns of pesticide producers. Especially where an indigenous biological control agent is being used, less stringent registration requirements may be appropriate for biological controls. Indeed there have been significant recent moves in that direction in the U.S.A. (Ruttan, 1982) which could lower registration cost for "biorationals" from around $10 million to around $100 000 and reduce the time taken for registration by 2½ years (Sundquist et al., 1982).

THE ROLE FOR GOVERNMENT

Spillovers from the searching ability or independent movement of classical biological weed control agents approximate a collective (or pure public) asset. Economists predict that as a rule these assets will not be supplied privately or will be supplied in insufficient quantities to maximise welfare (Samuelson, 1955). If economic welfare is to be maximised such goods must be supplied by the government, or private enterprise must be heavily subsidised by the government in order to supply them. Thus government involvement in classical biological control, and to some extent in augmentative, seems unavoidable on economic grounds; Oelhaf (1978) presents a strong case indicating that the U.S.A. government has underinvested in biological control. Since the effects of inundative biological controls are as a rule more localised, there is greater appropriation of gains by users of these. Hence, there are greater prospects for the commercial development and use of inundative biological controls by companies and private individuals.

Non-chemical pest control measures frequently involve the substitution of knowledge (by the farmer or by a government adviser) for the use of a chemical. This knowledge or information is a product which can be consumed without diminishing the consumption of others or its supply and is another public asset (Oelhaf, 1978). Yet this kind of knowledge — the biology/ecology of the plant/weed system-control methods other than herbicides — appears to be poorly distributed and undersupplied from society's point of view. The present system, with a heavy dependency on chemicals, is somewhat self-perpetuating in that the chemical companies have a vested interest in providing farmers with information relating only to chemical control methods, the source of the profits of the chemical companies. Consumer (i.e. farmer) ignorance makes advertising very effective and advertising by private enterprises continues until *its* marginal revenue product equals its marginal additional expense to these enterprisis (Oelhaf, 1978). As an example of the uneven distribution of knowledge, only 10% of insect control information of California cotton growers originates from the agricultural extension service, while 70% originates from chemical companies (Luck et al., 1977).

Given the comparative importance of beneficial spillovers and of 'public assets' involved in biological weed control (and deficient private appropriation from such control), the socially optimal supply of biological weed control requires a relatively large commitment by government to research into and development of such techniques. Furthermore, the government may need to implement directly or maintain biological control measures especially those of a classical or augmentative type. In addition, given the lack of industry advisory services and marketing of biological controls because of the appropriation problem, it is necessary to rely a great extent on government advisory and extension services for the implementation of biological controls.

If weed control methods are arranged in a spectrum as in Fig. 1, one can indicate heuristically the relative importance of expenditures by government, by the pest-control industry and by farmers in ensuring their implementation. Although there can be some integration of methods,

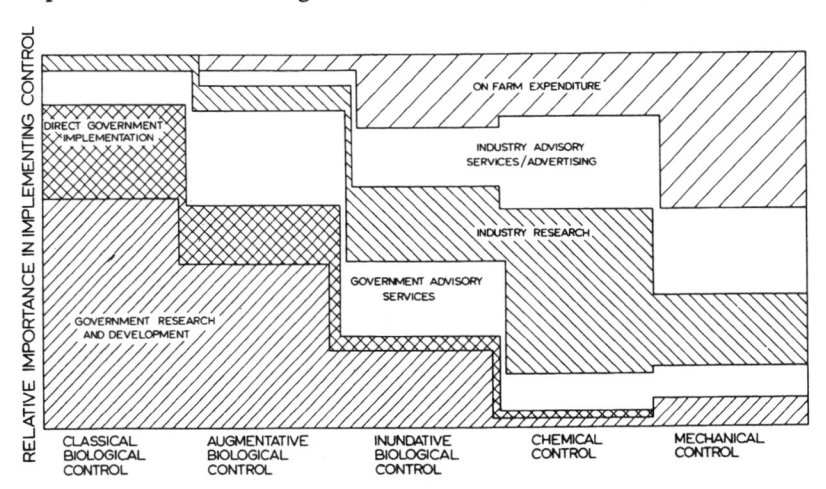

Fig. 1. Weed Control Spectrum. Conceptual framework showing relative importance of government, industry and on farm spending in various weed control methods. (Only forms of control which involve direct attack on the weed are considered here; indirect forms of control, e.g. competition from useful plants, could be integrated with any of these methods.)

each method requires a particular relative combination of inputs from government, industry and the farmer to achieve a socially optimal level of implementation. The government has a relatively larger role to play at the biological end of the spectrum in research and development, direct implementation, and advisory services (Fig. 1).

CONCLUSION

While it may be possible to reduce some of the constraints on the use of biological controls for weeds (e.g. institutional constraints), a number of constraints are inherent in this method of control and cannot be avoided. Inevitably governments will continue to play a large role in biological pest control. For this and other reasons there is likely to be a need for continuing cost-benefit analyses of these controls. Such analyses will require account to be taken of the type of factors outlined in this paper.

There appears to be a strong social case for greater government involvement in dissemination of information on weed control methods other than chemical control and in investment in research into biological weed control.

ACKNOWLEDGEMENT

The authors would like to thank Dr R.E. McFayden, Dr B.G. Johnston and the Editorial Board for their comments on the draft of this paper.

REFERENCES

Alcorn, J.L., 1976. Host range of *Puccinia xanthii*. Trans. Br. Mycol. Soc., 66: 365—367.
Andres, L.A., 1976. The economics of biological control of weeds. Aquat. Bot. 3: 111—123.
Andres, L.A., 1980. Conflicting interests and the biological control of weeds. In: Proc. 5th Int. Symp. Biol. Cont. Weeds, Brisbane, pp. 11—20.
Andres, L.A., Davis, C.U., Harris, P. and Wapshere, A.J., 1976. Biological control of weeds. In: C.B. Huffaker and P.S. Messenger (Editors), Theory and Practice of Biological Control. Academic Press, New York, NY, pp. 481—499.
Auld, B.A., Menz, K.M. and Monaghan, N.M., 1979. Dynamics of weed spread. Prot. Ecol., 1: 141—148.
Brett, R.B., 1982. Invited editorial. Aust. Weeds, 1: 2.
Bowers, R.C., 1982. Commercialisation of microbial biological control agents. In: R. Charudattan and H.L. Walker (Editors), Biological Control of Weeds with Plant Pathogens, John Wiley and Sons, New York, NY, pp. 157—173.
Charudattan, R., 1982. Regulation of microbial weed control agents. In: R. Charudattan and H.L. Walker (Editors), Biological Control of Weeds with Plant Pathogens, John Wiley and Sons. New York, NY, pp. 175—187.
Davis, D.E., Myers, K. and Hoy, J.B., 1976. Biological control among vertebrates. In: C.B. Huffaker and P.S. Messenger, (Editors), Theory and Practice of Biological Control. Academic Press, New York, NY, pp. 501—519.

Ecos, 1978. On the trail of the rabbit. Ecos, 18: 11—17.
Huffaker, C.B., Simmonds, F.J. and Laing, J.E., 1976. The theoretical and empirical basis of biological control. In: C.B. Huffaker and P.S. Messenger (Editors), Theory and Practice of Biological Control. Academic Press, New York, NY, pp. 41—78.
Luck, R.F., van der Bosch, R. and Garcia, R., 1977. Chemical insect control — A troubled pest management strategy. Bioscience, 27(9): 606—612.
Mandeville, T.D., Lamberton, D.M. and Bishop, E.J., 1982. Economic Effects of the Australian Patent System. Australian Government Publishing Service, Canberra, 228 pp.
Menz, K.M., Coote, B.G. and Auld, B.A., 1980. Spatial aspects of weed control. Agric. Syst., 6: 67—75.
Muir, W.R., 1978. Pest control — A perspective. In: E.H. Smith and D. Pimentel (Editors), Pest Control Strategies. Academic Press, New York, NY, pp. 3—7.
N.S.W. Gov. Printer, 1967. The prickly-pear pest in N.S.W. Government Printer, Sydney, N.S.W., 40 pp.
Oelhaf, R.C., 1978. Organic Agriculture. Economic and ecological comparisons with conventional methods. Allunheld, Osmun & Co., Montclair NJ, 271 pp.
Quimby, P.C., Jr. and Walker, H.L., 1982. Pathogens and mechanisms for integrated weed management. Weed Science, Suppl. 30: 30—34.
Ruttan, V.W., 1982. Agricultural Research Policy. University of Minnesota Press, Minneapolis, MN., 369 pp.
Samuelson, P.A., 1955. Diagrammatic exposition of the theory of public expenditure. Rev. Econ. Stat., 37: 350—356.
Sundquist, W.B., Menz, K.M. and Neumeyer, C.F., 1982. A Technology Assessment of Commercial U.S. Corn Production. Agricultural Experiment Station Bulletin 546, University of Minnesota, St. Paul, MN, 200 pp.
Templeton, G.E., TeBeest, D.O. and Smith, R.F., Jr., 1979. Biological weed control with mycoherbicides. Annu. Rev. Phytopathol., 17: 301—310.
Tisdell, C.A., 1982a. Wild Pigs: Environmental Pest or Economic Resource? Pergamon Press, Sydney, 445 pp.
Tisdell, C.A., 1982b. Exploitation of techniques that decline in effectiveness with use. Public Finance/Finances Publiques, 37: 428—437.
Tisdell, C.A., 1983. Thoughts on the patent system and the length of life of patents. Z. Betriebswirtsch., 53: 571—581.
Winch, D.M., 1971. Analytical Welfare Economics. Penguin, Harmondsworth, 208 pp.

[17]

Economic evaluation of biological weed control[1]

Summary

The existence of spillovers and new legislation makes it important to evaluate the economics of biologically controlling weeds in Australia before embarking on control. Classical biological weed control has been widely used in Australia with considerable success as measured in different ways, including high rates of return on R&D costs. Cost–benefit analysis has been used to evaluate the biological control programme for skeleton weed, *Chondrilla juncea*, and gave a benefit/cost ratio of 200:1. An inquiry by the Industries Assistance Commission (IAC) into the biological control of *Echium* species suggested a benefit/cost ratio of about 10:1. Room exists for improvement in the IAC's method of evaluation. Also, the CSIRO's (economic) test of whether a classical biological weed control programme is effective is found to have some deficiencies. Particular improvements are suggested.

Introduction

Most weeds of economic importance in Australia are introduced plants and are often amenable to biological control by natural predators or competitors from their country of origin (Groves and Cullen, 1981). But the importation of biological control agents can have a range of environmental, economic and social consequences, not all of which may be favourable or foreseen. While some members of the community may gain from introduction of a control agent, others may lose. Care is needed given the likely irreversible nature of introductions, the difficulties in containing populations once released and the possibility that a control agent may prove, on balance, to be a pest.

Prior to the *Biological Control Act* (1984), the public had little input into decisions about biological weed control in Australia. Evaluation tended to be the in-house business of the Department of Health administering the *Quarantine Act* (1908) and the CSIRO in consultation with the Australian Agricultural Council and other organizations importing biological control agents, for example, the Alan Fletcher Research Station and The Keith Turnbull Research Institute. The *Biological Control Act* (1984) sets out procedures to be followed for declaring target organisms and agent organisms for biological control. The Act establishes the Biological Control Authority. This Authority must inform the public of any proposal to declare a target organism or agent, and call for submissions. When, after examining submissions, it appears that there could be adverse environmental or personal effects from biological control, a public inquiry may be held. This may be undertaken by the Industries Assistance Commission or by other specified bodies. The first inquiry under the Act, the biological control of *Echium* species, was conducted by the Industries Assistance Commission (1985).

The report of the inquiry is then considered by the Biological Control Authority. The Authority may declare target organisms and agent organisms for biological control if it is satisfied on the basis of the report that the target organism is causing harm; is likely to be controlled by biological means; that 'control throughout Australia of the target organism would not cause significant harm to the environment and/or persons'; and any harm caused (losses or costs imposed) by biological control throughout Australia would be significantly less than the benefits obtained (Industries Assistance Commission, 1985). These conditions allow for considerable elasticity of interpretation, particularly terms such as 'significantly less than'. However, as a number of inquiries are completed, precedents will be established. The first inquiry, into the biological control of *Echium* species, used cost–benefit analysis as a means of providing guidance to the Biological Control Authority and this approach to evaluation will be considered here. In any case, the *Biological Control Act* means that greater emphasis must be placed on prior economic evaluation of biological weed control than in the past.

Types of biological control
Biological control of weeds may be of the classical type, augmentative or inundative (Tisdell et al., 1984a; Industries Assistance Commission, 1985). Particularly, in the case of classical control, government involvement in the release or non-release of the biocontrol agent is called for on economic grounds (Tisdell et al., 1984a; Menz et al., 1984). Primarily this is because of 'spillovers' from the introduction of a biocontrol agent. Persons not introducing the agent may benefit from or be damaged by the introduction because of the mobility and spread of the biocontrol agent. Such favourable or unfavourable spillovers will not be taken into account by individuals because they are likely to act in their own self-interest. No individual may find it personally worthwhile to introduce a biocontrol agent even though its introduction would be a net benefit to the community. Conversely, an individual may on occasion, benefit from his or her introduction of a control agent though, on balance, it damages the community. Thus there is a need for government evaluation and action.

In addition, decisions involving both economic and ecological factors have to be made about where to release a biocontrol agent, how to assist its establishment, whether to foster its spread and when to augment its population, if at all. These decisions may hinge to some extent on equilibria relationships between the weed (the host) and the biocontrol agent (the predator) (Auld and Tisdell, 1985). In some areas, for instance, it may be desirable to keep a reserve of the weed population so that its host can maintain a viable population.

Biological weed control in Australia and its value
Classical biological weed control has been widely and successfully used in Australia, and augmentative control is adopted for some weeds, e.g. for *Opuntia aurantiaca* Lindley, tiger pear, controlled by *Dactylopius austrinus* De Lotto, cochineal, in central western New South Wales (NSW Government Printer, 1967). Inundative biological control of weeds has not been used in Australia but fungal pathogens appear to be promising in this respect (Quimby and Walker, 1982; Bowers, 1982) and research into these is being undertaken in Australia.

Classical biological weed control in Australia has been successful since the programme was implemented in 1926 to control prickly pear, *Opuntia* spp. (see Haseler, 1980). Of 18 weeds in Australia targeted for biological control, only three (*Xanthium occidentale*, Noogoora burr, *Emex spinosa* and *Emex australis*) showed no response to biological control (Burdon and Marshall, 1981). At least five species were completely controlled and substantial control was obtained of at least another five, as interpreted by Burdon and Marshall (1981) from reports in the literature.

The success achieved with biological control may be measured in several ways, for example, by the reduction in the population of the targeted species or the decline in its rate of spread, by its increased susceptibility to control by non-biological means or by the level of increased economic returns achieved. Menz et al. (1984) have suggested that a biological weed control programme for the southern wheat–sheep zones of Australia costing between $1 m and $3 m could easily have a benefit/cost ratio of between 5/1 and 14/1. Marsden et al. (1980) estimated that on an outlay of $2.6 m for the biological control of *Chondrilla juncea* L., skeleton weed, the benefit/cost ratio was almost 200/1. This could, however, be an overestimate since some forms of *Chondrilla juncea* have proven to be resistant to released strains of the fungus *Puccinia chondrilla* (Groves and Cullen, 1981). On the other hand, even when the rust *P. chondrilla* did not kill skeleton weed, it reduced its vigour and hastened its decline when cultivated or subjected to competition from subterranean clover (Groves and Cullen, 1981). This raises the possibility that even where a biocontrol agent does not significantly reduce the population of a target weed but reduces its vigour, it may be of economic value by increasing the effectiveness of non-biological controls or reducing their cost (see Harris, 1980).

Cost–benefit analysis and the case of Echium *spp.*

The social cost–benefit analysis of techniques of weed control can be complex (Auld et al., 1987). Under some conditions, the net change in incomes received by farmers may be taken as a measure of the benefit, if it is assumed that no significant benefits are passed on to consumers through price reductions or quality improvements in products. If, after appropriate discounting for time-flows, increases in net incomes exceed additional costs of biological control, net benefits are positive and this gives some economic support for the control measure, the support being greater the larger the ratio of benefits to costs. This was the basic approach taken by the IAC in evaluating the biological control of *Echium* spp. (including Paterson's curse/salvation Jane). The Commission estimated a most likely benefit/cost ratio of 10.5/1 (Industries Assistance Commission, 1985).

The main discounted costs are seen by the IAC (1985) as loss in income from beekeeping ($10.4 m), loss of income to crop industries because of reduced crop pollination ($1.4 m) and research and implementation costs of CSIRO and state governments ($3.0 m) amounting to a total cost of $14.8 m. The main benefits were estimated as an increase in income in livestock industries of $131.9 m, increased income to crop growers of $19.1 m and savings in *Echium* control costs by public authorities of $4.5 m, making a total benefit of $155 m. However, it is unusual to evaluate the benefit/cost ratio in this way. It would be more usual for benefits to be calculated by deducting the reduced income of beekeepers, together with that from reduced crop pollination, from increases in net income elsewhere in the economy. If this is done, the relevant benefit/cost ratio becomes almost 48/1.

Some confusion appears to have crept into the IAC analysis. It ought to be concerned with evaluating the return on research and implementation costs of biological control, i.e. the investment in control measures (see Wise, 1977). There is no economic ground for including industry losses as part of the investment even though the procedure does bring attention to the potentially worst affected industries. This could, however, still be done without confusing the issue. Again, if net losses are imposed on graziers in low rainfall areas by *Echium*, why net out their losses from increased income for graziers in higher rainfall areas as the IAC does (IAC, 1985)? There is an internal inconsistency in its procedure. If the same approach were adopted as for the crop industries, the IAC's benefit/cost ratio would fall below 10.5.

Despite such issues, aggregate net benefit in relation to cost of biological control of *Echium* species is likely to be high. Nevertheless, we need an improvement in the evaluation procedures used by the IAC and better delineation by it between aggregate or overall impacts on incomes or returns, and income distributional consequences.

CSIRO's view of the effectiveness of a classical biological control programme
CSIRO considers biological control to be effective only when it reduces the density of the targeted weed to a level below the 'economic threshold' and so makes other forms of control unprofitable (IAC, 1985). While this is a necessary condition for economic gain from biological control in some circumstances, it is not a requirement for economic gain in other circumstances. A gain from the biological control may arise because it increases the kill rate of the non-biological weed control or lowers the cost of achieving a particular kill rate as, for example, suggested for skeleton weed by Groves and Cullen (1981). Moreover, when the cost of controlling a weed by non-biological means rises with its density, biological control may reduce densities from a level at which control is unprofitable to a level where non-biological control becomes profitable. In all these instances, non-biological control may continue to be profitable after a biological control programme is implemented. Biological control has an economic benefit despite the dictum of CSIRO (see Harris, 1980). For the traditional type of threshold model in which cost of non-biological control of weeds per unit area is constant (Tisdell et al., 1984b), the view of the CSIRO does seem well founded provided the cost of non-biological treatment is not reduced or its kill rate is raised by the biological control.

Conclusion
The economic evaluation of biological weed control using cost–benefit analysis is a complex task, which has assumed increased significance in Australia as a result of the *Biological Control Act* (1984). While substantial benefits in relation to costs have been obtained and are achievable from biological weed control programmes in Australia, economic evaluation methods and tests adopted by bodies such as the IAC and CSIRO require improvement. Ways to improve these have been outlined.

Acknowledgement
I wish to thank Dr Bruce Auld and Dr Ken Menz for bringing useful material to my attention for the preparation of this article.

Note

1. This article originally appeared in C.A. Tisdell (1987), 'Economic evaluation of biological weed control', *Plant Protection Quarterly*, **2**(1), 10–12. The content of this article is the same as in the original (except for minor changes) but its presentation has been improved. Consequently, pagination is no longer the same as in the original.

References

Auld, B.A., Menz, K.M. and Tisdell, C.A. (1987), *Weed Control Economics*, London and New York: Academic Press.

Auld, B.A. and Tisdell, C.A. (1985), 'Biological weed control – equilibria models', *Agriculture, Ecosystems and Environment*, **13**, 1–8.

Bowers, R.C. (1982), 'Commercialisation of microbial biological control agents', in R. Charudattan and H.L. Walker (eds), *Biological Control of Weeds with Plant Pathogens*, New York: John Wiley and Sons, pp. 157–73.

Burdon, J.J. and Marshall, D.R. (1981), 'Biological control and the reproductive mode of weeds', *Journal of Applied Ecology*, **18**, 649–58.

Groves, R.H. and Cullen, J.M. (1981), '*Chondrilla juncea*: the ecological control of a weed', in R.L. Kitching and R.E. Jones (eds), *The Ecology of Pests: Some Australian Case Histories*, Melbourne: CSIRO, pp. 7–17.

Harris, P. (1980), 'Stress as a strategy in the biological control of weeds', *Proceedings of the Fifth International Symposium on the Biological Control of Weeds*, p. 47.

Haseler, W.H. (1980), 'Lessons from early attempts at biological control of weeds in Queensland', *Proceedings of the Fifth International Symposium on the Biological Control of Weeds*, pp. 3–9.

Industries Assistance Commission (1985), *Draft Report on the Control of Echium Species (Including Paterson's Curse/Salvation Jane)*, Canberra: Australian Government Publishing Service.

Marsden, J.S., Martin, G.E., Parham, D.J., Risdill Smith, T.J. and Johnston, B.G. (1980), *Returns on Australian Agricultural Research*, Canberra: CSIRO.

Menz, K.M., Auld, B.A. and Tisdell, C.A. (1984), 'The role for biological weed control in Australia', *Search*, **15**, 208–10.

NSW Government Printer (1967), *The Prickly-pear Pest in N.S.W.*, Sydney Government Printer.

Quimby, P.C. Jr and Walker, H.L. (1982), 'Pathogens and mechanisms for integrated weed management', *Weed Science*, Suppl., **30**, 30–34.

Tisdell, C.A., Auld, B.A. and Menz, K.M. (1984a), 'On assessing the value of biological control of weeds', *Protection Ecology*, **6**, 169–79.

Tisdell, C.A., Auld, B.A. and Menz, K.M. (1984b), 'Crop loss elasticity in relation to weed density and control', *Agricultural Systems*, **13**, 161–6.

Wise, W.S. (1977), 'Cost–benefit analysis of agricultural research: hybrid maize reconsidered', *R & D Management*, **8**, 29–32.

[18]

Evaluation of biological control projects[1]
C.A. Tisdell and B.A. Auld[2]

Abstract
Methods for evaluation of biological control of weeds projects are discussed. The importance of deciding from what human and time perspective the evaluation is to be made is stressed. Social cost–benefit analysis (SCBA), an evaluation technique that focuses on society as a whole, is described. The application of SCBA to supply/demand analysis is illustrated and the specific case of SCBA for *Echium* spp. control in Australia is outlined. Another evaluation approach, 'multiple characteristics in assessment', is described. Evaluation at pre- and post-release is discussed, including a comparison of classical and inundative strategies. Any method of evaluation provides a limited perspective and guide, but some guidance is better than none.

Introduction
The nature of biological control of weeds extends within the range of possibilities from inundative to classical and this makes it difficult to generalize about the evaluation of biological control projects. This article highlights factors which should be taken into account in determining whether a biological control project is worth pursuing, and those which may favour inundative control compared with classical and vice versa.

Before evaluation can begin one has to consider from whose point of view and from what point of view it is to be done; this may dictate the method of assessment that should be used. In addition, the factors that need to be considered in assessment will vary with the stage of the project; for example, whether evaluation is being done prior to the project or after its implementation.

Criteria for evaluation

Background
An evaluation of biological control projects depends on the point of view from which the evaluation is to be undertaken. We shall suppose, in the first instance, that it is to be undertaken from an economic[3] point of view, and for this to be possible biological data are an essential prerequisite. Having decided on this, one needs to determine whose economic interests or objectives are to be taken into account. Headley (1985) discussed the use of cost–benefit analysis for researchers in biological control, principally from a farmer's point of view. However a biological control project may be assessed from the point of view:

(a) of society as a whole;

(b) of different interest groups within society; e.g. farmers or subgroups of farmers, manufacturers of an inundative biological control agent, government agencies such as the Department of Agriculture, consumers of products affected;
(c) of the interests of individuals; e.g. in their capacity as individual farmers, manufacturers, consumers, public servants etc., or in relation to an individual's interest as a whole taking into account his or her multiple roles in society.

Quite different evaluations and degrees of 'approval' of a biological control project may emerge depending on the reference group for assessment purposes. Though this is obvious, it is unfortunately often forgotten. Cost–benefit analysis is a measuring and comparison of costs and benefits associated with some action. Social cost–benefit analysis (SCBA) is an evaluation technique frequently used when the focus is on society as a whole, whereas individual farmers or firms may use private cost–benefit analysis to assess a biological control project from a private economic viewpoint.

We shall consider the evaluation of biological control projects from the point of view of society as a whole, concentrating first on social cost–benefit analysis and then on an alternative to it.

Social cost–benefit analysis (SCBA)
In evaluating weed control it is often assumed that benefits and costs are limited to farmers who may, for example, benefit through reduced costs. However consumers may benefit through lower prices for farm products or by an improvement in the quality of products purchased; e.g. less contamination of grain by weed seed; no herbicide residues. The benefits to producers are generally measured by changes in producers' surplus ('profits') and in consumers' surplus (see below and Auld et al., 1987).

The SCBA approach is based on the view that account should be taken of the total gains to society as a whole from a biological control project. Specifically, the Kaldor-Hicks criterion is used as an evaluator (Little, 1958; Mishan, 1971; Ng, 1979): If those gaining could compensate those losing (if any) from a biological control project and remain better off than before the project, the project would be judged to be a social improvement applying the Kaldor-Hicks test. Using this test, it is unnecessary for everyone to gain from the project and losers need not be compensated for their loss.

One limitation of the test is its lack of attention to the income distribution consequences of a project. If these are judged to be important from society's point of view an additional judgement must be made. If the income distribution consequences of a project are judged to be socially unsatisfactory, does this disadvantage outweigh the net benefits to be had if following the Kaldor-Hicks approach? If so, the change is judged socially unacceptable, otherwise it is socially acceptable (Ng, 1979).

Suppose a weed species primarily affects only one commodity and let the line marked DD in Figure 1 represent the demand for the commodity. Suppose that the line marked SS represents the supply of the commodity given the presence of the weed so that market equilibrium for the commodity occurs at E_1, which corresponds to a price of P_1 for the product and a quantity of production of X_1 for the commodity. In the absence of the weed, the supply curve of the product might be as indicated by line S_AS_A. Market equilibrium is established at E_2, the price of the product is lower at P_0 and production would be greater at X_2. If an effective biological control agent could be introduced to eliminate the weed, the (gross) benefits to society would be equal to the equivalent of the hatched area plus the dotted triangular area. The hatched area represents the reduction in total cost of producing existing total output, X_1. The triangular area represents the induced or stemming benefits flowing from extra sales of the product because its price is reduced. These areas when together provide an indirect measure of the total change in consumers' surplus plus producers' surplus. Note three things in relation to the size of this area: (1) the lower is S_AS_A in comparison to SS the greater are the benefits; (2) the further DD is to the right, that is, the larger the market for the product and the volume of production, the greater are the benefits; and (3) the less steep is the downward slope of DD (the more elastic is demand) the larger are stemming benefits.

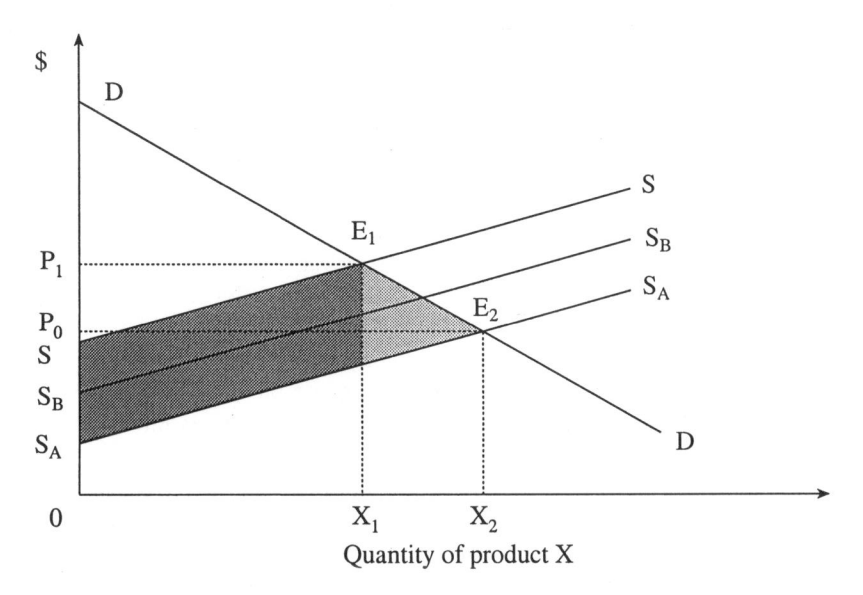

Figure 1 An illustration of how supply (S) and demand (D) of a product (X) establish equilibrium (E) prices (P). See text for further explanation

Even if the biological control were not completely effective in eliminating a weed, there could still be significant benefits (e.g. as shown by line S_BS_B: see discussion below). Moreover a new biological control project may be able to reduce the supply curve (costs) generated by existing methods of control.

A specific case
A specific case where SCBA has been applied to the evaluation of biological control is the *Echium* (Boraginaceae) species case in Australia. A government body, the Industries Assistance Commission (IAC, 1985) investigated costs and benefits to various groups. The IAC assumed that the control of *Echium* in Australia would not have a significant impact on the prices of farm products, therefore consumers would

be unaffected by the introduction of biological control agents. Social gain, it was assumed, would for practical purposes all be captured by farmers, some groups of whom would gain as a result of the biological control (e.g. wheat producers) and some of whom would lose (e.g. honey producers) as the weed is a source of nectar and pollen.

The IAC (1985) concluded that introduction of biological control agents for *Echium* in Australia would easily satisfy the Kaldor-Hicks test: farmers as a whole could expect large net benefits and there would be no reduction in consumers' welfare. While some primary producer groups were anticipated to be worse off as a result of the project, the gains of others were expected to be much larger than the losses of the disadvantaged. The judgement was made that net gains to the community were more than adequate to make up for the income distribution consequences. It was concluded that those disadvantaged could be compensated by 'once-off' direct cash payments from consolidated revenue rather than a levy on the beneficiaries.

As an exercise in SCBA the enquiry (IAC, 1985) has a number of limitations (Tisdell, 1987), even though it is useful in illustrating the application of the analysis to weed control projects. However SCBA itself has a number of limitations apart from its inadequacy in dealing with income distribution questions. For instance, the amount of information needed to apply it can be considerable and may only be obtained at considerable cost and effort, and even then considerable uncertainty may remain. It is also a single objective criterion in that it requires all relevant costs and benefits to be maximized. It is doubtful whether all relevant net benefits or outcomes from projects can be expressed solely in terms of money or by any other single measure, although progress has been made in the theory of valuation of non-market goods (Bromley, 1986). Even if this were possible, it may be less informative than identifying multiple components of the outcome such as Conway (1985a, b, 1987) has done. In addition assessment may involve multiple objectives rather than a single objective.

Multiple characteristics in assessment (Conway approach)
Conway (1985a, b, 1987) has proposed that 'agroecosystem technology packages' be assessed by Farming Systems Research (FSR). This requires alternative farming technologies such as the use of pesticides versus biological control of pests to be evaluated in terms of four factors: (1) their impact on levels of yields or income (of farmers); (2) consequences for instability of yields or incomes; (3) effect on the equitability of income distribution; and (4) their consequences for sustainability of yields or income.

In comparison to another technology, a technology that results in high income or yields, less instability of these, a more equitable distribution of income and greater sustainability of income or yield is to be preferred. While there are a number of difficulties in using Conway's approach (Tisdell, 1988), he has applied it with some success to evaluating alternative agricultural technology packages for farming in developing countries (Conway, 1985b).

In relation to biological control of a pest versus the use of pesticides, Conway (1987) has indicated that biological control is likely to be superior to the use of pesticides from the point of view of greater sustainability. This particularly concerned growing resistance of pests to pesticides and the possibility that this will induce unsustainability of yields. (However, we cannot rule out the long-term possibility of pests becoming increasingly resistant to some biological control agents (Charudattan, 1985).) Furthermore, in some cases biological control would also have a more favourable impact on farm incomes and on stability of yields. In addition, classical biological control can result in more equitable effects from an income distribution point of view. Once the classical biological control agent is released, both poor and rich farmers benefit without individual investment being required. In the case of pesticides, each individual farmer must purchase these for his own individual use. Poorer farmers may not be able to afford to do this or to obtain the required credit, unlike richer farmers, and so greater inequality of income is likely to emerge.

Some difficulties involved in using Conway's approach to evaluation are:

(a) It does not take account of consumers' surplus and therefore at this time is more applicable to subsistence economies than market-exchange economies. However, even in market economies it is relevant if farmers only are the focus of evaluation.
(b) In the case where one technique is superior to other techniques on the basis of one or more income characteristics and inferior on the basis of others, there is no clear indication of how to rank it in relation to the other techniques.
(c) Measurement of the characteristics poses some problems. For example, how are instability, equitability and sustainability of income to be exactly specified and measured (Tisdell, 1988)? In due course, some of these problems will probably be overcome. Despite these difficulties, Conway's approach offers a broadened perspective on evaluation of agricultural techniques, including techniques of weed control.

Pre-development of project and pre-release evaluation

Resources for biological control investigation are limited so it is desirable that they be allocated to projects with the greatest economic potential if an SCBA approach is taken. Factors that may be useful in identifying these projects are:

(a) The extra total cost imposed on production of commodities by the presence of a weed. This will depend on the cost added to each unit of production as well as the total volume of production affected.
(b) The potential for expanding sales of the commodities affected if costs of production are reduced. The greater the potential the more is control favoured.
(c) If methods of control of a weed already exist, would the biological control result in a substantial reduction in costs compared to these methods? If so this is favourable to the project.
(d) What is the likelihood from an ecological viewpoint that the biological control agent(s) will be effective in controlling the weed? The more likely that it will be effective the more propitious this will be for the project.
(e) What is the likelihood of the biological control agent attacking non-target species of value? The less likely is an attack the more favourable this is for the project.

(f) Will some groups of farmers have a substantial reduction in income if the biological control agent is introduced? If so, this may create social difficulties for its introduction.
(g) If the actual cost of the biological control project (the cost of research and evaluation of the agent) is low, this is favourable.

Inundative versus classical biological control

On occasion society may have to choose between trying to develop an augmentative/inundative biological control or classical/inoculative biological control. What factors may favour one over the other?

Classical biological control has the following disadvantages:

1. *Risks.* The biological control agent, when released, may attack non-target species; it is impractical to test the entire flora of a country.
2. *Low reversibility that adds to the riskiness.* Once the classical biological control agent is released and established, it may be difficult to eradicate if it has unforeseen consequences.
3. *Impact.* The impact of the biological control agent on weeds is independent of the wishes of individual farmers.

The following are advantages of classical biological control in comparison to inundative biological control:

1. *Benefit.* Occurs in perpetuity or at least for a long period.
2. *Extension or advisory costs are minimal.* However, costs may be involved in convincing the community that the release is worthwhile and socially beneficial, and in convincing suitable bodies such as Pasture Protection Boards or authorities to help in the initial spread of the biological control agent, or to undertake augmentative work.
3. *Basically only an initial cost or outlay is required rather than continuing expense.*
4. *Automatic density adjustment of the biological control agent.* The agent density tends to adjust automatically to the density of its weed-host. An automechanism (of adjustment) is involved and it 'searches out' its host (Tisdell et al., 1984a).
5. *Effective at low densities.* A classical biological control agent may even exert control at low densities of a weed where mechanical, chemical or inundative control would be uneconomic (Tisdell et al., 1984b).
6. *All farmers, poor and rich, may benefit from it.*

Inundative biological control has the following disadvantages in comparison to classical biological control:

1. *Limited area effects.* The biological control effects are usually limited to the area in which it is applied, and it does not search out its host on its own accord, but requires human intervention.
2. *No density adjustment.* An inundative biological control agent does not adjust its population automatically, as a rule, to the population of its host.

3. *Limited time effects.* Effects are usually limited in time as well as in area. It therefore requires continuing expenditure for it to have an effect.
4. *Uneconomic at low densities.* Use of inundative biological control agents is usually uneconomic in cases of low weed density or of low value of crop affected by the weed (Tisdell et al., 1984b).
5. *Higher extension and advertising costs.*

Advantages of inundative biological control include:

1. *Lower risk.* This is because of greater human control over the agent once it is released. If it has unwanted side-effects it can be withdrawn or applied only in cases where the side-effects will not occur or are acceptable. However, in cases where an exotic agent is introduced or a new agent is obtained by genetic engineering or induced mutation (Charudattan, 1985) it could be more persistent, and this risk would need to be guarded against.
2. *Less social conflict.* Inundative biological control agents may cause less social conflict in cases where some groups favour eradication of a weed and others favour its retention. Individuals wishing to control the weed can do so on their properties whereas those not wishing to do so can retain the weed.

In many cases, of course, there is not a choice between the two forms of biological control. It should also be noted that classical biological control falls into the category described by economists as a pure public good (Brown and Jackson, 1986; Tisdell, 1982). For it to be supplied, or supplied optimally, it usually has to be provided by the government. Its provision is subject to whatever political failures or shortcomings exist. On the other hand, inundative biological control falls into the category described by economists as a pure private good. It can be supplied by the market and by companies hoping to make a profit. Its provisions are subject to whatever market failures or shortcomings are present. Its acceptance will depend on the increased profits that farmers perceive that they will make from its use (Reichelderfer, 1985).

Post-release evaluation

Assessments of benefits can be made from three perspectives: (1) results achieved to date; (2) results expected in the future; and (3) past benefits or results achieved plus future benefits expected.

Evaluation after the event is of economic value as an information-providing exercise that may improve future decisions and provide useful guidelines for these. However, whether or not a biological control project should proceed or continue cannot be determined solely by reference to the past, in particular likely returns on past expenditures. Past outlays on the project are sunk costs and are irrecoverable and the future of the project should not be burdened by or be dependent on these (e.g. Heyne, 1976). All that counts from an economic viewpoint are the returns on any additional expenditure and whether these will be sufficiently high to make the additional expenditure worthwhile. For example, suppose that for a pre-release expenditure of $10 m, a substantial benefit of $2 m p.a. was expected originally from the release of a biological control agent, but that just prior to release this is revised downward as a result of experimental information to $0.2 m p.a., a return inadequate to justify the original project. An additional $0.1 m is needed to release the control agent effectively. Clearly the release should proceed from an economic viewpoint since the $0.2 m return must be related to the $0.1 m additional expenditure rather than to the $10 m outlay. The $10 m outlay is a sunk or irrecoverable cost.

One may wish to evaluate after release of a biological control to determine whether augmentation is necessary, whether some cultural practices or environmental conditions increase the effectiveness of the biological control or reduce this and whether or not it is having any unwanted side-effects. In the latter case if it is an inundative control, it can be withdrawn or conditions favourable to its use can be more accurately specified. Even when a project is unsuccessful it may still produce information of value for future research.

In the case of inundative control, companies will need to do market evaluation of the product. Its success depends not only on its biological effectiveness but also on its marketability. It has been suggested that a biological control agent must be completely effective in destroying its host for it to be of value, or at least sufficiently effective to reduce it below the economic threshold at which other measures are worthwhile applying (IAC, 1985; see 6.4.4).

However a biological control agent that is partially successful either in destroying or in weakening (reducing the vigour of) its host can be of considerable economic value (Harris, 1981; see also Figure 1).

Partial control can take many forms. For instance, it may involve:

1. control in some geographical areas only;
2. effective control only under certain environmental conditions;
3. control only *or* some members of the weed population in an area under stress; and
4. reduction in the vigour of some of the weed population.

Even if the biological control does not result in increased mortality of the weed population, it may cause morbidity. This may reduce the competitiveness of the weed with crops or make other forms of weed control or integrated pest management systems more effective.

Companies developing inundative biological control agents generally obtain a patent and a monopoly right for a limited period. They are therefore likely to charge a monopoly price for the agent or 'whatever the market will bear'. Nevertheless both consumers and farmers can be expected to benefit. Farmers will not use the inundative control technique if it is not more profitable to them than other available alternatives. Secondly, if the technique lowers costs of production, it is likely to result in lower prices of end products for consumers. There may, of course, be additional benefits to the community such as less use of dangerous chemicals with a switch from pesticides, or less soil erosion as a result of reduced use of mechanical means of weed control.

Conclusion

Biological control projects can be evaluated from a number of different perspectives.

This article emphasizes the importance of specifying clearly in advance the perspective from which an evaluation is to be undertaken. All available methods of evaluation provide a limited perspective and guide. This is true even of social cost–benefit analysis. Nevertheless, even limited guidance can be better than none at all, provided that absolute faith is not placed in it.

Acknowledgements

The authors thank P. Harris and E.J. Roberts for comments on the manuscript. B.A. Auld thanks the Australian Wool Corporation for financial assistance.

Notes

1. This article originally appeared in C.A. Tisdell and B.A. Auld (1988), 'Evaluation of biological control projects', in E.S. Delfosse (ed.), *Proceedings of the VII International Symposium on Biological Control of Weeds*, Federconsorzi Conference Hall, Rome, Italy, 6–11 March, pp. 93–100. The content of this article is the same as in the original (except for minor changes) but its presentation has been improved. Consequently, pagination is no longer the same as in the original.
2. Agricultural Research and Veterinary Centre, Forest Road, Orange, New South Wales, 2800, Australia.
3. It should be borne in mind that a *political* point of view is another approach that may override economic or ecological considerations (Menz and Auld, 1977).

References

Auld, B.A., Menz, K.M. and Tisdell, C.A. (1987), *Weed Control Economics*, London: Academic Press.
Bromley, D.W. (ed.) (1986), *Natural Resource Economics: Policy Problems and Contemporary Analysis*, Higham, MA: Kluwer-Nijhoff.
Brown, C. V. and Jackson, P.M. (1986), *Public Sector Economics*, 3rd edn, Oxford: Basil Blackwell.
Charudattan, R. (1985), 'The use of natural and genetically altered strains of pathogens for weed control', in M.A. Hoy and D.C. Herzog (eds), *Biological Control in Agricultural IPM Systems*, Orlando, Florida: Academic Press, pp. 347–72.
Conway, G.R. (1985a), 'Agroecosystems analysis', *Agricultural Administration*, 20, 31–55.
Conway, G.R. (1985b), 'Agricultural ecology and farming systems research', in J.V. Remenyi (ed.), *Agricultural Systems Research for Developing Countries*, Canberra: Australian Centre for International Agricultural Research, pp. 43–59.
Conway, G.R. (1987), 'Properties of agroecosystems', *Agricultural Systems*, 24, 95–117.
Harris, P. (1981), 'Stress as a strategy in the biological control of weeds', in E.S. Delfosse (ed.), *Proc. V Int. Symp. Biol. Contr. Weeds*, 22–27 July 1980, Brisbane, Australia, Melbourne: CSIRO, p. 47.
Headley, J.C. (1985), 'Cost–benefit analysis: defining research needs', in M.A. Hay and D.C. Herzog (eds), *Biological Control in Agricultural IPM Systems*, Orlando, Florida: Academic Press, pp. 53–73.
Heyne, P. (1976), *The Economic Way of Thinking*, 2nd edn, Chicago: Science Research Associates.
Industries Assistance Commission (IAC) (1985), 'Report on biological control of *Echium* species (including Paterson's Curse/Salvation Jane)', Canberra: Australian Government Publishing Service.
Little, I.M.D. (1958), *A Critique of Welfare Economics*, 2nd edn, London: Oxford University Press.
Menz, K.M. and Auld, B.A. (1977), 'Galvanised burr, control and public policy towards weeds', *Search*, 8, 281–7.
Mishan, E.J. (1971), *Cost–benefit Analysis*, London: George Allen and Unwin.
Ng, Y. (1979), *Welfare Economics*, London: Macmillan.
Reichelderfer, K. (1985), 'Factors affecting the economic feasibility of the biological control of weeds', in E.S. Delfosse (ed.), *Proc. V Int Symp. Biol. Contr. Weeds*, 19–25 August 1984, Vancouver, Canada. Ottawa: Agriculture Canada, pp. 135–44.
Tisdell, C.A. (1982), *Microeconomics of Markets*, Brisbane: John Wiley.
Tisdell, C.A. (1987), 'Economic evaluation of biological weed control', *Plant Protection Quarterly*, 2, 10–12.
Tisdell, C.A. (1988), 'Sustainable development: differing perspectives of ecologists and economists, and relevance to LDCs', *World Development*, 16, 373–84.
Tisdell C.A., Auld, B.A. and Menz, K.M. (1984a), 'On assessing the value of the biological control of weeds', *Protection Ecology*, 6, 169–79.
Tisdell, C.A., Auld, B.A. and Menz, K.M. (1984b), 'Crop loss elasticity in relation to weed density and control', *Agricultural Systems*, 13, 161–6.

[19]
Economic Impact of Biological Control of Weeds and Insects

CLEMENT A. TISDELL

Department of Economics, University of Queensland, St Lucia, Queensland, Australia

Introduction
Impact of classical, augmentative, and inundative control
Conventional indicators of biological control benefits
        Crude indicators—Cost savings as a measure of economic benefits—
        Profit increases or variations—Variations in land values—Benefits in
        terms of variations in producers' surplus and consumers' surplus
Two case studies
        Returns on biological control research by the Division of Entomology,
        CSIRO, Australia—Economic impact of control of *Echium* species
Possible economic impact of future biological control
Conclusions
References

Introduction

Use by humans of biological methods for the control of weeds and insects on a major scale appears to be relatively recent. The first major programme for biological control of a weed dates only from 1902, when several exotic organisms were released against *Lantana camara* L. (Verbenaceae) in Hawaii (Perkins and Swezey, 1924; Julien, Kern and Chan, 1984). This programme was pre-dated by the successful control of the cottony-cushion scale, *Icerya purchasi* Maskell (Coccoidea: Margarodidae), in California during the period 1888–89. Scale control was accomplished by the introduction from Australia of the parasite *Cryptochetum iceryae* (Will.) (Diptera: Cryptochetidae) and of a predator, the vedalia beetle *Rodolia cardinalis* Mulsant (Coleoptera: Coccinellidae) (Huffaker and Caltagirone, 1986). The latter state that,

Critical Issues in Biological Control
© Intercept Ltd, PO Box 716, Andover, Hants, SP10 1YG, UK

although no one has accurately measured the economic benefits from this project, they are vast and far exceed the estimated US $100 million in benefits of the biological control project against St John's wort, *Hypericum perforatum* L. (Guttiferae), in California.

Even though no concerted effort has been made to measure the global economic benefits of the biological control of weeds and insects, the total benefits must be reckoned in billions of dollars. For Australia alone they exceed A$ 1billion. For example, the economic benefits of the biological control of skeleton weed, *Chondrilla juncea* L. (Compositae), were estimated to be more than A$ 261 million based on 1975 prices and a discount rate of 10% (Marsden *et al.*, 1980). In present-day values, the economic benefits to Australia of having controlled prickly pear, *Opuntia* species (Cactaceae), by biological means would also be an enormous sum. However, from an economic point of view, it is not so much the absolute economic impact of biological control which is significant but its economic benefit relative to the resources used for bringing about this impact. The relative economic returns on biological control programmes in Australia have been high compared with the alternative possible returns on the resources used. For example, the internal rates of return on outlays for evaluated biological control projects of the CSIRO Division of Entomology were estimated to range between 13.9 and 14.1%, depending on the project involved (Marsden *et al.*, 1980, p.6). Overall, these returns are high compared with those available from alternative investment opportunities.

Julien, Kerr and Chan (1984) completed a comprehensive survey of all attempts at biological control of weeds with invertebrates and fungi prior to 1980. They reported that, of a total of 174 projects undertaken to control 101 species of weeds, 39% were considered successful. Success was defined as a degree of reduction to below a threshold appropriate to each situation. Their judgement on whether or not biological control was effective against these 48 weeds was essentially qualitative. Among the invertebrates and fungi used for biological control, the greatest reliance by far was placed on insects. They accounted for 488 of the 499 species of natural enemies released and were credited with 47 of the 51 effective cases of released insects, acarina, fungi and nematodes. The rate of effectiveness of insects was, therefore, slightly less than 10% in relation to the number of species released. But virtually nothing can be deduced from these data about the comparative economic benefits of the use of insects as control agents, especially because the overall returns from research projects are usually heavily influenced by the very large benefits of a small number of projects (e.g. Marsden *et al.*, 1980, p.5).

Impact of classical, augmentative, and inundative control

Biological methods of pest control are frequently classified into three categories: classical, augmentative, and inundative (Tisdell, Auld and Menz, 1984). Despite some slight differences in definition (Charudattan, 1985), the main features of these categories are broadly agreed. Most case studies of the economic impact of biological control are for classical methods, that is, the importation and release of exotic natural enemies against native or introduced pests.

To date, the major economic benefit of biological control of weeds and insects has come from classical methods. This is to be expected as few cases of inundative control are available for evaluation. However, since returns on several inundative control programmes have been sufficiently attractive, these agents are now commercially produced and marketed in the United States and elsewhere. For example, microbial herbicides are available in the United States to control milkweed vine, *Morrenia odorata* (Hook. & Arn.) Lindle (Asclepiadaceae), and northern jointvetch, *Aeschynomene virginica* (L.) B.S.P. (Leguminosae) (Charudattan, 1985). Various pathogenic preparations, including *Bacillus thuringiensis* Berliner (Bacillaceae), are also available to control insects. Prospects for the development of additional microbial pesticides, based on natural or genetically altered strains, for agricultural use are generally rated as good (Quimby and Walker, 1982; Templeton, 1982). As Charudattan (1985) pointed out, these control methods can be integrated with classical controls using, for example, arthropods.

As discussed elsewhere (Tisdell, Auld and Menz, 1984), classical biological control of pests is akin to the provision of a public or collective good. This means that providers of the good (in this case, those introducing the natural enemy or releasing it at a limited number of locations) can only appropriate a small fraction of the total economic benefits of their action. Therefore, classical biological control may not be provided at all or may be undersupplied if its supply is left to free market forces. By contrast, inundative controls, such as the application of microbial pesticides, are generally like chemical controls and are private goods. A large fraction of the economic benefits of providing these can be appropriated by the suppliers. Thus, if a profit (and a social gain) can be had from the provision of inundative controls, it is likely to be supplied by free markets (Tisdell, 1981).

Harris (1979) suggested that a complete biological control programme against a weed is likely to cost about 19–24 scientist years. Even so, a programme against a major weed is only likely to yield an economic benefit if it is possible to capitalize on studies done in other countries. In this respect, two comments seem in order. First, even in the case of chemical controls, companies prefer to concentrate on developing controls for major weeds because the latter provide the largest market. Secondly, an international externality or free-rider problem is highlighted in relation to biological control. A country carrying out research on the biological control of pests may provide valuable information to other countries at little or no cost to the latter. Consequently, in the absence of international co-operation, a pest that is of comparatively minor importance in many countries but of major importance world wide may be neglected by each country as a target for classical biological control, and considerable economic benefits may be foregone globally. International co-operation is needed to maximize the benefits and the economic impact of biological pest control.

Because the economic impact of biological control is usually multi-dimensional, its measurement can be a complex matter. A pest control technique normally has a range of economic effects, for example, on the level of economic returns, on the distribution of income (Mumford and Norton, 1987), on the variability of returns, and on the ability to sustain the benefits (DeBach, 1974; Wilson and Huffaker, 1976; Conway, 1987; Tisdell, 1988a). It is uncertain if these impacts can be satisfactorily reduced to a single measure even though economists seem to prefer a uni-dimensional measure.

In assessing the economic benefits of a change such as the introduction of a new technique, economists concentrate on gains and losses that can be measured in monetary terms. If the total value of the gains by gainers exceeds the sum of the losses by losers, then as a first approximation a social economic benefit is assumed. Social cost–benefit analysis tries to determine whether or not there is a net gain in this sense, but it ignores the way in which the gains and losses are distributed. This is not to say that the consequences of income distribution are unimportant. However, they are being introduced as a separate consideration by economists as the discussion below of the biological control of *Echium* species (Boraginaceae) will illustrate. Despite the fact that legitimate doubts have been expressed about the use of uni-dimensional measures for economic impact assessment, they are being widely used, and it is appropriate to review them.

Conventional indicators of biological control benefits

CRUDE INDICATORS

Even crude measures of the impact or the benefit of biological control may be better than no measures at all. These include indicators such as a reduction in the population size of the target pest, an increase in yields, an increase in production, and an increase in total farm revenue or receipts. In some circumstances, which can be specified (Tisdell, 1988b), social net benefits as defined by economists vary linearly, or nearly so, with these indicators. Nevertheless, crude indicators are not always adequate for the assessment of variations in economic benefits, and more sophisticated measures may be required.

COST SAVINGS AS A MEASURE OF ECONOMIC BENEFIT

Cost savings as a result of biological control have been used to measure its economic benefits. For example, Huffaker, Simmonds and Laing (1976) used this approach to measure the economic benefits of major successful biological control projects in California between 1928 and 1973. They estimated the total benefits up to 1973 as US$275 million in 1973 dollars. Although their estimate allowed for inflation, they did not consider the interest rate. If cost savings were accumulated to 1973 allowing for a realistic rate of interest, this would actually increase the size of the apparent benefits.

Huffaker, Simmonds and Laing (1976) estimated cost savings from biological control as being equal to the sum of:

1. The value of losses in production saved (i.e. the value of extra production no longer lost due to the presence of the pest);
2. Any savings over alternative pest controls, such as pesticides.

In relation to (1), it is as a rule assumed that the price of the product does not fall as a result of increased production. On this basis, benefits will be overestimated to the extent that extra production from losses saved also incur extra costs, for example, additional harvesting costs. These extra costs should be deducted. After this is done, however, benefits are likely to be underestimated because the costs per unit of production will decline. As a rule, this change makes it profitable for agriculturalists to expand their relative concentration on a particular product, that is, to expand the area sown to an affected crop.

PROFIT INCREASES OR VARIATIONS

When the price of a product does not vary with the volume of production, the benefits of biological control (at least in the short term) will be reflected in the variations of profits received by producers. In this case, producers capture the total economic benefits from biological control. In Australia, the Industries Assistance Commission (Anonymous, 1985) used discounted estimated profit increases and losses for various agricultural producer groups in order to assess the potential economic benefits from biological control of *Echium* species. While this method is likely to be more accurate than one based on cost savings, it will also probably be more costly to use because it requires more information.

VARIATIONS IN LAND VALUES

When a successful agricultural technique such as a method of biological pest control is introduced and raises the returns from agricultural land in various locations, the market value of that land can be expected to rise and so can the rent paid for its use. The total rise in land values can be an accurate indicator of any economic benefits. The increase reflects the capitalized value of extra profits that are available from the land as a result of the biological control programme.

Dodd (1940) indicated that the biological control of prickly pear in Australia resulted in a gain in value of A$0.50 per acre for 60 million acres of land, for a total increase in value of A$30 million at 1940 prices. Based on a cost of A$ 240 000 for the programme, a benefit–cost ratio in excess of 120 : 1 is implied.

Changes in land values after St John's wort had been successfully controlled in California were used by Huffaker and Caltagirone (1986) as an indicator of economic benefits. But they also added to this value any savings resulting from a reduced need for herbicide applications as well as the value of weight gains of livestock. However, double counting can be a problem with this method. For example, the increase in land values should reflect both the cost savings for herbicides and any additional profits available in the new situation.

This is not to say that variations in land values necessarily capture all economic benefits. The latter may be influenced by land speculation and by other developments in addition to the introduction of biological control. Also, consumers rather than producers may benefit from lower prices for agricultural products; these benefits will not increase land values.

BENEFITS IN TERMS OF VARIATIONS IN PRODUCERS' SURPLUS AND CONSUMERS' SURPLUS

The economic benefits that result from a new production technique can be shared between the consumers and the producers of a product. When the consumers capture some benefits, the increase in the (discounted) profits of producers (or more generally the rise in producers' surplus) or the rise in land values underestimates the total economic benefits of a biological control programme. When the price of a product declines (because pest control results in a greater supply of the product), consumers obtain an economic benefit which can be measured by the rise in consumers' surplus. This consumers' benefit should be added to the producers' benefit (Auld, Menz and Tisdell, 1987, ch. 8). Thus, the total economic benefit of a biological control programme that raises agricultural productivity is equal to the sum of the present discounted value of the producers' surplus plus the present discounted value of the consumers' surplus during the period in which the biological control is in place. If, for example, biological control reduces the cost of producing a product by $2 per unit and the extra supply causes the price to fall by $1 per unit, producers and consumers equally share the total economic benefit. As calculated by this approach, the total economic benefit in this case will be equal to twice the estimated value of benefits to producers.

Although most economists agree that the economic impact of a new technique can be measured more accurately if changes in consumers' surplus are taken into account, most studies of the economic impact of biological control have been limited to producers' benefits. However, several cost–benefit studies of new agricultural techniques have considered both factors (e.g. Hayami and Herdt, 1975; Alauddin and Tisdell, 1986). Using this approach, Vere, Sinden and Campbell (1980) estimated a social benefit–cost ratio of 11 : 1 for the control of serrated tussock, *Nassella trichotoma* (Nees) Arech. (Gramineae), by non-biological methods in New South Wales.

For several reasons, researchers undertaking cost–benefit analyses of biological control have been slow to allow for consumers' surplus. First, considerably more information and time is needed to estimate its variations. Secondly, in many cases, changes in prices and in consumers' surplus are generally believed to be small. Thirdly, where the product affected is principally an export item, any gains from a price reduction are mainly distributed among foreign consumers. Their gains will be unimportant if only national net benefits are being considered (e.g. Duncan and Tisdell, 1971; Edwards and Freebairn, 1982). For this reason, it is important to distinguish between an impact assessment based on national and one based on global economic benefits of biological controls. International economic impact from research into and application of biological control can occur via a number of mechanisms, several of which were discussed by Davis, Oram and Ryan (1987).

Two case studies

In Australia, two concerted efforts were made in recent years to estimate the economic impact of biological pest control. The first involved an assessment of returns on selected research projects undertaken by the CSIRO Division of Entomology (Marsden et al., 1980), and the second involved an estimation by the Industries Assistance Commission (Anonymous, 1985) of the possible economic impact of the biological control of *Echium* species. Each of these studies provided useful examples of the economic impact of biological control.

RETURNS ON BIOLOGICAL CONTROL RESEARCH BY THE DIVISION OF ENTOMOLOGY, CSIRO, AUSTRALIA

The joint study by the Industries Assistance Commission and the CSIRO of the benefits and costs of the CSIRO Division of Entomology (Marsden et al., 1980) marks a significant step in the systematic economic evaluation of benefits from research into pest control. Thirteen of the research projects conducted by the Division between 1960 and 1975 were evaluated, and discounted net benefits over the period 1960 to 2000 were estimated. Of the 13 projects, four were biological control projects involving the use of natural enemies of insects or weeds. The following research projects were evaluated: sheep blowfly control with diazinon; cattle tick research; scarab beetle research; locust investigations; subterranean clover stunt virus; oriental fruit moth control; fruit fly control; white wax scale control; orchard mite control; phasmatid research; sirex control; termite research; and skeleton weed control.

The study found that these projects (at a discount rate of 10%) had a measured economic benefit of A$350.4 million at 1975 prices and, except for termite research (which would not appreciably alter the result), had a cost of A$32.93 million. For these projects, the overall benefit–cost ratio was 10.64 : 1. This is an extremely favourable ratio because a project can be justified as efficiently using resources if its benefit–cost ratio exceeds unity. Furthermore, the study showed that, even if there had been benefits only from these projects, the existence and the work of the Division of Entomology could still be justified on economic grounds.

The total costs of the Division in the relevant period were A$154.3 million. Hence if a discount rate of 10% is used, the Division has a benefit–cost ratio of 2.27 : 1 on the basis of the above mentioned projects alone.

Table 1 Economic benefits and costs of research projects on biological and non-biological pest control by CSIRO Division of Entomology*

Project	Benefit (A$mio)	Cost (A$mio)	Benefit/Cost Ratio
Biological control			
Skeleton weed	261.2	2.33	112.1:1
Orchard mite	14.4	0.59	24.4:1
Sirex wasp	12.8	5.21	2.5:1
White wax scale	1.5	1.04	1.4:1
Total for biological control	289.9	9.17	31.6:1
Total for non-biological control	60.5	23.76	2.5:1
Total for all projects	350.4	32.93	10.6:1

* Data are from Tables 1.1 and 1.2 of Marsden *et al.* (1980). Returns are estimated for the period 1960–2000 and are expressed in constant (1975 $) net present values, based on an assumed 10% discount rate.

The main interest here, however, is in the economic impact and benefits of the Division's biological control projects. Relevant benefit and cost values are shown in *Table 1* for the following projects: skeleton weed (*C. juncea* L.); orchard mites (Acarina: Tetranychidae); *Sirex* wood wasps (Hymenoptera: Siricidae); and white wax scale (*Ceroplastes destructor* Newstead [Homoptera: Coccidae]). These four projects accounted for over 72% of the total benefits but used only 17.3% of the funds allocated to all 13 projects. The biological control projects provided an economic benefit of A$289.73 million at a benefit–cost ratio of 31.6 : 1. By comparison, the research projects on non-biological pest control provided an estimated benefit of A$60.5 million at a cost of A$23.76 million, giving a benefit–cost ratio of 2.5 : 1. Thus, both the absolute economic impact and the relative returns from resources used by the Division of Entomology for research on biological pest control were much greater than for projects on non-biological means of control.

All of the biological control projects of the Division of Entomology are interesting. It can be seen from *Table 1* that the major economic benefit from biological control was expected to come from the control of skeleton weed by the importation of several natural enemies, namely a rust fungus (*Puccinia chondrillina* Bubak & Sydenham), a gall mite (*Eriophyes chondrillae* Canestrini), and a gall midge (*Cystiphora schmidti* Rubsaamen). Benefits were estimated as the reduction in the costs of spraying plus the value of increased wheat production. However, production increases may have been somewhat overestimated because some forms of *C. juncea* have proven to be resistant to released strains of *P. chondrilla* (Groves and Cullen, 1981).

The wood wasp *Sirex noctilio* F. (Hymenoptera: Siricidae) was accidentally introduced into Australia, probably from New Zealand, and is a pest in plantations of *Pinus radiata* D. Don (Pinaceae). Wasp parasites of *Sirex* and a nematode have been introduced to control it. Economic benefits from its control were estimated in two parts: the value of softwood timber losses avoided in areas where *Sirex* is present, and the value of local timber losses avoided by the slower spread of *Sirex* (to new areas) because of the presence of introduced natural enemies.

Spider mites cause problems in Australian orchards of deciduous fruits (e.g. apples, pears and peaches) as well as with several other crops. The two-spotted spider mite (*Tetranychus urticae* Koch) and the European red mite (*Panonychus ulmi* L. [Acarina: Tetranychidae]), are the main pests. An organophosphate-resistant predatory mite, *Typhlodromus occidentalis* Nesbitt (Acarina: Tetranychidae), was imported from North America to control the orchard mites. *T. occidentalis* seems to control *T. urticae* effectively, but *P. ulmi* only partially. In cases where *T. urticae* is a problem but not *P. ulmi*, the introduced predator allows orchardists to reduce the frequency of acaricide applications. These savings in spraying costs were used as the basis for an assessment of the economic benefits from the biological control programme. Estimation of benefits is, however, complicated by the possibility that codling moth, *Cydia pomonella* L. (Lepidoptera: Tortricidae), will become resistant to organophosphate and that *T. occidentalis* will not be resistant to any pesticides replacing it. Furthermore, spider mites are likely to become increasingly resistant to the acaricides currently used for their control. These dynamic problems of increasing pest resistance complicate the estimation of economic benefits (see, for example, DeBach, 1974).

For the control of white wax scale, which is a significant pest in coastal citrus orchards in New South Wales and Queensland, natural enemies were imported from South Africa. One species achieved significant control in Queensland and another one some control in N.S.W. Again, the main economic advantage of biological control was seen as a reduction in spraying costs, that is, in the cost of chemical control; this change was used to estimate the benefit–cost ratio given in *Table 1*.

The above cases suggest that the net (social) economic benefits from research into and application of biological control methods are higher than for non-biological methods, at least in the public sector.

Furthermore, the relative economic returns obtained from the investment of funds in classical biological control programmes are very high when compared with any possible returns had the resources been allocated to other uses in the economy. If an average return of 10% in real terms can be obtained from investment elsewhere in the economy, the rates of return on all the biological control programmes mentioned above exceeded this figure. Internal rates of return on the resources used ranged between 13.9% for white wax scale control to 141% for skeleton weed control.

ECONOMIC IMPACT OF CONTROL OF *ECHIUM* SPECIES

The cost–benefit study of the CSIRO Division of Entomology (Marsden et al., 1980) paved the way for the next major study of the economic impact of biological control in Australia. This public enquiry into the benefits and costs of proceeding with the biological control programme for *Echium* species was conducted by the Industries Assistance Commission under the provision of the Biological Control Act 1984. The matter was contentious because one group of agriculturalists favoured the release of natural enemies, whereas another group opposed the scheme and had taken legal and political action to prevent it. The programme involved the introduction of several agents by the CSIRO, namely a leaf-mining moth (*Dialectica scalariella* Zeller [Lepidoptera: Gracillariidae]), a stem-boring beetle (*Phytoecia coerulescens* Scopoli [Coleoptera: Cerambycidae]), and two flea beetles (*Longitarsus aeneus* Kutsch and *L. echii* Koch [Coleoptera: Chrysomelidae]). As with the earlier CSIRO studies, it was assumed that product prices would not be affected by any increase in production resulting from weed control, so no allowance was made for increases in consumers' benefits. The study by the Industries Assistance Commission (Anonymous, 1985) highlights complications from changes in income distribution that may result from a successful biological control programme. The Commission's presentation also included a conceptual error, which caused the net economic benefits and the benefit–cost ratio to be underestimated (see below).

The costs and benefits from the biological control of *Echium* species, as estimated by the Industries Assistance Commission, are summarized in *Table 2*. All costs and benefits from the continuation and implementation of the programme over the 15-year period from 1985 (discounted by a 5% interest rate) are given in 1985 prices. The estimated costs include income losses to beekeepers (especially in South Australia, where bees use *Echium* species as a source for honey production); losses mostly to the

Table 2 Costs and benefits of the biological control of *Echium* species in Australia*

Costs and benefits	Discounted value (A$mio)
Costs	
Loss of income by apiarists	8.3
Loss of income due to reduced crop pollination	4.0
Reduced income due to lower stocking rates in semi-arid areas	1.1
Research and implementation costs of CSIRO and State governments	3.8
Total costs	17.2
Benefits	
Increased income from livestock in non-arid areas	115.2
Increased income from crops	20.2
Savings in control costs by public authorities	13.2
Total benefits	148.6
Benefit–cost ratio	8.6:1

* Data are from Table 7.1 of report by Industries Assistance Commission (Anonymous, 1985). Sum of returns for 15 years, expressed in constant (1985 $) prices and discounted at 5% interest rate.

fruit industry (as a result of reduced pollination by smaller bee populations); loss of income from grazing (in semi-arid areas where *Echium* species have value as forage during drought); and the costs of research and implementation incurred by the CSIRO and various State governments.

The major benefits from biological control of *Echium* species are higher incomes (or profits) from livestock grazing in other than semi-arid areas; greater incomes from crops such as wheat; and savings in control costs by public authorities. Total costs were estimated as A$17.2 million and total benefits as A$148.6 million, giving a benefit–cost ratio of 8.6 : 1. The apparent net benefits were A$131.4 million.

But, if normal practice for estimating cost–benefit ratios (using the Kaldor-Hicks principle) is followed, the loss in incomes by apiarists, by growers on account of reduced crop pollination, and by livestock producers in semi-arid areas (shown as cost items in *Table 2*) should be directly offset against increased incomes shown as benefits. When this is done, the only appropriate cost item (and resource-using item) is A$3.8 million as the cost for research and implementation by the CSIRO and State governments. After income losses have been offset against income gains, the total benefits are A$135.2 million, and the appropriate benefit–cost ratio is 35.6 : 1, not 8.6 : 1 as stated by the Industries Assistance Commission. Nevertheless, the main point is that the relative returns from implementation of the project (as well as its absolute impact) would be high. Thus, a decision about whether or not to implement the programme hinges on a value judgement as to whether those who will receive a reduced income should be 'sacrificed' so that the much larger benefits could be obtained by those who will gain. There is no 'scientific' economic answer to this question, but the type of economic information which is presented enables policy-makers to make an informed decision (see also Mumford and Norton, 1987).

Possible economic impact of future biological control

Even if discussion is limited to classical biological control, biological control has had a major favourable economic impact, and the average economic returns from research into and implementation of biological control programmes have been high. By contrast, expenditures on research into and development of chemical pesticides were about 8.5 times greater than those for biological control (Huffaker, Simmonds and Laing, 1976). This difference is a reflection of the greater ability of companies or business to appropriate benefits from toxicological research, rather than an indication that the economic benefits of chemical controls are more than eight times those of biological controls.

One concern regarding biological control can be summed up in this question: 'Will the degree of economic success achieved with classical biological control in the past be maintained in future efforts?' A number of biological control workers suggested that success rates should not be expected to taper off appreciably. For example, Wilson and Huffaker (1976) stated that 'enormous resources of natural enemies exist, and the research possibilities for [biological] pest control are virtually unlimited'. DeBach (1974, p. 195) claimed that 'despite the substantial progress that has been made in the field of biological control, there is no doubt that its application could be greatly increased'. By contrast, findings by Julien, Kern and Chan (1984) could suggest that the effectiveness of releases for the biological control of weeds is declining. These authors reported that effectiveness had decreased from 36% for releases made between 1940 and 1949 to 14% for releases made between 1970 and 1979, despite a marked increase in the number of attempts. It does not follow, however, that the average economic returns on research into biological controls have fallen; the opposite could well be true.

Control methods by augmentation and conservation of natural enemies or by inundation have been relatively little explored. Thus, the economic prospects of the use of, for example, pathogens, selective breeding, and genetic manipulation are good (Rabb, Stinner and van den Bosch, 1976;

Charudattan, 1985). If, as assumed, non-classical methods of biological control are likely to become relatively more important in the future, more economic research into these methods will be needed, including in particular the economics of mass-rearing of natural enemies (Mackauer, 1972) and of augmentation of r-selected natural enemies in annual crops (Ehler and van den Bosch, 1974).

Conclusions

Since the vedalia beetle was introduced in 1888 into California for the control of cottony-cushion scale, biological control has had a major economic impact globally. Economic methods have been developed to evaluate better its benefits and costs, but most economic studies have only been undertaken in the last decade and have, as a rule, been confined to classical biological control methods. These studies confirm that high levels of economic return have been achieved through the importation and release of exotic natural enemies for pest control. However, further economic research is needed to obtain a more accurate estimate of the global economic impact of biological controls and to provide an input into the development of non-classical biological control methods.

References

ALAUDDIN, M. AND TISDELL, C.A. (1986). Market analysis, technological change and income distribution in semi-subsistence agriculture: the case of Bangladesh. *Agricultural Economics* 1, 1–18.

ANONYMOUS (1985). *Report on the Biological Control of Echium Species (Including Paterson's Curse/Salvation Jane)*. Industries Assistance Commission. Australian Government Publishing Service, Canberra.

AULD, B.A., MENZ, K.M. AND TISDELL, C.A. (1987). *Weed Control Economics*. Academic Press, London.

CHARUDATTAN, R. (1985). The use of natural and genetically altered strains of pathogens for weed control. In *Biological Control in Agricultural IPM Systems* (M.A. Hoy and D.C. Herzog, Eds), pp. 347–372. Academic Press, Orlando, Florida.

CONWAY, G.R. (1987). The properties of agroecosystems. *Agricultural Systems* 24, 95–117.

DAVIS, J.S., ORAM, P.A. AND RYAN, J.G. (1987). *Assessment of Agricultural Research Priorities: An International Perspective*. Australian Centre for International Agricultural Research, Canberra.

DEBACH, P. (1974). *Biological Control by Natural Enemies*. Cambridge University Press, London.

DODD, A.P. (1940). *The Biological Campaign against Prickly Pear*. Commonwealth Prickly Pear Board, Brisbane.

DUNCAN, R.C. AND TISDELL, C.A. (1971). Research and technical progress—the returns to producers. *The Economic Record* **47**, 124–129.

EDWARDS, G.W. AND FREEBAIRN, J.W. (1982). The social benefits from an increase in production in part of an industry. *Review of Marketing and Agricultural Economics* **50**, 193–210.

EHLER, L.E. AND VAN DEN BOSCH, R. (1974). An analysis of the natural biological control of *Trichoplusia ni* (Lepidoptera: Noctuidae) on cotton in California. *Canadian Entomologist* **106**, 1067–1073.

GROVES, R.H. AND CULLEN, J.M. (1981). *Chondrilla juncea*: the ecological control of a weed. In *The Ecology of Pests: Some Australian Case Histories* (R.L. Kitching and R.E. Jones, Eds), pp. 7–17. CSIRO, Melbourne.

HARRIS, P. (1979). Cost of biological control of weeds by insects in Canada. *Weed Science* **27**, 242–250.

HAYAMI, Y. AND HERDT, R.W. (1975). Market price effects of technological change on income distribution in semi-subsistence agriculture. *American Journal of Agricultural Economics* **59**, 245–256.

HUFFAKER, C.B. AND CALTAGIRONE, L.E. (1986). The impact of biological control on the development of the Pacific. *Agriculture, Ecosystems and the Environment* **15**, 95–107.

HUFFAKER, C.B., SIMMONDS, F.J. AND LAING, J.E. (1976). The theoretical and empirical basis of biological control. In *Theory and Practice of Biological Control* (C.B. Huffaker and P.S. Messenger, Eds), pp. 41–78. Academic Press, New York.

JULIEN, M.H., KERN, J.D. AND CHAN, R.R. (1984). Biological control of weeds: an evaluation. *Protection Ecology* **7**, 3–25.

MACKAUER, M. (1972). Genetic aspects of insect production. *Entomophaga* **17**, 27–48.

MARSDEN, J.S., MARTIN, G.E., PARHAM, D.J., RISDILL SMITH, T.J. AND JOHNSTON, B.G. (1980). *Returns on Australian Agricultural Research*. CSIRO, Canberra.

MUMFORD, J.D. AND NORTON, G.A. (1987). Economic aspects of integrated pest management. In *Integrated Pest Management—Quo Vadis?* (V. Delucchi, Ed.), pp. 397–407. Parasitis 86, Geneva.

PERKINS, R.C.L. AND SWEZEY, O.H. (1924). The introduction into Hawaii of insects that attack *Lantana*. *Bulletin of the Hawaiian Sugar Planters' Association Experiment Station, Entomology Service* **16**, 1–83.

QUIMBY, P.C. AND WALKER, H.L. (1982). Pathogens as mechanisms for integrated weed management. *Weed Science Supplement* **30**, 30–34.

RABB, R.L., STINNER, R.E. AND VAN DEN BOSCH, R. (1976). Conservation and augmentation of natural enemies. In *Theory and Practice of Biological Control* (C.B. Huffaker and P.S. Messenger, Eds), pp. 233–254. Academic Press, New York.

TEMPLETON, G.E. (1982). Biological herbicides: discovery, development, deployment. *Weed Science* **30**, 430–433.

TISDELL, C.A. (1981). *Science and Technology Policy: Priorities of Governments*. Chapman and Hall, London.

TISDELL, C.A. (1988a). Sustainable development: differing perspectives of ecologists and economists, and relevance to LDCs. *World Development* **16**, 373–384.

TISDELL, C.A. (1988b). Biological control of pests: its economic nature and the evaluation of benefits of such control. Department of Economics, University of Newcastle, NSW, Australia (Unpublished manuscript).

TISDELL, C.A., AULD, B.A. AND MENZ, K.M. (1984). On assessing the value of biological control of weeds. *Protection Ecology* **6**, 169–179.

VERE, D.T., SINDEN, J.A. AND CAMPBELL, M.H. (1980). Social benefits of serrated tussock control in New South Wales. *Review of Marketing and Agricultural Economics* **48**, 123–138.

WILSON, F. AND HUFFAKER, C.B. (1976). The philosophy, scope and importance of biological control. In *Theory and Practice of Biological Control* (C.B. Huffaker and P.S. Messenger, Eds), pp. 3–15. Academic Press, New York.

The economic impacts of endemic diseases and disease control programmes

C.A. Tisdell [1], S.R. Harrison [1] & G.C. Ramsay [2]

(1) Department of Economics, University of Queensland, QLD 4072, Australia
(2) Secretariat for the Pacific Community, Private Mail Bag, Suva, Fiji

Summary
The authors discuss the evaluation of the economic impacts of endemic livestock diseases, and economic issues in control of these diseases. Particular attention is focused on helminths and on endemic vector-transmitted infections (particularly ticks and tick-borne diseases). Decisions relating to disease control have to be made by government and by the producer. Government requires information on the level of control to adopt, the extent of involvement needed, and how to fund animal health programmes (particularly how to share costs between taxpayers and livestock producers). Individual producers require information as to how much effort to invest in disease control, including information collection effort, and how to design control strategies. Economics can shed light on these issues. However, experience suggests that animal health policies are particularly difficult to evaluate from an economic viewpoint, with complex relationships between animal health, production impacts, market access, and non-production benefits of livestock. While little information is available concerning the cost of helminth diseases, many estimates have been made of the costs of ticks and tick-borne diseases at a regional and national level, sometimes demonstrating that eradication is warranted.

Keywords
Animal diseases – Disease control – Economics – Helminthiasis – Livestock – Tick-borne diseases.

Introduction

The term endemic diseases covers an extremely wide range and the characteristics of these diseases differ greatly. In this paper, therefore, the example of parasites as a source of disease will be highlighted, taking helminths as examples of endoparasites and ticks as examples of exoparasites. Even with such a focus, the number of parasite species and diseases is large, and there is, therefore, a need to concentrate on individual cases.

Endemic parasites are a major source of economic loss in animal husbandry, especially in tropical areas and developing countries, but as discussed later, the extent of those losses has yet to be accurately specified, and knowledge about the economics of treatment of these diseases is inadequate, mostly because the damage functions, and in addition, the response functions to treatment are imperfectly known. The response functions should be specified in relation to economic variables of importance (e.g. depending on the situation, weight gain, reproduction rates, lactation of mammalian livestock), rather than just clinical effects. Co-operation between veterinarians and economists is needed at an early stage if information of economic relevance is to be collected by veterinarians.

Extra cost is incurred in the collection of extra veterinary data of economic relevance, however, without this additional step, little progress will be made in the economic evaluation of diseases. This point cannot be over emphasised, and is supported in the conclusion of a recent paper (36).

Economic impacts of parasites of livestock may be divided into two groups – direct and indirect impacts. Direct impacts are those attributable purely to the presence of the parasite, e.g. reduced economic productivity of the livestock or economic loss due to mortality. Indirect impacts occur because some parasites, such as ticks, are also disease vectors, and the vector-borne diseases, if transmitted to a livestock host, often result in much greater economic loss than the presence of the parasite. Furthermore, by weakening the host the presence of a parasite may make the animal much more susceptible to infection and adverse impact from other diseases, or to environmental stresses such as food deficits (18).

The range of parasites of livestock is very large. Helminth parasites belong to three phyla of invertebrates: platyhelminths (flatworms), acanthocephalans (spiny-headed worms) and nematodes (roundworms).

In addition to ticks, ectoparasites include mites, lice, fleas, leaches, mosquitoes and flies of various kinds. However, as mentioned earlier, to deal with the full range of these parasites in this article would be impossible, and the focus here is limited to helminths and ticks. This article first considers general issues involved in the evaluation of the economic impact of endemic diseases and the economics of disease control. Subsequently, these issues are considered specifically for helminths and then for ticks and tick-borne diseases, and general conclusions are drawn.

Evaluation of the economic impact of endemic diseases and the economics of disease control

Evaluation of the economic impact of endemic diseases

Endemic diseases cause large economic losses but few precise estimates of the magnitude of these losses have been made, mainly because the amount of information required is large and the cost of obtaining the information is high. Furthermore, as mentioned earlier, veterinary research effort is not always sensitive to the data needs of economic evaluation. Economists can only make progress with economic evaluation if veterinary data is provided in a required form so that it can be combined with economic data and the appropriate economic analysis can be undertaken (30).

Economic evaluations can be undertaken for a variety of target groups, or geographical areas, for example:
- an individual livestock holder
- a region
- a nation
- for the global situation.

Different types and amounts of information are required for these different levels of coverage.

In collecting information on economic impact, the coverage required must be determined (this depends on the target group), but in addition, the purpose for which the information is to be used must be taken into account. For example, a common objective in relation to disease control is to estimate the *total economic loss* caused by the prevalence of disease. But what is the purpose of this? True, it provides an indication of the economic benefits to be had if an *inexpensive* method could be found to reduce substantially the incidence of the disease. However, such relative losses from diseases are a poor guide to the allocation of research funds, because the likelihood and cost of making new discoveries concerning the control of a disease and the economics of remedies discovered must also be considered (30).

Information regarding whether the full economic potential from remedying a disease is being obtained, given existing knowledge about disease control, is useful from several points of view. It is important to determine the scope which exists for increasing economic benefits from the use of existing veterinary knowledge. For an individual livestock holder, the scope for this will be equal to the difference between the economic gain using the current control practice and the economic gain using an optimal economic control strategy. As discussed later, such an approach has been adopted by several researchers (1, 27, 30). Similar estimates can be made regionally, for a nation or globally. However, accurate estimates cannot always be obtained by merely aggregating economic gains for individual livestock owners considered in isolation. For example, this will be true if an increase in supply of livestock products, as a result of disease control, reduces the market price of these products (18), or if a favourable environmental health spillover (externality) is generated by the disease control measures adopted by individual farmers. In some cases, measures taken by individual farmers to control livestock diseases also reduce the likelihood that the livestock of other farms will contract the disease. Where a difference exists between potential and actual net economic benefits from disease control practices, stockholders may be encouraged by government extension services to change control strategies.

Optimal economic strategies are not always the same for all stock owners and may vary with environmental conditions. This provides challenges for government extension services, which cannot, as a rule, target individual clients, and must focus on broad groups of clients. Therefore, government advice is generally based on average situations, and one has to decide how finely to structure this advice for target clients in different regional and environmental situations. To some extent this involves economic choice – the extra economic benefit for more structured and specifically directed advice must be weighed against the additional cost (4, 44). Improvement of advice is only socially desirable up to the point where the extra economic benefits generated equal the extra cost.

Economics of control of endemic diseases

The economics of control of endemic diseases is complex since the epidemiology of most is complicated and disease prevalence often changes radically with alterations in environmental conditions. In particular, the occurrence of parasitism seems to be greatly influenced by environmental conditions. This means that simple models of the economics of control of diseases can often only be applied to particular endemic diseases if considerable modification and further development is undertaken.

Livestock holders often fail to control endemic diseases in an economically optimal manner and convincing individuals to change this behaviour can be difficult. For example, particularly in developing countries, the presence of such diseases is often unrecognised or considered 'normal', and is thus fatalistically accepted by livestock holders. For example, although the occurrence of bovine fasciolosis is a major problem in Indonesia (42), control attempts are uncommon 'because the extent of the problem is largely unrecognised by farmers (the disease is common and unspectacular, and the main clinical signs of failure to thrive and reduced exercised tolerance are similar to those of poor nutrition or regarded as "normal"), modern anthelmintics are expensive and there is no reliable information on the benefits of control' (42).

This raises a further aspect, namely that the economics of control of a disease may vary from country to country and from region to region. The occurrence of endemic diseases can be greatly influenced by cultural practices – both agricultural practices and wider practices. For example, attitudes to the keeping and handling of dogs influence the occurrence of hydatidosis in humans. Treatment of animal manure by humans can influence whether the life-cycle of some helminths is assisted or thwarted. If cow dung is composted at high temperatures, before being returned to the field, or burnt for fuel, as in parts of the Indian subcontinent, the eggs of some species of helminths will be destroyed, thereby reducing these populations (37).

Control strategies for endemic diseases which are economically feasible in developing countries often differ from those which are economically optimal in more developed countries. This is because farmers in developing countries usually only each keep a small number of livestock, and often use communal water and feed sources, capital is usually in short supply, and agricultural activities differ from those in developed countries, with specialised farming being more the exception than the rule. In addition, cultural differences between countries can influence practical control policies. To be relevant, economic advice must be varied to take account of all such considerations. In addition, the objective of keeping livestock may vary from society to society and from group to group: the purpose is not always to earn the largest stream of net cash income, as seems to be the basic motivation in developed countries. Economic advice needs to be formulated by considering the differing objectives of target groups and the different cultural practices and habits of these groups, especially those relevant to the farming and grazing activities of livestock holders. Nevertheless, fundamental factors can be identified which are relevant to most analyses of the control of endemic livestock diseases.

For example, one needs to know the relationship between possible disease control strategies and the benefit expected by livestock holders or other targeted groups. It is necessary to determine what attributes of the livestock are valued and how much value is placed on each to obtain an objective function. The problem is a holistic one and calls for an interdisciplinary approach.

Ideally, the aim should be to recommend a specific disease control strategy which maximises the objectives of the client. Mathematically, the problem can be considered as a constrained maximisation problem – to maximise the objective or utility function of the client, subject to the resource and other constraints faced by the client and taking into account the possibilities generated by the available disease control strategies.

A simple form of this approach (which is particularly relevant to developed countries and increasingly to less developed countries as they become more market-oriented) is as follows:

a) the objective of the livestock holder is to maximise net monetary income (or net discounted cash flow)

b) this depends on the quantity and quality of saleable products produced by or attributable to having livestock in the farming enterprise as well as the prices of these products

c) allowance is made for financial outlays on account of these livestock, which include any outlays on disease control.

In considering this problem, one of the important points for attention is the relationship between strategies for control of a disease and the impact of these strategies on the quantity and quality of the saleable livestock products. In essence this means that a production function response or relationship needs to be estimated (30).

To take a simple case, suppose that a farmer wishes to maximise annual net income and that, for livestock, the only attribute of relevance is the weight of meat of the livestock. Suppose that preventive treatment for the relevant livestock disease is available which can be administered with varying intensities. Intensity, for example, may be the number of times per year that the stock are drenched to control helminth infestations. Additions to meat weight may increase with the number of preventive doses per year, but at a decreasing rate. In stationary conditions, a functional relationship exists between the number of doses and meat weight gains, which in principle can be determined scientifically.

The following procedure can be adopted to determine the economically optimal number of annual doses of the treatment to administer: increase the number of doses as long as the additional revenue (which equals extra weight of meat in kilograms multiplied by price paid per kilogram for this meat) exceeds the cost of an extra dose of the treatment, e.g. additional drenching. As a rule, the net income maximising dose rate will be the highest dose rate for which the above condition is satisfied. The optimal number of doses can be expected to vary with factors such as the price of meat and the cost of dosing the animals.

In practice, the situation will be much more complicated. More than one livestock product may be affected by a disease, and the economics of treatment may vary with environmental conditions, which may be relatively unpredictable. Thus, the economic decision to control a disease may have to be made in uncertain and variable conditions. Therefore, attitudes to risk and instability must be taken into account. In animal health economics, sensitivity analysis is usually used to measure these variables (10, 30). Sensitivity analyses specify outcomes for different possible scenarios or states of nature. The likelihood of these may also be indicated. Depending upon the attitude of the decision maker towards the bearing of risks and income variability, he or she can adopt a particular strategy. Essentially, this was the approach adopted by Meek (27) in analysing the economics of alternative strategies for the control of ovine fascioliasis in Australia.

The problem becomes more complicated when alternative techniques of parasite control exist, since these need to be compared to decide which are the most economic, including possible combinations of these techniques (43). More detailed analysis of animal health economics is provided by Dijkhuizen and Morris (10).

The main purpose of this sub-section is to emphasise that the economics of control of endemic diseases is especially complicated. This is particularly true of diseases related to parasitism because of the high degree of sensitivity of these (and of the economics of control) to variable environmental conditions. Furthermore, especially in the case of helminths, cultural factors and the often complicated helminth life-cycles add extra dimensions to difficulties of determining the economics of control.

Helminthiasis: economic impact and economics of control in livestock

Economic impact of helminthiasis

Many qualitative statements can be found in the literature which indicate that the economic impacts of helminth infections are substantial, but few quantitative estimates of these impacts exist. The opening section of a major FAO-sponsored review of helminth diseases states: 'Helminthiasis infections of food animals cause significant economic losses. The effect of the infection is determined by a combination of factors of which the varying susceptibility of the host species, the pathogenicity of the parasite species, the host/parasite interaction, and the infective dose are the most important. The economic losses are closely associated with the extent to which the pathogenic effect of helminth infections influences the production of the individual host. This may vary considerably from clinical disease including mortality to chronic production losses which may appear as, e.g., reduced growth rate, weight loss and/or reduced fecundity or it may go unnoticed' (35).

A further cost arises in that some helminths can be transmitted to man. Many parasitic helminths can in fact be transmitted between vertebrate animals and man (i.e. these helminths are zoonoses) and 'about 20 species are of public health importance causing severe or fatal infections. In many parts of Africa parasitic helminths are responsible for enormous economic losses, hampering rural development programmes and reducing the pace of economic growth' (25).

Bain and Urquhart estimated that the economic loss from stomach worm in British cattle in the mid-1980s would have been approximately £45 million annually if parasitic gastroenteritis had been completely uncontrolled (2). This potential loss would show a several fold increase if the figure were to be scaled up to cover the whole of Europe. The research was mostly concerned with the economic impact of the most pathogenic of the stomach worms in cattle in Europe, *Ostertagia ostertagi*. Estimates of financial loss are based on mortalities in young cattle and the failure of cattle to reach potential normal weights (weight loss). These are assumed to vary according to the severity of infection. In the absence of parasitic control, it is assumed that 5% of cattle will be severely infected, 20% moderately so, 45% will have sub-clinical infections and the remainder will not be affected. Only the severely infected group is assumed to suffer mortality; 24% of this group are supposed to die annually, involving a loss of £200 per dead animal on average. Weight loss per calf is assumed to decline as the degree of infection declines, being 35 kg for severely infected animals, 20 kg for the moderately infected and 10 kg for sub-clinical cases. Each kilogram loss is supposed to reduce the value of a calf by £1.20. The estimated annual production of calves in Britain was 3.2 million over the period in question and consequently the estimate of a £45 million annual loss can be easily obtained.

In this analysis, no account has been taken of any possible impact on prices of livestock nor of the possibility of a reduction in the milk yields of older cattle. In addition, no estimates have been made of profits forgone by not adopting the most economic methods of control of stomach worms of cattle, that is economic gains which could be made by adopting superior available methods of control of stomach worms. Therefore, this exercise appears to be of limited value. Nevertheless, it resulted in Zinsstag et al. (45) proclaiming that 'In Africa such estimations are not available, but losses are expected to be even higher due to poor nutrition, which substantially enhances the pathogenic effect of parasites (15)'.

In an early study of the economics of control of helminthiasis in weaned lambs (1), it was estimated on the basis of experimental evidence in Victoria, Australia, that if all young sheep in the high rainfall area of Australia were given the critical treatment scheme (anthelmintics at the most appropriate time) rather than being treated for clinical parasitism only, the additional net benefit would be approximately AUS$6.7 million annually at 1970-1971 wool prices. In comparison to traditional drenching schedules, the extra net benefit to wool growers from the critical scheme would amount to AUS$5.4 million. These economic gains increase with wool prices. The precise figures are not very important for this article; what is more important is that owners of sheep could have considerably increased their incomes by adopting a superior *available* strategy for helminth control.

Unfortunately, the review by Over et al. (35) of helminth diseases of livestock in developing countries sheds little light on the economic impact of these diseases. Data such as infection rates of livestock and rejection rates of offal at abattoirs due to helminth infections provide a restricted basis for estimating economic impacts. Furthermore, no indication is given of economic benefits which are unrealised due to failure to adopt appropriate *available* control techniques.

Economics of control of helminthiasis

Social and private perspectives on control effort

The economics of helminth control may be viewed from a government or social perspective, or from the perspective of the individual producer. The government needs to take account of externalities, such as trade impacts and effects on public health. The individual producer is concerned with maximising private goals of revenue, stock quality and other objectives, within the resource limitations and regulatory framework.

Government wishes to determine the optimal control expenditure, including public and private sector expenditure. The social cost-benefit analysis framework generally accepted for evaluation of animal health programmes involves making estimates of all socially relevant costs and benefits, including both market and non-market items, and hence deriving 'incremental cash flows'. However, when expenditure levels vary between alternative disease control policies, it has been argued that benefit-cost criteria do not rank alternatives correctly, and that a loss-expenditure tradeoff curve provides a better indication of optimal control effort (21, 22, 23). This takes the form of a concave tradeoff curve between disease cost (C) and control expenditure (E) as illustrated in Figure 1. The points along the loss-expenditure frontier may be taken to represent the capitalised value, or discounted sum, of disease and expenditure costs over a number of years. This curve represents the set of choices available to a country in terms of effort on disease control. With no control expenditure, disease cost, c_1, is high. As control expenditure is increased, disease cost first decreases rapidly, the C-E tradeoff curve being almost vertical. But with increasing expenditure the curve flattens as the marginal rate of improvement with respect to cost declines. If the disease can be eradicated, control expenditure may fall to zero or to some low amount representing the cost of preventing new outbreaks.

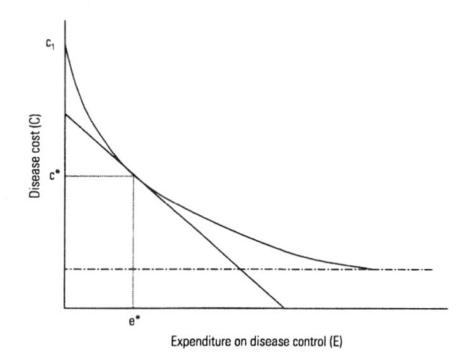

Fig. 1
The McInerney loss-expenditure tradeoff model

Since both axes are expressed in dollars terms, and a dollar in disease cost is regarded as equivalent to a dollar in control costs, the line with slope -1 represents combinations of equal cost to the country, i.e. an isocost line. One such line is drawn in Figure 1; this is the line which is tangent to the C-E curve. Any C-E combination to the right of this line would represent greater overall cost; any point to the left is not achievable. Hence e^* is the optimal expenditure level and is associated with a disease cost c^* (total cost $c^* + e^*$). For this formulation, the disease cost variable would need to include all relevant items including non-market costs (e.g. environmental impacts, animal welfare changes).

The distribution of costs between government and livestock producers is a further issue. There is increasing emphasis on 'user pays' policies, and this could be applied to animal health research, disease control, inspection for export and so on. However, it is notable that measures which reduce livestock producer costs may lead to greater benefits for consumers than for producers. On the other hand, measures which enhance export markets may impose a cost on consumers, to the benefit of producers and traders (middlemen).

Issues in evaluating control economics

The economics of control of helminthiasis tends to be complicated because the life-cycles of helminths can be complex and are dependent on environmental conditions which can vary considerably. Furthermore, in most cases, the life-cycles of helminths depend on multiple hosts. Schwabe (40) suggests that helminths show unusual biological complexity and points out that 'broadly adapted to causing disease in other species, parasitic nematodes, cestodes and trematodes undergo successive stages of development, often as quite dissimilar forms, including in their feeding and other habits'. Four types of zoonosis cycle exist:

– direct zoonoses can be perpetuated by a single vertebrate reservoir, e.g. trichinosis

– cyclo-zoonoses require more than one vertebrate species to complete a life-cycle, e.g. echinococcosis

– meta-zoonoses require both vertebrate and invertebrate host species, e.g. schistosomiasis

– sapro-zoonoses require a non-animal reservoir or development site to complete a life-cycle, e.g. some species of nematodes (40).

Knowledge of such cycles is essential for the sound economic and epidemiological control of helminths. For instance, as Zinsstag et al. (45) point out: 'The precise knowledge of the biology and seasonability of the (gastrointestinal) parasite and the groups at risk in the various agro-ecological zones are prerequisites for any economically and epidemiologically sound approach to the control of gastrointestinal parasites'.

Even in the case of direct zoonoses involving a single vertebrate reservoir, the question is raised of what is the best stage in the life-cycle of the parasites to target for economic control (e.g. the eggs or a later stage in the life-cycle), and at what time to target these stages. Some anthelmintic treatments are ineffective if the parasites enter a dormant stage (2). If a parasite has more than one host (e.g. liver flukes depend on snails as hosts as well as vertebrates), the question arises of whether it is economic to control this reservoir in order to reduce infection in the end-host. Furthermore, can livestock be economically excluded from sources of environmental contamination with parasites or excluded from these at critical times? To this end, and for the control of particular helminth diseases, strategic rotational grazing of livestock or strategic cropping and grazing cycles can be practised.

Usually, a wide range of *possible* strategies are available for the control of helminths, but only a few of these alternative strategies are explored from an economics point of view. Some may be excluded on *a priori* grounds because they are clearly uneconomical. Those explored often reflect the particular expertise of the researcher. Some of the available studies of the economics of control of helminthiasis are considered below.

Anderson et al. (1) and Morris et al. (29) explored the economics of anthelmintic drenching of sheep, based on the results of experimental work in Victoria. The main helminths of concern were *Ostertagia* and *Trichostrongylus* spp. The economics of four alternative control strategies were compared:

– treatment 1: no preventive treatment

– treatment 2: traditional preventive treatment as indicated by surveys of farmers

– treatment 3: varied prophylactic treatment based on critical periods in the life of the helminth species involved

– treatment 4: fortnightly treatment.

The studies found that both treatments 2 and 3 were economic, with treatment 3 being the most profitable when administered either to weaned lambs (1) or to breeding ewes (29). Fortnightly treatment was more profitable than traditional treatment.

These results were based on partial budgeting. Sensitivity analysis was applied and the profitability of the fortnightly treatment was shown to be particularly sensitive to wool prices. In the case of breeding ewes, it was concluded on the basis of the economic evidence presented that 'there is no merit for the farmers in considering the option of either the 'traditional' treatment scheme or 'two-weekly' treatment scheme in preference to the 'critical' treatment scheme, since the latter was the most financially rewarding under all circumstances evaluated' (29).

While the above contribution is a significant one, the study is based only on a small experimental sample on one farm, of limited numbers of sheep and results over one year. Further testing at other locations and other replications would be desirable to test the robustness of the results. When experimental research on the control of animal diseases, including field experiments, is performed, some doubt must arise regarding the extent to which the results can be generalised to a wider population. An interesting social science question raised by this research (but not explored) is why farmers used the traditional method rather than the more profitable one? Is it because of lack of information or for some other reason?

In northern Victoria, Australia, an analysis of the economics of control of ovine fascioliasis was undertaken in sheep grazing on irrigated and non-irrigated pasture (27, 30). A model was developed to analyse infections of sheep by *Fasciola hepatica*, taking into account environmental factors, the life-cycle of *F. hepatica*, and the impact of ovine fascioliasis on wool production (and value of wool sold) and value of sheep sold. Parameters in the model were specified using scientific literature, expert opinion and two experiments designed to

supplement this information. Using simulation techniques, the economics of control were examined for the following:
- five alternative anthelmintic strategies
- the use of a molluscide to destroy snail hosts
- rotational grazing.

Although most attention is given to anthelmintic strategies, the economic study is more comprehensive than most because other control possibilities are also considered. The economic benefit from anthelmintic treatment is shown to depend on stocking rates and percentage of the paddock (used by sheep) occupied by snail habitats. The most intensive strategy for anthelmintic treatment (drenching approximately every eight weeks) was shown to give the highest financial return. The research provides a valuable holistic framework for studying the economics of control of ovine fascioliasis (27). In this regard, this study is a relatively unique contribution to the veterinary economics of the control of helminthiasis.

The case studies mentioned above concern a relatively developed country. Similar case studies for developing countries are scarce, but the need for economic evaluation is often crucial, as is evident from recent research on strategic gastrointestinal nematode control in cattle in The Gambia (45). Significant weight loss from nematodes was demonstrated in cattle, but financial analysis of the recommended control measures was not completed.

An interesting economic evaluation of the economic returns from vaccinating cattle in the Sudan against *Schistosoma bovis* has been completed (18). *Schistosoma bovis* vaccine has proven to be effective in experiments in the White Nile province of the Sudan in reducing mortality and weight loss (or growth delay) from *S. bovis* infection. The results of these experiments are used together with other data to estimate the economic return to be expected from an *S. bovis* vaccination campaign in the Sudan extending over a five-year period.

For the purpose of estimation, the Sudan is divided into three areas – central, western and southern. 'The area divisions reflect judgements on the probabilities of infection and vaccine coverage that could be reasonably expected under the conditions in the different provinces' (18). Returns for vaccination in each of the areas are estimated for high and low levels of production losses avoided as a result of *S. bovis* vaccination. The estimated benefit-cost ratios and internal rates of return are set out in Table I. Benefits exceed costs in all areas except southern Sudan where low levels of production losses are avoided. For central Sudan, estimated economic benefits are especially high, being well in excess of costs in all circumstances considered, with internal rates of return varying from a low of 43% to a high of 328%, depending on whether low or high productivity losses are avoided.

The methods used to estimate bovine schistosomiasis losses in the Sudan and the costs of the vaccination programme, and to

Table I
Benefit-cost ratios and internal rates of return for a five-year vaccination programme against bovine schistosomiasis in the Sudan

Area and coverage assumed	High production loss avoided B/C[a]	High production loss avoided IRR[b]	Low production loss avoided B/C	Low production loss avoided IRR
Central (90%)	22.7	328	1.5	43
Western (50%)	5.7	209	1.2	17
South-east (50%)	4.0	157	0.7	−12

Source: based on McCauley *et al.* (18), p. 738
a) Benefit-cost ratio based on net present values of benefits and costs
b) Internal rate of return

estimate the value of losses avoided are clearly described in the research (18). For the purpose of the exercise, the study concentrates on sales of meat and takes into specific account two factors: mortality from *S. bovis* and growth delay weight loss. It is recognised that additional factors could be considered, such as losses in production from reproduction inefficiency and condemnation of livers at abattoirs. However, reliable data was not available concerning increased reproduction inefficiency, and losses due to liver condemnation are small.

While the estimates in this study rely on various assumptions, the systematic framework provides a solid basis for economic evaluation. The model can be easily reworked for changed assumptions or conditions. In addition, scope exists for adapting the model to estimate the economic returns from control of other animal diseases. However, the model evaluates one method of control only, whereas comparative economic analysis of different methods and techniques of control would be desirable.

The level of subsidy or the extent to which farmers would have to pay for vaccination of cattle in the Sudan against *S. bovis* is unclear. Even when returns from an investment are high, individual farmers in less developed countries (LDCs) may fail to undertake this investment because of lack of finance or because of suspicion of new techniques. Farmers in LDCs often face different economic constraints to those in developed countries. As highlighted by Roberts and Suhardono (38), control measures for animal diseases which are economic in developed countries may be uneconomic in less developed countries or prevented from being adopted by cultural practices. Therefore, the development of different strategies for controlling helminths in LDCs may be necessary. In relation to the control of fasciolis in ruminants in LDCs, the following observations have been made: 'Anthelmintics are not affordable. Recent observation of a major fasciola resistance gene with substantial dominance in Indonesian Thin Tail sheep infected with *Fasciola gigantica* suggest that parasite control by breed substitution, or cross-breeding and selection, are feasible. Such control should be inexpensive to implement and sustainable' (38).

Breed substitution is in fact often used as a mechanism for coping with parasites; even in developed countries, including Australia, *Bos indicus* cattle are often substituted for *Bos taurus* in tick-infested areas because the former species has superior resistance to ticks. However, such substitution can involve economic costs, for example if the market value of products from the substituted breed is less than the alternative. Roberts and Suhardono also suggest that some simple changes to agricultural cultural practices in LDCs may entail low cost and provide some control of fasciolosis, however, an economic study was not undertaken (38). The study is nevertheless significant in that the importance of relating research on the control of animal diseases to the social and economic context in which this research is to be applied is emphasised. To complicate matters, it might also be added that social and economic conditions are rarely stationary. Changes in the direction of research and in the type of advice given to client groups are generally required as circumstances change.

It should be noted that in undertaking social evaluation of methods of disease control, as opposed to the calculation of private economic benefits, factors additional to those mentioned above should be taken into account. For example, the use of molluscicides to destroy snails which are hosts for a helminth may destroy other biota of value. All externalities and environmental spillovers should be taken into account in a social evaluation. Furthermore, more frequent administration of anthelmintics may accelerate resistance of helminths to these chemicals, so the more immediate economic benefits must be balanced against loss of economic benefits in the longer term. In addition, some anthelmintics (and other chemical means of controlling parasites) result in chemical residues in meat or milk which can limit the marketability and suitability of the product for human consumption. This restricts the economic and social value of the use of such controls.

Thus, it can be seen that the economics of control of helminthiasis is complex, partly because of the complex epidemiology involved. A sound knowledge of both biological and economic relationships is needed to evaluate disease control strategies involving helminths. That few in-depth economic studies of the control of helminthiasis have been completed is therefore not surprising. However, the authors are surprised to find that no substantial studies in the area seem to have been published since 1984.

Tick and tick-borne diseases: economic impact and control economics

Ticks and tick-borne diseases are accepted as major causes of economic loss to livestock industries in tropical and sub-tropical regions of the world. In particular, these losses occur in the ruminant industries. According to McLeod and Kristjanson (24): 'Tick-borne diseases severely constrain cattle production in Asia, Africa and Australia. The diseases theileriasis, babesiosis, anaplasmosis and heartwater can cause mortality, reduced milk and beef production, depressed manure production and reduced animal draft power. The distribution of these diseases is dictated by the presence of specific tick vectors for each of the diseases. Anaplasmosis and babesiosis are primarily transmitted by ticks of the genus *Boophilus*, found throughout Africa, Asia and Australia. Heartwater (cowdriosis) is vectored by *Amblyomma* spp. in Africa. Theileriosis, called tropical theileriosis in Asia and northern Africa is vectored by ticks of the genus *Hyalomma*. It is known as East Coast fever (*Theileria parva*) in Central and Eastern Africa where it is transmitted by the tick *Rhipicephalus appendiculatus*.'

The effects of these agents can be devastating as evidenced by the outbreak of East Coast fever in southern Africa at the beginning of this century, which decimated the cattle industry. In addition, the consequences to an individual producer can be high when naïve cattle are introduced into an area where these diseases occur, or where the vector transmitting a disease is introduced to a new area. The latter may occur either as a result of changes in seasonal conditions (for example warmer, wetter conditions, favouring tick development and reproduction) or due to changes in tick control methods. Therefore, the problem confronting decision makers is not whether the diseases are important, but when the disease should be controlled and which method or combination of methods is most appropriate.

Tick-borne diseases provide many difficulties in determining the most economically efficient approach to control. Problems arise due to the complexity of the system, including both the large number of tick species present and of diseases transmitted by the ticks, and the interactions between animals, ticks and disease agents.

Disease control involves a number of activities in various combinations and is not an all-or-nothing event. In addition, disease or vector control activities often affect other vectors or diseases, providing benefits (and in some cases costs) not directly related to the original disease control programme. In addition, control methods for ticks and tick-borne diseases can be difficult to use and are not always effective. For example, vaccines are usually live and although of low virulence, can have side effects requiring close monitoring of livestock for signs of vaccine reactions and need for treatment.

Tick control can have additional side effects: the most commonly applied method (the use of chemical acaricides) produces residues in meat if animals are slaughtered within the prescribed withholding period. Environmental contamination and erosion of areas around dip tanks can also occur.

The factors outlined above, in association with the interactions that occur between vector ticks and the diseases produced, make decisions about tick and tick-borne disease-control complex on a technical and economic basis. For example, treatments to control one species of tick will affect other species of ticks, and the introduction of one programme such as dipping, for tick control, may require introduction of another programme involving vaccination against tick-borne disease (due to reduced immunity). These flow-on effects need to be considered in the analysis. The complexity of animal health systems, in particular, the large number of interactions, calls for the development of simplified models to enable the important factors in the system to be determined and considered in the analysis.

Economic impacts

The effects of disease on livestock production

Disease has a variety of biological effects on animals that are exhibited as production losses. Disease affects the ability of an animal to survive, grow and reproduce. In addition to the effects of disease on individual animals, herd effects are also seen, including adverse modification of the herd structure (26).

Close clinical observation, physical measurement and laboratory examination of specimens is often required to determine the effect of a disease on the productivity of an animal (30). The effects of a disease on animal production are difficult to estimate because of the large number of variables that are affected, such as age, breed, production status and condition of the host, pathogenicity of the disease-causing organism and environmental factors (16). Because information is limited on the effects of diseases on production, estimates must usually be made using a combination of published data and expert opinion (11, 33).

The effects of parasitic infestation (which can produce a chronic disease) on animal production have been widely reported (1, 14, 27, 30). A system has been developed to outline the information needed to determine the effects of a disease on livestock production (30). The system uses a combination of experimental studies and expert opinion to determine the effects of a disease on animal production.

In the case of sporadic or exotic diseases there is much less certainty about disease occurrence. To perform a field experiment or observational study would either involve artificial infection of a group of animals or the use of a large number of sites, which would be expensive. In addition, the large number of variables which could not be controlled in an observational study would lead to the need for unacceptably large sample sizes (17). Under these conditions, the use of a modelling approach is appropriate to estimate the effect of the disease on production. This is especially true under extensive grazing conditions where the effects of disease on individual animals are difficult to measure. However, for modelling to be carried out successfully, an understanding of the potential effects of disease on the productivity of affected animals is essential.

Although systems for the assessment of the effects of chronic diseases, in particular internal parasitism, are well documented, techniques to assess the effects of acute infectious diseases on animal production under grazing conditions are not. Extensive production of livestock differs significantly from intensive production, for example, animals are rarely observed closely, animal health information is therefore scarce, inputs are considerably lower per head and feed intake is difficult to measure.

Few studies have been carried out to assess the effect of disease on animal production in extensive areas in Australia. The costs of outbreaks of bovine ephemeral fever have been assessed using crude estimates of disease incidence and the effect of disease on production (39). In addition, the benefits of using an improved vaccine against tick-borne diseases in Australia (3) have been estimated, as well as the costs of bovine pestivirus infection in extensively grazed cattle (20). However, none of these studies examined in detail the loss in production due to the disease compared with the loss avoided when the disease was controlled. No field studies comparing the effect of various disease control measures on animal production have been carried out in extensively grazed areas of Australia.

Effects of disease on livestock productivity

Morris and Meek divide the effects of chronic disease on the productive performance of livestock into two categories, namely, apparent alterations in efficiency and real reductions in efficiency (30). This conceptual framework is expanded by Morris and Marsh, who have defined apparent alterations in efficiency as changes in production caused by animals eating less food (31). In some cases, appetite suppression may be due to a direct and specific effect on appetite, while in others the effect may be indirect due to the reluctance of the animal to forage due to pain or discomfort associated with movement or prehension, caused by the disease.

Real reductions in efficiency are defined by Morris and Meek as being due to depression of feed digestibility or of feed conversion efficiency (30). The estimation of real reductions in efficiency is complicated by interactions between these two factors because the level of feed intake can affect the efficiency with which feed is used.

The differentiation between a reduction in feed intake and a true reduction in productive efficiency is important because if a reduction in feed intake is the factor causing the lost production, an increase in stocking rate will increase production as an alternative to controlling the disease (30). Often, the dividing line between apparent and real effects is not clear, and if feed intake cannot be measured, as occurs in extensively-raised cattle, the two effects cannot be differentiated (31).

The effects of acute disease contrast with those of chronic disease because acute disease is short-lived and affected animals usually either recover rapidly or die. In addition, animals that have recovered from acute infectious diseases are often not susceptible to a second attack of that disease, whereas in the case of diseases such as mastitis and internal parasitism, recurrent infections or infestations occur.

Categories of livestock production affected by disease

While the effects of a disease on animals are extremely variable, a simplified approach must be taken to enable the examination of these effects. The production loss in grazing animals due to diseases can be divided into the following categories:

- death
- weight loss
- reproductive loss
- lactation effects.

A description of each of these factors is given below.

Production loss due to death

The death of animals due to disease can have several effects on herd production. Deaths result in a reduced number of animals available for sale and a modification in the herd structure. In extensively grazed animals, production loss due to death is difficult to assess.

Production loss due to weight loss

The final effect of weight loss, due to disease, on production will depend on several factors, the most important of which are:

- the amount of weight lost due to the disease

- the composition of that weight loss (i.e. body fluid, gut content, muscle or fat)

- the rate at which the weight is recovered (this is affected by compensatory growth, the level of nutrition and the composition of the weight loss).

Considerable information is available on the effects of the restriction of nutrition on the subsequent growth and development of cattle (34). Much of this information is contradictory but there is general agreement that several factors are important in determining if compensatory growth occurs and if it does, how much compensatory growth occurs. These factors are as follows:

- breed
- age, liveweight and maturity
- stage of growth and condition (ratio of fat:lean meat:bone)
- severity and duration of restriction
- type of feed
- level of nutrients in the feed (34).

Although clinical disease does cause weight loss, little work has been done to measure the amount and variation in the amount of weight lost, the type of weight lost and the ability of animals under various conditions to recover the weight lost.

Production loss due to product quality

Ticks and tick-borne disease can cause downgrading of live animals at sales, and of meat, offal, and hides.

Production loss due to effects on lactation

Disease can vary both the quantity and the quality of the milk produced. The effects on quantity can vary from a mild temporary reduction to a total cessation of milk production. The effects vary depending on the stage of lactation at which the disease occurs and the severity of disease. In beef cattle, the main effect of a reduction in lactation is on the growth and survival of calves.

Production loss due to reproductive loss

Diseases can have several effects on reproduction. These effects on reproduction are firstly examined for females and then for males. The effects of disease on female reproduction vary according to the time that the disease occurs in relation to the reproductive cycle of the individual. The system of management, either controlled seasonal breeding or continuous breeding, will influence the proportion of females at each stage of the reproductive cycle at different times of the year, and therefore the effect of a disease on reproduction in the herd.

Diseases can affect reproductive efficiency in the following ways:

- silent oestrus periods

- prevention of fertilisation

- early embryonic loss

- loss in mid gestation

- abortion in the last trimester of pregnancy

- birth of dead, weak, or deformed calves which die soon postpartum

- delays in heifers breeding due to body weight and condition being below optimal.

The effects on male reproduction are more restricted and relate to the ability of the male to seek, mate with and fertilise receptive females. The effects of a disease in males can be summarised as follows:

- reduced mobility, with the consequence that affected animals are not able to seek and mate with receptive females

- reduced libido

- temporary or permanent infertility due to direct effects of a disease on spermatogenesis

- temporary infertility due to effects on spermatogenesis and sperm survival due to pyrexia associated with a disease.

Temporary effects are especially important if a disease outbreak occurs during or just before the breeding season.

Indirect production effects and non-production effects

Disease can mask genetic differences, making herd improvement difficult, and can be an impediment to introducing more intensive systems of livestock production (12, 28).

Apart from impacts on livestock products, tick-borne diseases can impose costs in terms of reduced consumer surplus, adverse impacts on human health, reduced draught and transport services in countries where cattle and buffalo are used for these purposes, and impose adverse effects on animal welfare. Reduced reproduction can lead to lower stock numbers, where stock ownership is a measure of wealth, a status symbol, source of creditworthiness, or provides other social values. In general, the literature overlooks these costs, with disease costs defined in terms of production and control costs only, as described, for example, by Mukhebi (32).

Costs of ticks and tick-borne diseases

There is no shortage of reported estimates of regional and national costs of ticks and tick-borne diseases. Often these estimates are made to assist in evaluation of tick eradication proposals. A number evaluations in various countries have been reviewed by Davis (8, 9), some examples of which follow:

a) The Bureau of Agricultural Economics (BAE) estimated the annual cost of cattle tick to producers in Australia (including production loss, chemical purchases, capital works, mortality and hide damage) at £9.5 million (about AUS$90 million at current prices) (6).

b) The cost of *Boophilus microplus* to beef producers and government in Australia (including research, loss of production, tick fever vaccinations, labour, dip maintenance and chemicals) was estimated by the Cattle Tick Control Commission to be AUS$41.27 million (over AUS$250 million at current prices) (7). Indirect costs due to tick fever deaths or illness were approximated by assuming 1% mortality. No estimates were made for the dairy industry. The estimated cost to Queensland in this study was approximately AUS$140 million at current prices, considerably higher than found in the BAE study (6).

c) McLeod and Kristjanson noted that 120 scientists have been involved in preliminary estimation of tick-borne disease costs in South Africa, Kenya, Tanzania, Zimbabwe, Thailand, China, India, Nepal, Indonesia, the Philippines and Australia (24). As an example, the cost estimates for Australia of US$37 million in 1997 (approximately US$38 million at current prices), are shown in Table II.

d) In the United States of America (USA), an eradication campaign for *B. microplus* and *B. annulatus* was successfully completed in 1947, at a cost of US$53.5 million (at 1953

Table II
Annual cost of ticks and tick-borne diseases, Australia, 1997 (US$ million) (24)

Cost category		Tick worry	*Anaplasma* and *Babesia*
Control costs	– chemicals	4.2	0.0
	– labour	0.0	0.8
	– vaccines	12.0	0.0
Productivity loss	– mortality	0.0	15.1
	– milk production	0.7	0.06
	– live weight	3.9	0.0
Total		20.8	15.9

prices). A retrospective economic analysis (5) estimated a benefit-cost ratio of the eradication of 140:1.

e) In Argentina, the cost of tick infestation was estimated at US$154.6 million for weight loss, death and hide damage, and US$34.9 million for the cost of tick control and tick fever vaccine (41).

f) The global cost of ticks and tick-borne diseases was estimated in 1979 at US$7 billion (over AUS$26 billion at current prices) (19).

The above studies suggest there are major variations in cost estimates, e.g. AUS$90 million and AUS$140 million for Queensland, and AUS$250 million and AUS$60 million for Australia. In addition, the cost of ticks and tick-borne diseases in Australia is generally agreed to have fallen sharply with the wider use of *Bos indicus* cattle.

Control economics
Control economics from a social perspective

The optimal level control effort for ticks and tick-borne diseases may be viewed in terms of the loss-expenditure tradeoff curve. In this context, Mukhebi (32) views immunisation in terms of the tradeoff curve developed by McInerney (21, 22). However, often the decision faced will not be to decide the optimal amount of control, but rather to decide whether to continue current practice or introduce an eradication campaign, on a local, regional or national basis. In fact, for developed countries engaging in intensive livestock production, tick eradication may be the optimal policy. However, this will be true only for a very few countries (such as those in the Caribbean where the USA wishes to prevent the introduction of *Amblyomma* into the USA). Often eradication will be favoured where the process is moderately easy technically, e.g. areas of low rainfall or low temperatures. In some cases, eradication will proceed incrementally, as individual districts work towards and achieve eradication. The Cattle Tick Control Commission noted essential conditions for successful eradication: ability to muster all tick hosts; adequate treatment facilities; effective industry co-operation; reasonable prospects of avoiding reinfection; efficient weapons of control; and adequate funding (7).

Where livestock production systems are more extensive (such as in Australia, South America or Africa), the economics of eradication becomes questionable. Here the payoff from eradication is lower, ticks do not represent an acute problem, managing a full muster may be difficult, and more wild hosts are likely to be present, making eradication more difficult.

If eradication is warranted, the issue arises as to how the cost is shared between government and private sector. Davis suggests that this may be through consolidated revenue, charges on producers, or joint funding, noting that in voluntary tick eradication programmes in Queensland, government met 50% of the cost, and had representation on a corporatised local management committee (9).

It has been argued (13) that in the case of disease (or vector) eradications, a different form of tradeoff curve is relevant, as illustrated in Figure 2. Substantial overhead costs (e_1) may be incurred in setting up a tick eradication programme, which in themselves do not achieve any reduction in disease cost. In addition, even at a moderate level of variable expenditure ($e_2 - e_1$), little progress may be achieved towards eradication. At a high level of control expenditure, collapse of the tick population may take place. If the C-E curve is of this shape, then the choice is one of either eradication or non-eradication, and not an intermediate point.

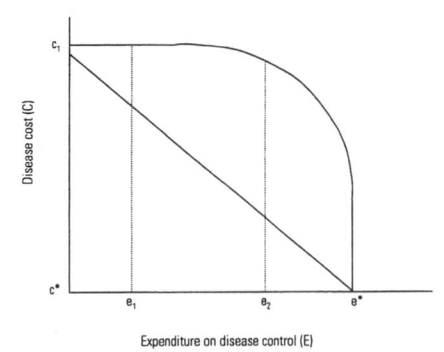

Fig. 2
Disease cost-control expenditure tradeoff model for an infectious disease which requires a threshold protection level

Control economics from a private perspective

As noted above, in intensive livestock production systems, the optimal policy is likely to be tick eradication. In extensive systems and in developing countries, choice of optimal control measures is more difficult, and strategies need to be compared in terms of expected profit or cost to the individual producer. The complexity of the economic evaluation may necessitate a form of modelling and simulation. From the viewpoint of the livestock owner, several questions arise with respect to disease control actions, as follows:

– does the control method work (what is the chance of the method failing)?

– will production be greater after the control programme is implemented?

– will the benefits derived exceed the costs (including opportunity costs)?

– can the control programme be put into practice within the constraints in which the livestock owner operates?

At the private producer level, decisions arise concerning the following options:

– to live with the disease, with little or no control measures

– to continue current practice, or standard practice in the district, e.g. spraying for ticks when numbers climb

– to adopt a regular vaccination schedule

– to increase control incidence, perhaps supported by information collection

– to attempt to achieve a disease-free herd or flock, with respect to particular diseases.

Control measures

Davis (8) categorises control measures for the cattle tick species *B. microplus*, as follows:

– chemical control
– use of tick-resistant (*B. indicus*) cattle
– pasture rotation
– use of TickGARD, a cattle tick vaccine.

Davies notes that tick fever – *Babesia bovis*, the major indirect effect of the cattle tick – can be controlled by reducing cattle tick populations, selecting cattle which have natural resistance to tick fever (*B. indicus* cattle), and using tick fever vaccines. With constant tick populations, cattle will have an acquired resistance; vaccination is more critical where outbreaks occur only occasionally.

Estimating the production loss avoided

In the simplest form, the production loss avoided due to disease control equals the reduction in the number of cases of disease due to disease control multiplied by the production loss per case. However, the production loss avoided will vary with the age and sex class of the animals and the severity of the disease. The calculation can be made more accurate if the production loss avoided is estimated as the sum of the production loss of each age and sex class of the animals in the herd and is weighted according to the different severities of disease.

Most control studies, models and programmes involve a combination of experimental studies, expert opinion, field studies and modelling and simulation. The complexity of the system means the level of knowledge about the biology of a disease is limited. However, perfect knowledge of all interactions is not required to make an appropriate decision about control of the disease. Attempts to collect all information about the disease and control method before control is instituted will bear a cost and may be uneconomic. It is necessary that any analysis considers the likely consequences of action and inaction. Lack of information on the likely effects of disease control does not prevent an effective and detailed analysis being carried out.

The private use of animal health information

Information provides support for private decision makers. Any decision in livestock disease control involves uncertainty because the aim is to predict what may occur in the absence of a disease control programme and the effect that a disease control programme may have on disease occurrence. Decisions are therefore made using the imperfect knowledge that is available to the decision maker (in this case knowledge is defined as the sum of available information). However, perfect knowledge is not required for rational decision-making and the optimal decision-making usually involves imperfect knowledge and hence imperfect information (4).

The collection of additional data, which is processed into information, is one way to decrease uncertainty. However, the production of information requires the use of resources, and hence bears a cost. Because of these costs, it is often uneconomic to decrease uncertainty through the gathering of additional information. Alternatively, the effects of uncertain events can be reduced by, for example, maintaining flexible policies that enable a rapid response to change in animal health status. Examples are: preparation for immediate vaccination if cases of a disease occur in the district, or reduction of the hazard, as in the case of farming disease-resistant *B. indicus* cattle where there is a risk of disease caused by *B. bovis* infection. However, maintaining flexible policies will also bear a cost.

The effect on a decision of the collection of additional information, leading to increased knowledge, can be to change the decision, decrease the uncertainty in making the decision, or decrease the need for flexible policies.

The benefit gained by a farmer from collecting additional information will depend on the current level of knowledge. If, when using the information currently available, the level of knowledge of the producer is at a low level, then the collection of additional information will most probably bring a large increase in information for a relatively small expenditure. However, if the level of knowledge is high, the benefit from collection of additional information is much less per unit of expenditure. A farmer with a high level of knowledge would be expected to gain less benefit from the same expenditure on information collection than a farmer with a lower level of knowledge. A farmer with a high level of knowledge may gain little information from additional expenditure and the expenditure could exceed the value of the information.

Collection of information

Although collecting information is regarded as an essential part of economic analysis, the economics of obtaining additional information is not usually assessed when designing animal health programmes. Often, additional data must be collected and analysed before the presence of a disease problem and the possible benefits of controlling a disease are recognised (i.e. the efficiency gap is detected). The value of information will depend on the attitude of the farmer to using information and the degree of confidence the farmer has in the accuracy of the information. Disease control is often carried out because a severe disease outbreak in the past has stimulated producers to avoid similar losses in the future. Ellis and James (11) note the lack of use of available information on disease control by farmers, commenting on the slowness of farmers to commence new disease control measures which have been shown to produce financial rewards. Furthermore, it is suggested that this is due to the reluctance of farmers to spend money and effort to obtain a benefit that has not previously been obtained.

If a producer is already making the most appropriate decisions then the private benefit received from the improved animal health information will be small, and will depend on the value placed on the decreased uncertainty in relation to decisions and the reduction in the cost of maintaining flexible policies due to the decreased uncertainty. It is also possible that additional information will change a specific decision from the appropriate one to an inappropriate one, resulting in a negative pay-off from the additional information (though not in an expected pay-off sense). It will not be worthwhile for a producer to collect additional information on animal health unless the benefit gained from using that information exceeds the costs of collection.

The relationship between cost and value of additional information in private decisions

If a farmer decides to gather additional information on the occurrence of disease on the farm, there are many methods of collecting that information. While some of these methods will not be feasible, or will be prohibitively expensive, the farmer will generally have a choice. To select the method to be used, the farmer can compare the cost of the method and the value of the information likely to be produced. In most cases the method will not be an all-or-nothing method, and by increasing expenditure, the farmer will obtain increasingly accurate information. However, the relationship between the cost of collecting information and the value of that information to a private decision maker is almost certainly not a linear relationship and several possibilities exist for that relationship.

Specific scenarios for animal health decisions

The following sections examine the issues confronting individual producers in a variety of situations, with some examples of analysis for different production systems. A relatively simple situation will be examined first with more complex decisions considered in subsequent examples.

Scenario 1: control of ticks and tick-borne diseases for an extensive beef producer in Australia

Here, the major tick-borne disease is caused by *B. bovis* and tick control is often not practised on the property. The producer faces the decision of whether to vaccinate. The questions are whether the benefits from vaccinating exceed the cost, and how the net pay-off compares with alternative uses of the money, such as depositing it in a bank account or investing in shares. In addition to these financial concerns, the producer may wish to be perceived by other cattle producers as a good manager who maintains healthy stock, and by the community as someone who is concerned about the welfare of animals. Therefore, the producer will also want information on the effect of vaccination on morbidity and mortality in the herd.

How can this information be provided to the producer? Discounted cash flow analysis, including partial budgeting, can provide information on expected returns from vaccination. A variety of information is required to produce annual partial budgets. How extensive does the partial budget need to be, and how accurate need the budget data be? In its simplest form, the calculations might resemble the following:

a) 500 cattle are to be vaccinated each year

b) the vaccine costs AUS$1.50 per dose and administration costs AUS$0.50 per head (mainly for labour)

c) the vaccine is effective and all animals that require vaccination are vaccinated

d) cattle only need to be vaccinated once in a lifetime

e) in the absence of vaccination, three cattle die as a result of *B. bovis* infection each year and each animal that dies has a value of AUS$400.

From these data, the partial budget would take the following form:

Vaccine costs: 500 × AUS$1.5 = AUS$750
Administration costs: 500 × AUS$0.50 = AUS$250
Total costs: AUS$1,000.

Benefits from preventing stock deaths:
AUS$400 × 3 = AUS$1,200
Annual net cash benefit: AUS$200.

The farmer can see that if the estimate of three deaths is accurate, the gains will exceed the expenditure on disease prevention (by AUS$200/year), provided the vaccine is effective.

Many livestock producers carry out analyses similar to the one described above, to assist in decision making.

In some cases, a limited partial budget may provide the farmer with sufficient information to meet decision-making needs. However, the disease may also have effects such as reducing cattle growth rates and reducing reproductive efficiency. An approach to estimating the production loss avoided and financial benefit gained due to a vaccination programme to control *B. bovis* is provided by Ramsay (37) and a brief description is presented here. The approach illustrates several areas in which additional information may be of use and others where information may be of limited use to a decision maker.

The production loss is divided into loss due to deaths, weight lost and not regained by the time of sale, and reproductive loss. A simulation model is developed which is used to determine the production loss avoided for two vaccination programmes, for each of three herds containing cattle that are resistant, susceptible and of intermediate susceptibility to disease, following infection with *B. bovis* at high, medium and low levels of incidence of infection.

The model developed provides an effective method of examining the production benefits gained from adopting a vaccination programme. Simulations indicate that the major production benefit from *B. bovis* vaccination arises from deaths avoided. Considerably less benefit is predicted to be gained in the form of weight loss avoided due to prevention of clinical disease from which animals recover. The effect of *B. bovis* vaccination on reproductive efficiency is also predicted to be small.

As expected, the production benefits of *B. bovis* vaccination are predicted to be greatest in the susceptible cattle and least in disease-resistant cattle. The incidence risk of infection also affects the production benefits of *B. bovis* vaccination, with the loss avoided being highest where the incidence risk of infection is medium and least where the incidence risk of infection is high, as in these cases, herd resistance to *B. bovis* is also high.

Livestock producers in extensive grazing areas of Australia suffer from a shortage of animal health information. However, a rational decision can be made without perfect information and it will not be worthwhile for a producer to collect additional information on animal health unless the benefit of that information exceeds the cost of collecting the information. A clear understanding of the interactions between the disease incidence and production effects will enable the livestock producer to consider the decision to gather additional animal health information systematically and examine the potential returns. However, the benefits from collection of additional information need to be calculated for each individual situation.

The information above with respect to the effect of B. bovis on production demonstrates that collection of additional information to determine more precisely the weight loss due to disease caused by B. bovis will not be of great benefit to the decision maker, nor will be additional information on the reproductive effects of the disease. However, additional information on the number of animals likely to die following infection and on the incidence of risk of infection could be of use to the decision maker.

Before deciding whether to collect additional information, the livestock owner needs to determine if the benefit from collecting the information is sufficient to justify the costs of collection. The collection of additional information does not always result in a net gain. When the gain from collection of additional information is small, it is possible that the farmer would gain more by using the money elsewhere rather than collecting additional information. In the case of B. bovis, vaccine is relatively inexpensive in comparison to the cost of the collection of additional information, and provides long-lasting protection. Therefore, it is possible that money may be more effectively spent on the purchase of vaccine rather than on gathering additional information. If vaccination must be carried out annually, or a more expensive and less effective vaccine is used, the benefit from collecting additional information would be expected to be greater.

Scenario 2: cattle producer in a country in Africa where multiple ticks and tick-borne diseases are present

In the case of a large-scale beef producer the question to be asked may be 'How should I control ticks and tick-borne disease on my property?' The producer faces a number of decisions which can interact producing flow-on effects. For example, dipping practised to control ticks will also affect the exposure of stock to agents causing tick-borne diseases. If a producer is considering reducing dipping intensity what benefits will be obtained? The livestock might enter a higher priced market for chemical-free beef, and chemical and labour costs may be reduced. Environmental contamination on the property could be reduced since the producer is no longer using dip wash. In addition, the farmer may wish to protect the environment and farm in an ecologically sustainable manner. However, by reducing dip use, production could be reduced and greater morbidity and mortalities experienced.

In the case of a small-scale dairy or beef producer, the effect of losing a single animal may be much larger than for a large-scale beef producer. For example, a producer with three cows will lose a third of productive capacity if one cow dies. Therefore, this producer is more interested in reducing risk of loss of an animal rather than the average long-term effects of disease control. The producer may be prepared to spend relatively more on control of ticks and tick-borne disease because although the risk of loss of an animal to tick-borne disease may be low, the loss would have major economic consequences.

Conclusion

Economic study of endemic diseases and disease control is a complex area, with a large number of disease vectors and diseases, and complex relationships between treatment, environment and livestock performance to consider. While considerable technical study of these diseases has been performed, economic analysis of the effects and the control of the diseases remains a relatively neglected field.

Decision-making about livestock diseases needs to be viewed from both social and private perspectives. In making social decisions it is necessary to take a broader view than in making private decisions, and consider externality costs, optimal overall control effort, and sharing of costs between taxpayers and livestock owners. Individual producers face complex decisions about disease control effort, including whether to invest in obtaining further information before making decisions.

Major differences in livestock systems and in optimal disease control programmes arise in developed and developing countries, between intensive and extensive production systems, and even between properties, due to differences in environment, resources and objectives of livestock owners. Non-production costs of diseases can be important, but typically are not adequately taken into account in disease cost studies.

While little information is available on the cost of helminth diseases, many estimates have been made of the costs of ticks and tick-borne diseases at regional and national level, sometimes demonstrating that eradication is warranted. Introduction of more tick-resistant cattle appears to have substantially reduced costs associated with ticks and tick-borne diseases.

References

1. Anderson N., Morris R.S. & McTaggart I.K. (1976). – An economic analysis of two schemes for the anthelmintic control of helminthiasis in weaned lambs. *Aust. vet. J.*, **52**, 174-180.

2. Bain R.K. & Urquhart G.M. (1986). – The significance and control of stomach worms in British cattle. *Outlook Agric.*, **15** (1), 10-14.

3. Bartholomew R.B. & Callow L.L. (1979). – A benefit-cost study (ex-post) of the development and introduction of a new vaccine against bovine babesiosis. *In* Proc. 2nd International Symposium on veterinary epidemiology and economics (W.A. Geering, R.T. Roe & L.A. Chapman, eds), 7-11 May, Canberra. Australian Government Publishing Service, Canberra, 448-457.

4. Baumol W.J. & Quandt R.E. (1964). – Rules of thumb and optimally imperfect decisions. *Am. Econ. Rev.*, **54**, 23-46.

5. Bram R.A. & Gray J.H. (1979). – Eradication – an alternative to tick and tick-borne disease control. *Wld Anim. Rev.*, **30**, 30-35.

6. Bureau of Agricultural Economics (BAE) (1959). – The economic impact of cattle tick in Australia. Australian Government Publishing Service, Canberra, 85 pp.

7. Cattle Tick Control Commission (1975). – Cattle tick in Australia: inquiry report. Australian Government Publishing Service, Canberra, 108 pp.

8. Davis R. (1996). – An overview of the status of cattle tick *Boophilus microplus* in Queensland. Research Papers and Reports in Animal Health Economics, No. 14. Department of Economics, University of Queensland, Brisbane, 21 pp.

9. Davis R. (1997). – A review of economic evaluations of government policies for the control of cattle tick. Research Papers and Reports in Animal Health Economics, No. 32, Department of Economics, University of Queensland, Brisbane, 46 pp.

10. Dijkhuizen A.A. & Morris R.S. (eds) (1997). – Animal health economics: principles and applications. Post Graduate Foundation in Veterinary Science, University of Sydney, 306 pp.

11. Ellis P.R. & James A.D. (1979). – The economics of animal health. 1. Major disease control programs. *Vet. Rec.*, **105**, 504-506.

12. Harrison S.R. & Tisdell C.A. (1997). – Animal health programs in sustainable economic development: some observations from Thailand. *In* Development that lasts (K.C. Roy, H.C. Blomqvist & I. Hossein, eds). New Age International, New Delhi, 201-214.

13. Harrison S.R. & Tisdell C.A. (1999). – Economic analysis of foot and mouth disease control and eradication in Thailand. Australian Centre for International Agricultural Research, Canberra, 95 pp.

14. Hawkins C.D. (1977). – The effect of disease on productivity: a study of liver fluke in sheep. PhD dissertation, University of Melbourne, 209 pp.

15. Holmes P.H. & Coop R.L. (1994). – Workshop summary: pathophysiology of gastrointestinal parasites. *Vet. Parasitol.*, **54**, 299-303.

16. James A.D. & Ellis P.R. (1978). – Benefit-cost analysis in foot and mouth disease control programmes. *Br. vet. J.*, **134**, 47-52.

continued over

17. James A.D. & Ellis P.R. (1979). – The evaluation of production and economic effects of disease. In Proc. 2nd International Symposium on veterinary epidemiology and economics (W.A. Geering, R.T. Roe & L.A. Chapman, eds), 7-11 May, Canberra. Australian Government Publishing Service, Canberra, 363-372.

18. McCauley R.H., Majid A.A. & Tayer A. (1984). – Economic evaluation of the production impact of bovine schistosomiasis and vaccination in the Sudan. Prev. vet. Med., 2, 735-754.

19. McCosker P.J. (1979). – Global aspects of the management and control of ticks of veterinary importance. Recent Adv. Acarol., 11, 43-53.

20. McGowan M.R., Baldock F.C., Kirkland P.D., Ward M.P. & Holroyd R.G. (1992). – A preliminary estimate of the economic impact of bovine pestivirus infection in beef herds in Central Queensland. In Proc. Australian Association of Cattle Veterinarians, 'New Horizons', 10-15 May, Adelaide. Australian Association of Cattle Veterinarians, Adelaide, 129-132.

21. McInerney J.P. (1988). – The economic analysis of livestock disease: the developing framework. In Proc. 5th International Symposium on veterinary epidemiology and economics (P. Willeberg, J.F. Agger & H.P. Riemann, eds), 25-29 July, Copenhagen. Acta vet. scand., 84 (Suppl.), 66-74.

22. McInerney J.P. (1991). – Cost-benefit analysis of livestock diseases: a simplified look at its economic foundations. In Proc. International Symposium on veterinary epidemiology and economics (S.W. Martin, ed.), 12-16 August. Department of Population Medicine, Ontario Veterinary College, University of Guelph, 149-153.

23. McInerney J.P., Howe K.S. & Schepers J.A. (1992). – A framework for the economic analysis of disease in farm livestock. Prev. vet. Med., 13, 137-154.

24. McLeod R. & Kristjanson P. (1999). – The economic impact of selected tick-borne diseases on cattle in Africa, Asia and Australia. eSYS Development, Sydney, and ILRI, Nairobi (in press).

25. Macpherson C.N.L. & Craig P.S. (1991). – Parasitic helminths and zoonoses in Africa. Unwin Hyman, London, 281 pp.

26. Matthewman R.W. & Perry B.D. (1985). – Measuring the benefits of disease control: relationship between herd structure, productivity and health. Trop. Anim. Hlth Prod., 17, 39-51.

27. Meek A.H. (1977). – Economically optimal control strategies for ovine fascioliasis. PhD thesis, Faculty of Veterinary Science, University of Melbourne, Melbourne, 178 pp.

28. Morris R.S. (1990). – Animal health economics, principles and applications: how economically important is animal disease and why? International Training Centre, Wageningen Agricultural University, Wageningen, 10 pp.

29. Morris R.S., Anderson N. & McTaggart I.K. (1977). – An economic analysis of two schemes for the control of helminthiasis in breeding ewes. Vet. Parasitol., 3, 349-363.

30. Morris R.S. & Meek A.H. (1980). – Measurement and evaluation of the economic effects of parasitic disease. Vet. Parasitol., 6, 165-184.

31. Morris R.S. & Marsh W.E. (1992). – The relationship between infections, diseases and their economic effects. In Modelling vector-borne and other parasitic diseases (B.D. Perry & J.W. Hansen, eds). Proc. Workshop held at the International Laboratory for Research on Animal Diseases (ILRAD), 23-27 November, Nairobi. ILRAD, Nairobi, 199-213.

32. Mukhebi A.W. (1996). – Assessing economic impact of tick-borne diseases and their control: the case of theileriosis immunisation. In Epidemiology of ticks and tick-borne diseases in Eastern, Central and Southern Africa (A.D. Irvin, J.J. McDermott & B.D. Perry, eds). Proc. Workshop, 12-13 March, Harare. International Livestock Research Institute (ILRI), Nairobi, 174 pp.

33. Mukhebi A.W., Perry B.D. & Kruska R.L. (1992). – Estimated economics of theileriosis control in Africa. Prev. vet. Med., 12, 73-85.

34. O'Donovan P.B. (1984). – Compensatory gain in cattle and sheep. Nutr. Abstr. Rev., B, 54, 389-410.

35. Over H.J., Jansen J. & van Olm P.W. (1992). – Distribution and impact of helminth diseases of livestock in developing countries. FAO Animal Production and Health Paper No. 96. Food and Agriculture Organization, Rome, 221 pp.

36. Perry B.D. & Randolph T.F. (1999). – Improving the assessment of the economic impact of parasitic diseases and their control in production animals. Vet. Parasitol. (in press).

37. Ramsay G.C. (1997). – Setting animal health priorities: a veterinary and economic analysis with special reference to the control of Babesia bovis in central Queensland. PhD thesis, Department of Economics, University of Queensland, Brisbane, 325 pp.

38. Roberts J.A. & Suhardono (1996). – Approaches to the control of fasciolosis in ruminants. Int. J. Parasitol., 36, 971-981.

39. St George T.D. (1986). – The epidemiology of bovine ephemeral fever in Australia and its economic effect. In Proc. 4th Symposium on arbovirus research in Australia, Brisbane. Commonwealth Scientific and Industrial Research Organisation (CSIRO), Division of Animal Health, Brisbane, 281-286.

40. Schwabe C.W. (1991). – Helminth zoonoses in African perspective. In Parasitic helminths and zoonoses in Africa (C.N.L. Macpherson & P.S. Craig, eds). Unwin Hyman, London, 1-24.

41. Spath E.J.A., Gugielmone A.A., Signorini A.R. & Mangold A.J. (1992). – An estimation of the economic costs caused by the tick Boophilus microplus and tick-borne disease in Argentina. Report (in Spanish). Instituto Nacional de Tecnología Agropecuaria, Balcarce, Argentina.

42. Suhardono, Roberts J.A., Copland J.W. & Copeman D.B. (1997). – Control of bovine fasciolosis in Indonesia. *Epidémiol. Santé anim.*, **31-32**, 0.2.19.1-0.2.19.3.

43. Tisdell C.A. (1995). – Assessing the approach to cost-benefit analysis of controlling livestock diseases of McInerney and others. Research Papers and Reports in Animal Health Economics, No. 3. Department of Economics, University of Queensland, Brisbane, 22 pp.

44. Tisdell C.A. (1996). – Bounded rationality and economic evolution: a contribution to decision making, economics and management. Edward Elgar, Cheltenham, 336 pp.

45. Zinsstag J., Ankers Ph., Itty P., Njie M., Kaufmann J., Pandey V.S. & Pfister K. (1997). – Effect of strategic gastrointestinal nematode control on productivity of N'Dama cattle in the Gambia. *Epidémiol. Santé anim.*, **31-32**, 02.10.1-02.10.3.

[21]

Genetically modified (transgenic) crops and pest control economics

C.A. Tisdell and C. Wilson

Background

Crops have been genetically modified or engineered mainly in two different ways to promote the control of pests. Some crops have been modified to make them more resistant to particular types of herbicides. Herbicide-resistant varieties have been developed, for example, for corn, cotton and soybeans. Crops have also been genetically engineered to make them toxic to various insect pests.

'Transgenic crops are proprietary, developed almost exclusively by the private sector in industrial countries, with the majority of the global transgenic crop area to-date grown in the countries of the North' (James, 1998, p. 5). At the beginning of the 21st century, the USA accounted for 74 per cent of the global area sown to transgenic crops, Argentina 15 per cent and Canada 10 per cent, with smaller areas in another five countries or so accounting for the remainder (James, 1998).

The principal transgenic crops in 1998 in terms of area grown are, in descending order: soybean, corn, maize, cotton, canola/rape seed and potatoes. Soybean accounted for 52 per cent of the global transgenic area and corn 30 per cent (James, 1998, p. 3).

Use of transgenic herbicide-resistant crops is dominant. At the beginning of this century, they accounted for about 71 per cent of the total global area sown to transgenic crops, and insect-resistant crops constituted around 28 per cent of this area. Crops genetically modified for both insect resistance and herbicide tolerance accounted for approximately 1 per cent of the transgenic crop area globally, and those with genes altered to influence the quality attributes of the final product amounted to much less than 1 per cent of this area. Thus transgenic modification for pest control accounts for almost all the area planted globally to transgenic crops, with herbicide-resistant crops comprising the major portion and insect-resistant crops most of the remainder. Although some scientific progress has been made in the genetic engineering of animals, most progress has been with plants (Moffat, 1998).

In this chapter, economic factors liable to influence choice by individual farmers of herbicide-resistant crop varieties are discussed initially using threshold models. Industry-wide influences on the economics of the development and use of these transgenic crops are then considered, and subsequently, the economics of using crops modified to be toxic to insects is discussed. The discussion concludes with a consideration of general social and ecological concerns about the genetic engineering of crops as a means of pest control.

Economics of private choice of farmers of herbicide-resistant crops

Considerable controversy exists about the economic benefits to individual farmers of using herbicide-resistant crops (HRCs). Pimentel and Ali (1998, p. 247) suggest that in some instances HRCs are ineffective in increasing crop yields compared to traditional varieties and cost more overall so that farmers actually lose financially by choosing HRCs. They suggest, for example, that use of herbicide-resistant corn may reduce a farmer's profit by US$143 per hectare. On the other hand, there are also counter claims, particularly by companies such as Monsanto, which are active in developing and promoting HRCs.

However, the aim here is not to indicate whether a particular HRC is likely to be the superior economic choice by a farmer compared to the best non-engineered variety of the same crop. Rather the purpose here is to specify general economic factors that need to be taken into account by individual farmers in making this choice. The types of threshold models introduced in Chapters 11 and 12 will be adapted to do this. This analysis also enables an assessment to be made of whether the adoption of HRCs increases or decreases the cost of herbicide use.

One of the main advantages claimed for HRCs is that they raise crop yields compared to conventional varieties because the latter experience significant growth retardation when herbicides are applied (see James, 1998, p. 4). However, as far as soybeans are concerned, Duffy (2001) found that yields on average from herbicide-tolerant soybeans were slightly lower than for non-tolerant soybeans, and at the same time, seed costs per acre were significantly higher. On the other hand, weed management costs per acre were considerably less for the herbicide-tolerant soybeans (Duffy, 2001). Nevertheless, taking all these factors into account, Duffy (2001) concludes that there is virtually no difference in the profitability from using herbicide tolerant soybeans or non-tolerant ones. It is also possible for the price received for produce for transgenic crops to be less than for non-modified crops depending on consumer acceptance. This, however, does not explain the rapid expansion in area sown to herbicide-tolerant soybeans in the USA. He suggests advertising pressure and desire for 'clean fields' (that is, fields clear of weeds) are the major driving forces for the adoption of herbicide-tolerant soybeans.

Using economic threshold models, consider some alternative scenarios involving a choice between an HR variety of a crop and a conventional one. In Figure 1, 0AB presents the net gain in income per hectare if a farmer plants an HRC, without considering weed treatment costs and the extra cost of the seed for the HRC. It is a function of weed density in the absence of weed management. Curve 0CD represents a similar relationship for the conventional variety. If the cost of weed control is 0K per hectare for both the HRC variety and the traditional variety and their seed costs are the same, the cultivation of HRC is the more profitable alternative. Furthermore, it will be profitable to engage in weed control at a lower density of weeds than with the conventional crop. If weed management in either case involved a similar cost of herbicide application, treatment of weeds in an HRC would be profitable if weed density happened to be w_1, or higher, but in the case of the conventional crop, it needs to be w_2, or higher, to be profitable. Furthermore, if the cost of weed management of the HRC is higher than for the conventional variety, it can be up to KM higher before treatment at lower densities than for the conventional crop would become

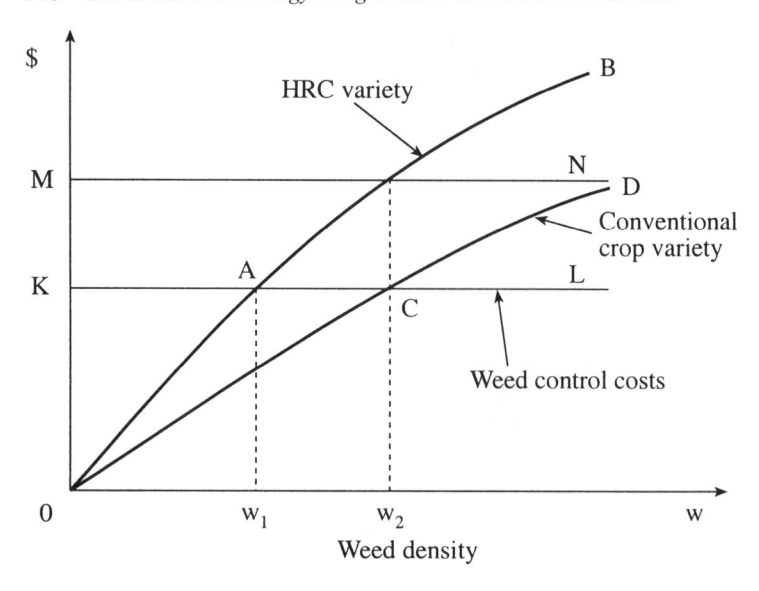

Figure 1 A case in which an HRC variety enables greater economic gains to be obtained from weed control before weed management and seed costs are taken into account

uneconomical. Even at a much higher cost for weed control than for the conventional variety, HRC may be a more profitable alternative where high weed densities would occur in the absence of weed management.

On the other hand, Duffy (2001) found that weed management costs are lower with herbicide-tolerant soybeans. This may be because farmers are more likely to use glyphosphate herbicide rather than other herbicides as a means of weed control when they plant HRCs. Herbicide tolerance is chiefly to glyphosphate. Monsanto is the major producer of it and the main supplier of HRC seed. If weed management costs are lower with HRCs, then even if their yields are not higher than with traditional varieties, it will be economic to control weeds at lower weed densities than in the case of conventional varieties.

For example, in Figure 2, suppose that the net gain in income from weed treatment is the same from conventional and HRC varieties and is as shown by curve 0AB. If the cost of weed management with the conventional crop is 0K and with the HRC variety 0P per hectare, the HRC is more profitable when weed control is economic. Furthermore, control becomes economic for a lower weed density, w_0, if the HRC variety is planted, than in the case of the conventional variety where control only becomes economic at a weed density of w_1 or more.

Consequently, since HRC varieties as a rule involve the application of glyphosphate as a herbicide, greater net income from weed management or lower weed management costs when HRC seed is planted rather than conventional seed, results in more widespread application of glyphosphate. This boosts glyphosphate sales. Thus a company such as Monsanto gains both from the sale of HR seeds and from extra sales of glyphosphate. Monsanto has a monopoly on glyphosphate supplies due to patent protection.

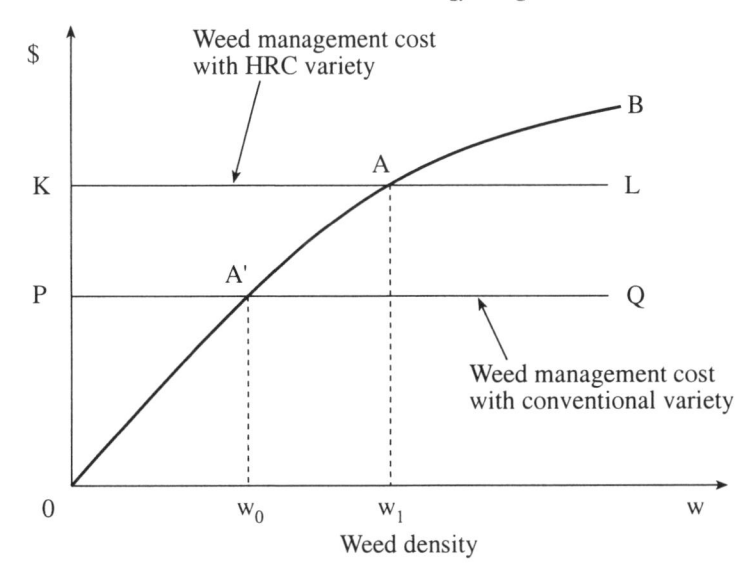

Figure 2 A case in which weed management costs are lower for an HRC variety than a conventional one

It is also possible that risk-averse farmers will find planting HRC seed attractive given the possibilities illustrated in Figure 3. At low weed densities (densities below that at which treatment is economic), there is no private economic advantage in planting HRC seed if its price is the same or higher than conventional seed. But if there is a chance that weed densities will rise above the economic treatment threshold, the extra expected benefits to a farmer from HRC plantings can exceed that from conventional plantings because of reduced weed control costs and/or extra net gain from higher yields or because of a harvest with less contamination by seeds of weeds. The planting of HRC seed may also be a maximin gain strategy or one that involves minimax regret in terms of potential profit forgone by the farmer by not planting HRC seed.

Industry-wide considerations in relation to herbicide-resistant crops
To the extent that transgenic herbicide-resistant crops lower the cost of production from crops, they tend to reduce the market price received for crop yields. If we assume that output from genetically modified crops is a perfect substitute for that from non-modified ones, the supply curve of output from crops that are to an increasing extent genetically modified will then shift to the right. Given normal supply/demand curves, this will lower the equilibrium price of the produce involved. Consumers' surplus should, therefore, rise. Whether or not producers' surplus increases depends on the nature of the shift in the supply curve. A parallel shift to the right in the supply curve will increase producers' surplus (unless the demand curve is perfectly inelastic, but if the slope of the supply curve merely declines, this surplus may fall (see Duncan and Tisdell, 1971; Hayami and Herdt, 1977; Lindner and Jarrett, 1978, 1980; Alauddin and Tisdell, 1991, Ch. 6).

However, private companies with proprietary rights to HRC seeds are in a position to exert some control on the surplus earned by farmers from HRC seeds. Companies like Monsanto often appear to have a monopoly-like position. They are suppliers both of transgenic seed and of pesticides. According to Pimentel and Ali (1998), agrochemical companies such as Monsanto are increasingly reducing their research emphasis on pesticides and increasing that on developing transgenic crops resistant to weeds, insect pests and plant pathogens. The costs of developing new transgenic crops and having them approved appears to be of the order of around one-fifth of that for new pesticides, based on estimates in Pimentel and Ali (1998, p. 246). Companies such as Monsanto find that such development can be very profitable in its own right and, in some cases, extra economic benefit is obtained by selling extra amounts of herbicides such as glyphosphate-based herbicide.

The simple monopoly model might be adapted to assess the situation that a company such as Monsanto faces in pricing transgenic seed. Monsanto has a monopoly or virtual monopoly on the supply of transgenic seed for several important crops, including HRC soybean (Nelson, 1999), in view of its patent rights. In addition, it has a monopoly in the herbicide glyphosphate, although its patent rights on this are expiring (Nelson, 1999).

In Figure 3, let line AD represent the demand for HRC seed, e.g. soybean, from a company like Monsanto. Then AMR is the corresponding marginal revenue curve. Its per unit cost of supplying HRC seed is shown by line BC. Then if the monopolist wants only to maximize its profits from the sale of HRC seed, it does so by charging a price P_1 for its seed and selling a quantity X_1. However, in Monsanto's case, the sale of HRC seed is complementary to the sale of its herbicide based on glyphosphate on which it can earn above normal profit. Suppose that for every unit of the type of HRC seed sold, Monsanto earns an additional BF in extra profit from the sale of more herbicides. When this profit-spillover benefit to the company is taken into account, the real per unit cost to the company of selling its HRC seed is only 0F. Therefore, its effective per unit cost of supplying HRC seed is shown by line FG. Consequently, the most profitable strategy of the monopolist will be to charge only a price of P_0 per unit of its HRC seed and sell an additional $X_2 - X_1$ units of it. The greater is the added profit from complementary herbicide sales obtained by the monopolist because of extra sales of HRC seed, the lower is the profit-maximizing price for that seed, other things equal.

In order to protect its monopoly proprietary rights, a company selling transgenic seed, such as Monsanto, makes agreements with farmers that they will not save this seed or trade in it. The agreements also permit the company to test crops on the farms for up to three years after sale of transgenic seed to determine whether such seed has been saved. In addition to paying the per unit cost of seed, a buyer must pay an upfront technology access fee.

The transaction costs for Monsanto involved in such arrangements can be high, as pointed out by Barnett and Gibson (1999). Given such high market transaction costs, it clearly will be in Monsanto's private interest to introduce a terminator gene into its seed, if it can. From its point of view, the ideal would be for its seed to contain its special characteristics only for the initial sowing. This effect would be similar to the loss of vigour of hybrid corn when the seed is saved by a farmer and resown. One

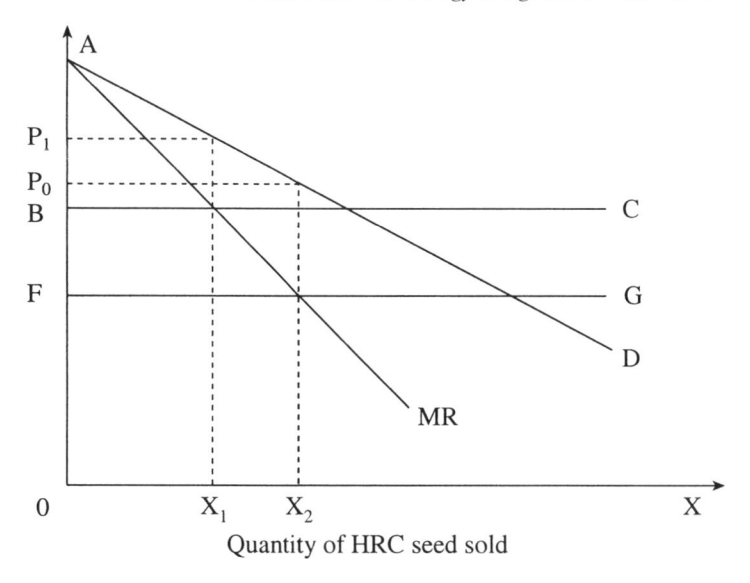

Figure 3 Microeconomics of pricing of HRC seed by a monopolist producing a complementary product

could imagine that the development of such seed would be a high priority for proprietors of transgenic seed.

As suggested in the previous section on theoretical grounds, one would expect the advent of glyphosphate-tolerant crops to increase the use of glyphosphate-based herbicide. In fact, the use of glyphosphate has increased in recent years (USDA, 1999). However, in the same period, the use of other synthetic chemical herbicides has decreased in the USA by a greater amount than the increase for glyphosphate used (USDA, 1999). Whether that is, however, beneficial environmentally is unclear because aggregate weight or volume of chemicals used is unlikely to be an accurate indicator of the environmental impacts of their use.

It might also be observed that an industry-wide effect occurred because of the substitution of glyphosphate for other herbicides. Nelson (1999) reports that since the introduction of glyphosphate-resistant technology, the prices of nearly all non-glyphosphate herbicides have fallen, as might be expected on the basis of economic theory. This tends to reduce the comparative economic attractiveness to farmers of glyphosphate-resistant transgenic crops.

Transgenic crops for the control of insect pests – private returns

Currently, the most widely adopted transgenic crops for the control of insect pests are those that contain insecticidal proteins from the bacterium *Bacillus thuringiensis*, Bt. This soil bacterium produces proteins that are toxic to the larvae (caterpillars) of the lepidoptera order, represented by butterflies. The Bt toxic protein, after ingestion, creates a hole in the caterpillar's digestive tract (Barnett and Gibson, 1999). In transgenic crops, this protein may be present in different parts of the crop depending on the type of genetic engineering involved. For example, in the case of Bt corn, the protein may be expressed in the leaves or in the cob and the tassel, and thereby target different insect pests. Moreover, the protein varies naturally and, as a result, in its toxicity to various members of the populations of lepidoptera. In general, only partial control of the target species is obtained.

Divergent views exist about private returns from the adoption of Bt crops in the USA. Duffy (2001) found, on the basis of empirical evidence, that farmers' returns from Bt corn are essentially equal to those from non-Bt corn. He suggests, however, that 'many farmers plant Bt corn as a sort of insurance policy'. On the other hand, Marra et al. (1998) found that there is a slight private gain from growing Bt corn, and this arises mainly from an increase in average yield per acre. By contrast, there is strong evidence that the private rate of return from growing Bt cotton is quite high. However, Pimentel and Ali (1998) suggest that the cotton bollworm may rapidly develop genetic resistance to Bt as a result of diffusion of Bt crops.

Marra et al. (1998) suggest that a farmer, in assessing whether it is profitable to adopt a Bt variety of crop rather than a conventional one, should consider whether the increase in total revenue exceeds pesticide application savings plus pesticide materials savings plus the additional seed cost plus the technical fee. If there is excess, the Bt variety maximizes profit. In this assessment, the increase in total revenue may be replaced by the increase in net revenue estimation on the basis that pesticide savings, the additional seed costs, plus the technical fee are ignored.

The decision about whether a Bt variety of a crop is more profitable than a non-transgenic one can be considered using an economic threshold type of analysis. This is done in Figure 4. Here, variable x represents the expected (average level) of cotton bollworm or European corn borer infestation, and curve 0AB represents the farmer's expected extra net revenue from planting a Bt variety, plus the savings in the farmer's insecticide (or more generally pest management) costs, all compared to the alternative of the conventional variety. 0K represents the extra cost of Bt seed plus the technical access fee. In this case, if the expected infestation exceeds x_1, adopting the Bt variety of the crop rather than a conventional variety increases profit.

As a rule, Bt varieties make a greater addition to total yield and to net revenue in areas where the potential yield of the crop is greatest. So, while in a high yielding region or farm, curve 0AB may apply, the corresponding additional net benefit curve in a less productive region or farm may be as indicated by curve 0CD. Therefore, one would expect Bt to be adopted more slowly in regions where potential yields tend to be lower, other things equal. Where potential yields are lower, the degree of infestation by the target pest must be higher than in high yielding regions before Bt adoption becomes economic. In the case illustrated in Figure 4, in the higher yielding region a farmer finds it profitable to adopt the Bt variety if the infestation of the target insect exceeds x_1, but it must exceed x_2 in the lower yielding region before a Bt variety is the most profitable choice. Given similar levels of infestation of the target insect by regions, this theory predicts that adoption of Bt varieties will be more widespread in areas with higher potential crop yields than in those with lower potential yields. The pattern of Bt cotton adoption when compared between North Carolina, South Carolina, Georgia and Alabama using the data supplied by Marra et al. (1998, see their Tables 8 and 9), seems to be consistent with this prediction, given that cotton yields in the USA rise on average as one moves south.

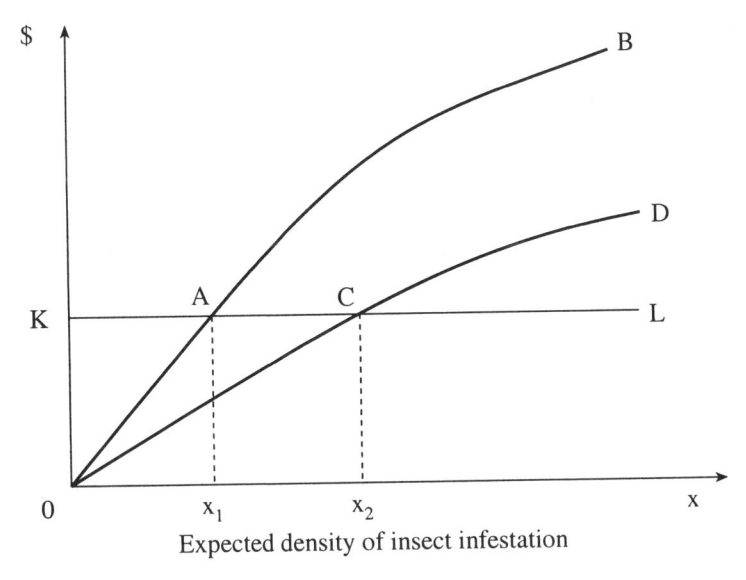

Figure 4 Economic threshold analysis as a guide to whether it is more profitable to adopt a Bt crop variety rather than a non-transgenic one

Environmental externalities and social issues
The environmental, health and social consequences of the introduction of transgenic crops remain controversial. Some see these overall impacts as slightly positive whereas others fear that they are likely to be quite negative. Shelton et al. (2002) take a positive stance and argue that generally Bt crops have shown positive economic benefits to growers and have reduced the use of synthetic chemical insecticides. They further believe that the risks of adverse 'potential ecological and human health consequences of Bt plants, including effects on non-target organisms, food safety, and the development of resistant insect populations' are lower than with current or alternative technologies (Shelton et al., 2002, p. 845). On the other hand, Paoletti and Pimentel (2000, 1996) are more wary of the environmental risks associated with the widespread introduction of genetically engineered crops. However, their main message appears to be that both conventional pest control reliant on pesticides and that dependent on new biotechnology involve significant environmental risk and safety concerns (Paoletti and Pimentel, 2000, p. 279).

A major problem in assessing the environmental risks posed by alternative pest control strategies involving biotechnology and conventional methods, such as synthetic chemical pesticides, is that scientists' knowledge of these risks is incomplete. Even where adequate knowledge exists and risks are known, social conflict may arise when risk is thrust upon individuals externally because individuals often differ in their willingness to bear risk. This is one reason, for example, why some consumers demand that genetically modified foods be labelled so that they can make their free choice about whether to consume these foods and adjust to any perceived risks associated with them.

Collective environmental risks associated with the introduction of transgenic crops can be quite high for farmers. For example, producers of organic food that rely on Bt sprays may find that target insects quickly become resistant to Bt spray as a result of the widespread use of transgenic Bt crops. In the case of some crops, cross-pollination occurs between the transgenic and conventional crops when they are grown nearby. In this way, seed saved from a GMO free crop may become 'polluted' by a transgenic crop. In the case of crops genetically modified to be herbicide-tolerant, this tolerance might be transferred by cross-pollination to weedy relatives of the crops or if use of a particular herbicide, such as glyphosphate, becomes more widespread, natural selection may result in increasing evolutionary selection of resistant weeds.

Table 1 identifies some possible external benefits and costs of glyphosphate-resistant crops and of Bt transgenic crops, but is not definitive. It is influenced by the summary of Marra et al. (1998).

Table 1   External costs and benefits from some transgenic crops

Glyphosphate (herbicide) tolerant crops
External costs: • genetic tolerance transmitted to weedy relatives   • glyphosphate-resistant weeds and evolutionary selected crops   • in some cases, cross pollination with non-genetically modified crops   • possibly, in some cases, more herbicide use with greater external effect   *External benefits*:   • in some cases, less herbicide use and use of herbicides possibly with fewer external effects   • may result in less tillage to reduce weeds and thereby reduce soil erosion
Bt transgenic crops
External costs: • increased potential for Bt resistance in targeted insects   • destruction of non-target insects, including beneficial ones   • reduced effectiveness of Bt sprays used by organic farmers due to growing resistance of target species   *External benefits*:   • in many cases reduced use of synthetic chemical insecticides with reduced adverse external effects

While part of the social opposition to GMOs is based on health and environmental concerns, some of the opposition seems to arise from monopoly rights held by larger companies in transgenic organisms, and their procedures to maximize their profit from the development of these organisms. In the case of transgenic crops, farmers are not allowed under their contract to save seed. They may also have to comply with other provisions. In the case, for example, of Bt corn a proportion of a farmer's crop must be planted with non-Bt corn. This is intended to provide refuge for insects and reduce the rate at which the insect population may develop resistance to Bt. The Environmental Protection Agency in the USA requires such a policy to be followed, and it may be economically beneficial to growers collectively as well as to suppliers of Bt corn seed by prolonging the demand for this seed. However, while it is a US requirement in relation to Bt corn, it is not one for Bt cotton (Barnett and Gibson, 1999). This seems rather inconsistent since the European corn borer and the cotton bollworm are the same insects with a different common name. The difference is probably historical. Bt cotton was introduced earlier than Bt corn and, at the early time, the resistance issue may not have been so prominent. Nevertheless, up to a point, both suppliers of Bt seed and growers have an interest in controlling the rate of increase in resistance of target insects, but collective action is needed to do this given the common property nature of the problem.

Concluding comments

The marketing of transgenic food and produce is still in its infancy, having commenced only around the mid-1990s. While the adoption of transgenic crops has spread rapidly in the USA and Canada, internationally adoption is uneven, and in some countries considerable consumer resistance to GMO food has emerged, for example, in the European Union and Japan. Some proponents of the biotech revolution believe that it has strong parallels with the green revolution. For instance, Nelson (1999, p. 7) suggests that the parallels between the biotech and green revolutions are strong.

> Both create commercial opportunities for the sale of inputs and thus 'modernize' agriculture, making it part of an industrial process rather than a self-contained traditional rural activity. Both have a tendency to reward adopters and those who are already more integrated with other sectors. Concerns with the thinning of the gene pool were also expressed at the time of the earlier green revolutions, and efforts were made to preserve genetic diversity.

However, there are also key differences. Most consumers did not distinguish between production from high yielding varieties developed as a result of the green revolution and that from traditional varieties. Secondly, green revolution technologies were mostly developed by the public sector and the intellectual property was available as a public good. On the other hand, the current biotech revolution is driven by the private sector.

> The fact that the current biotech revolution is driven by involvement of the private sector in the development and marketing of new varieties of crops and entails a change in consumer perceptions of the products, sets it apart from earlier green revolutions. (Nelson, 1999, p. 7)

While transgenic technologies hold great promise for increasing crop production for human end-use, increased yields from such crops may be less sustainable than originally hoped for.

References

Alauddin, M. and Tisdell, C.A. (1991), *The 'Green Revolution' and Economic Development*, London: Macmillan.
Barnett, B.J. and Gibson, B.O. (1999), 'Economic challenges of transgenic crops: the case of Bt cotton', *Journal of Economic Issues*, **33**, 647–60.
Duffy, M. (2001), 'Who benefits from biotechnology?', Paper presented at American Seed Trade Association Meeting, 5–7 December, Chicago, IL, <http://www.leopold.iastate.edu/pubinfo/papersspeeches/biotech.html>
Duncan, R.C. and Tisdell, C.A. (1971), 'Research and technical progress: the returns to producers', *The Economic Record*, **47**, 124–9.

Hayami, Y. and Herdt, R.W. (1977), 'Market price effects of technological change in income in semi-subsistence agriculture', *American Journal of Agricultural Economics*, **59**, 245–56.

James, C. (1998), *Global Review of Commercialised Transgenic Crops: 1998*, ISAAA Briefs No. 8, ISAAA, Ithaca, NY, <http://www.isaaa.org/publications/briefs/Brief8.htm>

Linder, R.K. and Jarrett, F.G. (1978), 'Supply shifts and the size of research benefits', *American Journal of Agricultural Economics*, **60**, 48–58.

Linder, R.K. and Jarrett, F.G. (1980), 'Supply shifts and the size of research benefits: a reply', *American Journal of Agricultural Economics*, **62**, 841–2.

Marra, M., Carlson, G. and Hubbell, B. (1998), *Economic Impacts of the First Crop Biotechnologies*, North Carolina State University, Department of Agricultural and Resource Economics, 1998, <http://www.ag-econ.ncsu.edu/faculty/marra/FirstCrop/sld001.htm>

Moffat, A.S. (1998), 'Totting up the early harvest of transgenic plants', *Science*, **18**, 2176–8, <http://www.sciencemag.org/cgi/content/full/282/5397/2167>

Nelson, G.C. (ed.) (1999), *The Economics and Politics of Genetically Modified Organism in Agriculture: Implications for WTO 2000*, Bulletin 809, The College of Agricultural, Consumer and Environmental Sciences, University of Illinois at Urbana-Champaign. See for Executive Summary: <http://web.aces.uiuc.edu/wf/GMO/execsummary.html>

Paoletti, M.G. and Pimentel, D. (1996), 'Genetic engineering in agriculture and the environment: assessing risks and benefits', *Bioscience*, **46**, 665–73.

Paoletti, M.G. and Pimentel, D. (2000), 'Environmental risks of pesticides versus genetic engineering for agricultural pest control', *Journal of Agricultural and Environmental Ethics*, **12**, 279–303.

Pimentel, D. and Ali, M.S. (1998), 'An economic and environmental assessment of herbicide-resistant and insect/part-resistant crops', *Indian Journal of Applied Economics*, **7**, 241–52.

Shelton, A.M., Zhao, J.-Z. and Roush, R.T. (2002), 'Environmental risk of pesticides versus genetic engineering for agricultural pest control', *Annual Review of Entomology*, **47**, 845–81.

USDA (1999), 'Impacts of adopting genetically engineered crops in the U.S. preliminary results', Economic Research Institute, USDA, <http://www.monsanto.com/monsanto/biotechnology/backgroundinformation/99july20USDA.html>

PART IV

MARINE PRODUCTION – BIOECONOMIC ASPECTS

[22]

Economic problems in managing Australia's marine resources[1]

Abstract

While this chapter pays most attention to policies adopted for management of commercial fisheries in Australia, it emphasizes that management of Australia's marine resources involves wider issues and interests. Government regulation of the domestic commercial fisheries and its control of foreign fishing in Australian waters are discussed. Regulation of the domestic fisheries involves producer protection against rent-dissipation and consumer protection, for instance from excessive mercury content in fish. However, government management may also have dissipating effects. The southern bluefin tuna (SBT) industry is given considerable attention since it involves a migratory species, a transborder one, shared by different Australian states and by Japanese vessels in international waters. While a management plan has now been adopted for the Australian SBT industry, Japan and Australia have yet to reach a negotiated solution on their joint catch rate. The Nash extended bargaining game model is used to consider a possible settlement between Australia and Japan. A number of gaps and shortcomings in Australian research on marine resource economics are noted, such as the relative lack of attention to such matters as recreational fishing, recreational demands on marine areas, demands for preservation of marine species and the economics of the management of marine parks.

Introduction

The Australian exclusive economic marine zone (the 200 nautical mile zone) is the second largest in the world (Gopalakrishnan, 1980, p. 42) but Australian production of fish and related products is small by world standards, partly because the biological primary productivity of the surrounding seas appears to be comparatively low (Firth, 1969, p. 55). While Australia imports a considerable amount of fish (is a net importer of fish), it is a net exporter of edible marine products when its exports of crustaceans and molluscs are taken into account. In 1983–84, Australia's most important exports of marine edible products in order of value were prawns or shrimps ($140m) closely followed by rock lobster ($131.5m) then abalone ($51.5m) and scallops ($28.4m), and of much less value ($6.6m) tuna (*Australian Fisheries*, Oct. 1984, p. 57). Commercial fisheries are important to a number of Australian communities and contribute to decentralization of population; a factor to be taken into account by Australian governments, especially in politically marginal electorates.

However, there are many other claims on Australia's marine resources apart from commercial fishing. These include recreational fishing and other recreational uses of the sea, for instance, boating, transportation (navigation lanes), sea-bed and other mineral exploitation, marine-based tourism, species and natural phenomena preservation, waste disposal and so on. The management of Australia's marine resources requires that consideration be given to all claims since sometimes they are in conflict with one another and there is often a social need for the government to impose rules and assist in conflict resolution.

Management problems raised by each of the uses mentioned above will be selectively considered in turn. General policies for the management of Australian commercial fisheries are discussed first, followed by foreign fishing in Australian waters and the Australianization goal. Particular consideration is given to the southern bluefin tuna fishery because it involves conflict between Australia and Japan over a migratory species. Then recreational fishing and other aspects of marine resource-use and management are addressed.

Australian commercial fisheries general management problems and questions
The Australian Constitution empowers the Commonwealth Government to make laws governing the 'fisheries in Australian waters beyond territorial limits' but state governments retain the residual power of regulation within the three-mile territorial limit. Commonwealth involvement in fisheries management began in 1968 when the 12-mile Declared Fishing Zone was established and became of increasing importance in 1979 when the 200 nautical mile Australian Fishing Zone became effective. Until recently, state rivalries and lack of co-ordination have hampered the overall management of Australian fisheries. However, the *Fisheries Amendment Act* 1980 made provision for the setting up of fisheries joint authorities by the Commonwealth in negotiation with the states.

> The authorities will make and implement decisions under the law applying to a particular fishery on such matters as closed seasons or areas; permitted or prohibited types of fishing methods or gear; catch quotas; controls on fishing effort; criteria for entry to limited entry fisheries, conditions of licences, and other measures necessary for the effective management of the fishery. (Senate Standing Committee on Trade and Commerce, 1982, p. 50)

There is also the possibility of agreement being reached between a state and the Commonwealth for the application of the law of one or the other to a particular fishery in waters adjacent to that state. The Senate Standing Committee on Trade and Commerce (1982, p. 51) reported

> It is expected that fisheries peculiar to a State or Territory will be the responsibility of that State or Territory and managed in accordance with that State or Territory's law. Fisheries falling within this category will include the lobster, scallop and abalone fisheries. Fisheries that are migratory or are common to a number of States (for example, the northern prawn fishery and the Australian salmon fishery) where a unified approach to management is desirable are likely to be managed by joint authorities. Fisheries of national and international importance or undeveloped fisheries are likely to be managed by joint authorities.

The southern bluefin tuna industry is an example of the latter.

Commonwealth involvement in the management of Australian fisheries has expanded considerably since 1982 and a number of the recommendations made in the report on *Development of the Australian Fishing Industry* (Senate Standing Committee on Trade and Commerce, 1982) have been implemented or are in the process of being adopted. Until recently, it is said, 'a common view in Government and industry was that as long as fish stocks were protected from severe depletion or extinction little else needed to be done' (Bain, 1984, p. 15). It seems possible that this was the political strategy of least resistance, given conflict and bargaining because of overlapping federal–state jurisdiction and lack of agreement among fishing industry members on desirable management action. Minimum regulation has much to be said for it from an economic point of view when the level of exploitation of a fishery is low and account is taken of the enforcement and administrative costs associated with regulation. However, low levels of exploitation are no longer the case in many of Australia's fisheries.

Bain (1984, p. 15) (then First Assistant Secretary, Fisheries Division, Department of Primary Industry) claims that in Australia

> the body of resource management theory which indicates that unregulated expansion of fishing effort on common property stocks will lead to a socially undesirable level of resource exploitation and a lowering of national economic welfare, did not receive wide recognition or acceptance until the last year or so.

Changed attitudes seem to have come about because available data 'clearly point to declining catch per unit effort in many major fisheries, excess fishing capacity and, in a few cases, biological pressure on resources' (Bain, 1984, p. 14).

Management plans, in which the Commonwealth has been a principal moving force, have been drawn up for the Northern Prawn Fishery and the Southern Bluefin Tuna Fishery and are in the process of being implemented. Interim or draft management plans have been devised for East Coast Prawns, Bass Strait Scallops and the South East Trawl and discussions have been commenced by an industry–government task force concerning the management of the shark fishery.

The management plan for the Northern Prawn Fishery involves a limitation on vessel numbers and boat capacity units. Boat units based on under deck volume and maximum continuous engine rating are to be initially allocated to each licensee and are intended to limit catching effort. However, boat units and licences are to be transferable (*Australian Fisheries*, June 1984). In order to reduce excess capacity in the Northern Prawn Fishery, a scheme to buy back boat units and withdraw these from the fishery is to be implemented by the government. The voluntary licence entitlement buy-back scheme (which applies to both boat units and licence entitlements) is to be funded by an annual levy on those holding boat units for the fishery. It has been anticipated that rents obtained by those remaining in the fishery will rise as boat units are bought back and excess capacity is reduced. The extent, however, of the gain by those remaining in the industry will depend on the prices paid for boat units bought back. If the price is sufficiently high the rent remaining after buy-back for continuing operators could be small. It is too early yet to say what the supply of boat units for buy-back is likely to be and what prices are likely to prevail. Observe too that even with boat unit restrictions an operator might increase effort without buying additional boat units, for example by working a licensed boat for a greater number of days or hours in any season. In the case of the Western Rock Lobster industry, days fished per month have tended to increase with quotas on lobster pots and restrictions on boat replacement (Rogers, 1982). However, it is unrealistic to expect economic perfection in these matters: possibly one has to be guided by Simon's administrative satisficing man rather than maximizing economic man in this respect (Simon, 1957).

It should not of course be thought that the above-mentioned fisheries are the only managed ones in Australia. State governments do have management strategies for several of their fisheries. Managed fisheries include the abalone fisheries of Victoria (Stanistreet, 1982), the South Australian prawn fishery (Byrne, 1982) and the Victorian scallop fisheries (Sturgess et al., 1982). Although the number of licensed operators is limited in these fisheries and regulations on effort apply, regulations have been circumvented to some extent, factors of production not subject to limitation in use have been substituted for regulated ones and fishing effort has tended to rise.

The main objectives of Australian fisheries management are far from concise. *The Fisheries Act* 1952 (Sec. 5b) indicates that the Australian government is responsible for

(a) ensuring through proper conservation and management measures that the living resources of the Australian fishing zone are not endangered by over-exploitation; and
(b) achieving the optimum utilization of the living resources of the Australian fishing zone.

This responsibility is capable of wide interpretation. However, it is clear that marine species are to be preserved from extinction and that (optimum) utilization is to be promoted. Nevertheless, the objective function to be optimized is not clear. Bain (1984, p. 16) indicates a number of priorities that he feels may be implied. These include avoiding depletion of living resources below their *long-term maximum sustainable yield* (MSY) and fostering expansion of the Australian fishing industry in the less developed fisheries. Note, however, that the MSY strategy is not necessary as a rule for species preservation. The maximum economic yield (MEY) strategy *may* be compatible with species preservation and usually involves retention of a larger fishing stock than the MSY strategy. Yet it can be optimal from a collective economic viewpoint to exploit some fish species on such a scale that they are driven to extinction (Clark, 1976). Presumably clause (a) above rules this strategy out and it could be argued that the Act tends to favour conservation over economic exploitation. However, it is only by studying the management of particular fisheries under commonwealth control that one can find out how priorities are established in practice.

Australian fisheries management policy at the present time (1983) is aimed at restricting effort in established fisheries by limitations on entry and other measures. To the extent that this is successful, rents should be generated in the fisheries. This raises the following questions. Who should obtain the rent? Should it go to those who obtain a licence? Who should be given access? Currently there is no bidding system for licences or rights. It could be argued that all rights of exploitation should be held by the government, that a market should be established in these and all revenues (less transaction costs) from trade in these *rights* should go to the Commonwealth. A suitable worked-out system would ensure that all rents go to the Commonwealth and that economic efficiency is promoted. Otherwise, those established historically in the fisheries are likely to appropriate a good deal of the rent. In practice, complex formulae are sometimes used to determine access to a fishery. For example, in the Victorian abalone fishery, age, waiting time and whether or not an applicant has a relative in the industry are all taken into account in determining eligibility for access (Stanistreet, 1982). A further possible consideration in management could in some instances be aboriginal claims. For example, an application has been made to the Northern Land Council on behalf of aboriginal interests to close seas in the vicinity of the north coast of Groote Eylandt (*Australian Fisheries*, Vol. 43, No. 4, p. 6).

In 1984, the federal government passed legislation to 'provide basic authority to raise revenue to support fisheries management measures and costs of associated research for effective management of fisheries'. These levies should result in the Commonwealth appropriating some of the rent from managed fisheries and will also in effect act as a tax on entry or continued presence in managed fisheries. To some extent, the levies may encourage the employment of public servants in lieu of rent-dissipating employment in the fisheries. One is only looking at one side of the picture by pointing to excess employment in the fisheries. Management also uses resources and increased government management or regulation is only justified up to the point where the additional benefits equal the additional costs. Furthermore, managerial groups can differ in their efficiency. To some extent, managerial groups within government enjoy a protected position so particular attention needs to be given to measures to ensure the efficiency of these managers and to management formulation itself. Otherwise, the rents may be dissipated in X-inefficiency and excessive employment of public servants, rather than excessive entry into the fisheries. Consequently, one distortion could replace another without any rents being left to contribute to general government revenue. Industry consultation will not adequately guard against this possibility. Indeed, if rents are to be appropriated from the fishery, members of the industry may prefer to see these spent on management of, and research for, the fishing industry rather than added to general government revenue.

Producer protection as well as consumer protection involves costs. An interesting example of the latter involves mercury content in fish. In 1981, the National Health and Medical Research Council recommended that the mean mercury level in a consignment of fish should not exceed 0.5 mg/kg and that it should not exceed 1.5 mg/kg in any individual sample. If this standard happened to be applied, a considerable proportion of Australian fish would be rejected and sampling could result in delays in processing and problems of storage: the school shark and gemfish industries could be significantly affected and some losses could be expected for tiger flathead, snapper and other catches (Senate Standing Committee on Trade and Commerce, 1982, pp. 85, 86). It seems that the mercury standards adopted by the states are not uniform and that shark may be shipped from a state where different standards apply. Large sharks for example may be cut into fillets in South Australia and shipped to Victoria. South Australia and Tasmania have in fact adopted a mean maximum mercury content for consignment of 1.0 mg/kg.

The mercury content of shark seems to be closely related to its length (Lyle, 1984) and it has been suggested that a legal maximum length might be used to protect consumers. In northern Australian waters, adoption of the NHMRC standard would lead to rejection of over 40 per cent of the shark catch (Lyle, 1984, p. 450). The standard to be adopted needs to be carefully considered in the light of the fish eating habits of Australians, variations in these, the costs of enforcement and the actual health risks from what appears to be largely mercury intake from natural sources. It is an area in which economists and social scientists as well as natural scientists need to make a contribution.

Foreign fishing in Australian waters and Australianization of the fisheries

> Australia is in the position of a privileged trustee of the marine resources of the Australian Fishing Zone rather than an absolute owner. It has the right to determine the terms and conditions under which it will permit foreign boats to fish in the Zone but it also has the obligation to permit such fishing to take any surplus of the total allowable catch that Australian fishermen are unable to take. Foreign fishermen are required to be licensed to fish within the Australian Fishing Zone and must comply with the terms and conditions of access determined by the Australian Government. (Senate Standing Committee on Trade and Commerce, 1982, pp. 11–12)

At the present time (1983), Japanese, Taiwanese and Korean vessels are licensed to operate in the AFZ (Australian Fishing Zone) and initially Polish vessels were also licensed to operate in the AFZ. (For general background information on foreign fishing in the AFZ, see Senate Standing Committee on Trade and Commerce, 1982, Ch. 10.)

Economic benefits to Australia from foreign fishing in the AFZ are claimed to include or have included:

(a) Royalty payments. For instance, in 1982 Japan paid a $1.44 million access fee and Taiwan has paid an access fee of around $1 million p.a.
(b) Income obtained as a result of the use of Australian port facilities.
(c) Information on fishing stocks. This is pertinent to assessing the size of the stocks and whether or not an Australian fishery based on the stocks would be economically viable.
(d) Development of new Australian fisheries.

Benefit (d) is achieved not only because initial risks are taken by foreign vessels and information provided but also in some cases by arranging for the systematic and phased introduction of Australian vessels into the new fishery. For example, a joint-venture agreement has been signed with Taiwanese interests operating in the northern AFZ that requires the phasing in of Australian vessels.

> Apart from the vessel replacement program the joint venture company will be required to purchase annually 300t of pelagic or dermersal fish from Australian fishermen and to make maximum use of Australian shore facilities. In addition it will be required to spend in Australia, on Australian goods and services, an amount of not less than 10 per cent of the catch taken by the foreign vessels licensed under the agreement. (Branford, 1984, pp. 14–15)

Perceived disadvantages or costs associated with foreign fishing in the AFZ include:

(a) Possible competition with Australian fishermen for (migratory) common stock. This was a complaint against Japanese longliners operating in the AFZ for

southern bluefin tuna. A similar complaint was made about the impact of Japanese longliners operating off the Queensland coast on the black marlin fishery.
(b) The costs of enforcing regulations and seeing that adequate information is in fact obtained. Australian monitors board foreign ships for this purpose.
(c) The possible crashing of populations before they can be exploited by Australian fishing vessels. This crashing might come about because of initial over-exploitation by foreign fishing vessels, or permanent change in species composition might come about due to high initial levels of exploitation of stocks of particular species.
(d) Incidental take of species of high conservation value such as dugong and porpoises/dolphins can occur.
(e) Where fishing grounds of Australian and foreign fishermen overlap, obstruction may occur and damage to Australian fishing gear (as well as foreign gear) may result.
(f) Given the interdependence of nature, the foreign exploitation of a previously unutilized fish stock may have adverse consequences for the population of a utilized species via the food chain.
(g) Information collected may be of limited value because catch rates on commencement of a fishery may exceed sustainable rates. It takes some time for possible equilibrium or sustainable rates to be established, or determined, and for reliable information to be obtained.

An in-depth analysis of the economic benefits and costs of foreign fishing is required. Some items that appear to be a benefit may not be a collective benefit. For example, the forcing of the purchase of Australian goods and services or the purchase of Australian-caught fish on foreign vessels could benefit special Australian interests but at the same time lead to a reduction in the general licence or royalty rate obtained from foreigners if this policy reduces the profitability to foreigners of fishing in Australian waters.

Australianization of the fisheries seems to have been an important goal for the commonwealth government (see Senate Standing Committee on Trade and Commerce, 1982). The following two requirements are among those that have been adopted in recent years:

(a) Foreign vessels fishing in the AFZ should land a proportion of their fish in Australia for processing or sale (Senate Standing Committee on Trade and Commerce, 1982, pp. 149–50).
(b) Phased replacement of foreign fishing vessels in the AFZ by Australian ones.

The requirement to land fish in Australia would, of course, be an interference with the free market, and may impose a cost on foreign operators that could eventually be reflected in lower access fees. In some cases too, there may not be an Australian market for the product, or the cost of meeting Australian standards may be high. For example, the landing of shark from northern Australian waters by Taiwanese vessels may prove to be relatively uneconomic if the maximum mercury content requirements for Australia are much lower than for Taiwan. It may also be uneconomic to establish processing facilities in remote areas unless sufficient volume and regularity of local fish supply are available. In some cases, however, it is possible that landings might help to develop local tastes and demands and could possibly result in some import substitution. If processing for export is being considered the question needs to be considered of how competitive the locally processed product is likely to be. Each case needs to be considered on its own merits. Economic loss to Australia can result from rigid adherence to a local landing requirement even though remote communities might hope to benefit from it.

As for the replacement of Australian vessels by foreign ones, much depends on the relative economics of operation of Australian vessels. In some cases, replacement might add to costs (thereby lowering rents) and if carried out completely could result in lack of access to original overseas markets for a product that cannot be easily marketed in Australia or elsewhere. While the replacement policy may increase employment in some remote areas of Australia, it should be asked whether more cost-effective methods of increasing employment in remote areas are available, and whether these are politically possible to adopt.

Australianization of the fisheries can involve economic costs and these need to be considered carefully. Is Australianization in accordance with Australia's comparative advantage or in conformity with its likely comparative advantage after learning and social adjustment have taken place? Is there some other substantial reason for all-Australian operation of fisheries in its waters when this does not accord with its comparative advantage? Exclusive fishing in the AFZ by Australian fishermen is unlikely to be in the general economic interest of Australia, even in the longer term.

Southern bluefin tuna – international conflict over a migratory species

The southern bluefin tuna *Thunnus maccoyi* (Castlenau) spawns in the Indian Ocean in an area between Indonesia and north-west Western Australia (WA). The young migrate slowly down the Western Australian coast around the Great Australian Bight and into waters off New South Wales (NSW). At various times some leave the main body to move out into the southern Indian Ocean generally, and eventually all appear to do so, except those suffering mortality. They mature in the southern Indian Ocean and return for spawning to the grounds off the north-west shelf of Australia. The stock available for spawning influences the flow of migrants to the AFZ, the catch in the AFZ and beyond, and recruitment to mature stock for spawning. An interesting interdependence problem therefore exists. The general pattern of migration and ages of the migratory stock in Australian waters can be seen from Figure 1.

The migratory southern bluefin tuna (SBT) stock is most heavily fished by Australian and Japanese fishermen with catches by New Zealand being very small. The Japanese catch of SBT by volume has shown a downward trend over the last 20 years, whereas the trend for Australia is the reverse. The Australian catch at over 21 000 tonnes in 1982–83 exceeded the Japanese catch for the first time but the value of the Japanese catch is much higher than the Australian because a large proportion of it supplies the sashimi (raw fish) market. There is conflict or competition between Australia and Japan for SBT stocks. There is also interstate rivalry within Australia. The largest volume of catch within Australia is made by South Australia (SA), with the smaller

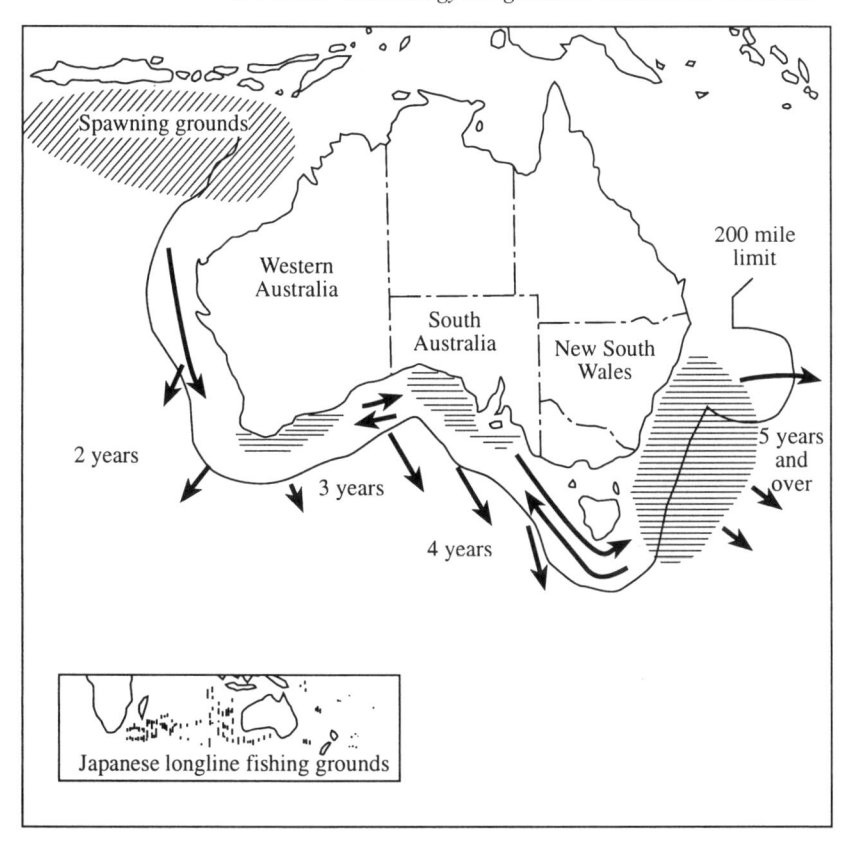

Figure 1 Southern bluefin spawning area, migration pattern about Australia and location of the Japanese longline fishery after Majowski et al. (1984) and IAC (1984)

catch of WA and NSW being of similar volume in recent years. However, the number of fish caught by WA far exceeds those caught in NSW since most of the WA catch consists of fish of 2–3 years of age whereas the NSW catch consists of juveniles of 5 years and over. The Japanese catch is mainly of adults caught by longliners. Catches early in the migratory path tend to lower stock later along the path. Thus catches in WA tend to reduce the stock available to SA. Catches in SA tend to lower the stock available to NSW and eventually to Japan. In these circumstances co-operative management appears to be rational. There is a need for co-operative management within Australian waters and in international exploitation of the species. An overall management plan for the AFZ has been devised but no agreement has been reached on international management.

In November 1983, the management of the SBT fishery and other issues such as adjustment assistance to the SBT processing sector and arrangements for the import of tuna for processing were referred to the Industries Assistance Commission (IAC) and it reported on 28 June 1984 (Industries Assistance Commission, 1984). The Report notes

268 Economics and Ecology in Agriculture and Marine Production

... some trends over the last 10 to 15 years are apparent. The catches from the Japanese and New South Wales fisheries, based on larger, more mature fish have declined, while the South Australian and Western Australian fisheries, based on smaller fish have expanded. As a result, the world catch in terms of weight has fallen even though the number of fish caught has been increasing.

In 1979, CSIRO scientists monitoring the SBT fishery concluded it was fully exploited. Any further increase in exploitation by Australian fishermen 'would reduce the Japanese catch' and any significant increase in the Japanese catches might reduce the Australian catch . . .

Scientists considered that breeding stock had been reduced to a level at which recruitment might be adversely affected if the parental biomass were reduced much further. (Industries Assistance Commission, 1984, pp. 10–11)

It has been argued that a spawning stock of around 220 000 tonnes is needed if tuna recruitment rates are to be maintained and the risk of extinction of SBT stock due to natural disaster is to be kept low (Industries Assistance Commission, 1984, pp. 109–10). Using this level of stock as a desirable goal, the IAC estimated the trade-off curve between the volume of Japanese and Australian catches per annum needed to achieve this goal. The suggested trade-off function put forward by the IAC is indicated in Figure 2.

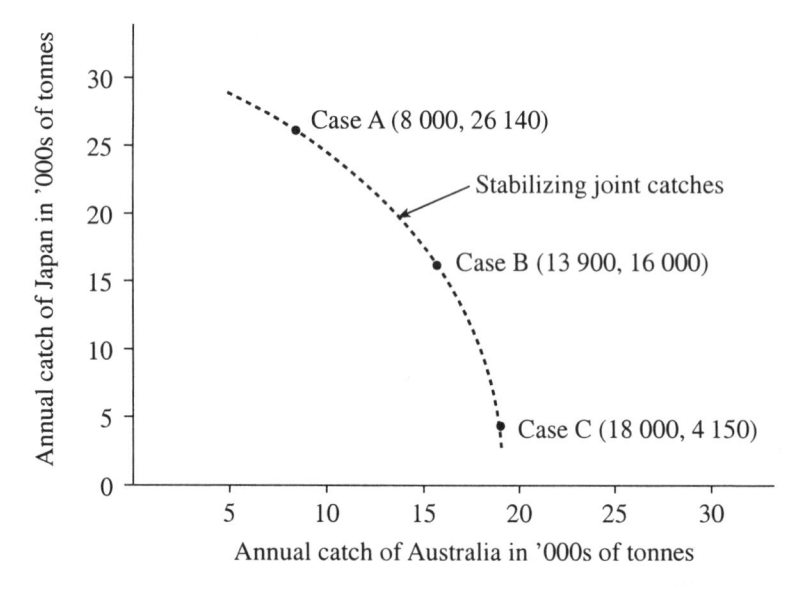

Figure 2 IAC (1984) estimates of joint SBT catches by Australia and Japan which stabilize the parental biomass at the 'safe' level of approximately 220 000 tonnes

The results suggest that the world catch has been well above sustainable levels and that the *maximum volume* sustainable catch would be achieved if Japanese longliners were to dominate the industry, namely a world catch of approximately 34 140 tonnes.

The returns for three different catch combinations were investigated in detail by the IAC and these combinations are indicated as cases A, B and C in Figure 2. The IAC argued that discounted joint net returns in the industry would be maximized for case B, calculations being based on the then estimated operating costs in the fishery (IAC, 1984, p. 141), but details of these estimates are not given. The specific assumptions behind the model need to be carefully considered but at least it provides an interesting start on the interdependence problem.

The implications of the Kennedy and Watkins (1984) model are rather different because they suggest that the Pareto optimal solution would be for Japan to take over all harvesting. They argue from the dynamic optimizing of their model that

> the elimination of Australian harvesting of fish under four years of age is in the overall interests of Australia and Japan . . . The results further suggest that Japan could profit by more than compensating Australia for agreeing to reduce harvesting and even agreeing to cease harvesting altogether. (Kennedy and Watkins, 1984, pp. 24, 25)

The Kennedy and Watkins approach is based on the criterion of maximizing economic welfare in terms of discounted consumers' surplus plus producers' surplus. The demand curves for Australia and Japan are specified separately and it is assumed that Australian caught fish will not be supplied to the high valued sashimi market.

An SBT management plan commenced in the AFZ on 1 October 1984 and will apply for the 1984–85 season. Japanese vessels have been effectively excluded from the SBT fishery in Australian waters (*Australian Fisheries*, Vol. 43, No. 12). A national quota of 14 500 t will apply and individual transferable quotas have been allocated to eligible boats using a formula incorporating an allowance for past catches and market value of boats and gear. For details, see Bowerman (1984). No size limits will apply to tuna but there will be a two-month seasonal closure of grounds in WA plus a permanent closure north of 34° south off the west coast of WA (*Australian Fisheries*, Vol. 43, No.2). Note that transferable quotas lead to greater economic efficiency than non-transferable ones and from this point of view are welcome. It is also interesting to note that the national quota of 14 500 t is only slightly in excess of that recommended by the IAC, namely 14 000 t (IAC, 1984, p. 38).

Whether or not this Australian quota is maintained may depend on Japanese reactions. If Japan increases its volume of catch above recent levels, Australia may retaliate by increasing its catch. It is not inconceivable that Japan could try to retaliate in response to its effective exclusion from SBT in the AFZ. Australia is likely to put continuing political pressure on Japan to reduce its catch to complement Australian restrictions. On the other hand, there does not appear to be any major reason for Japan to reduce its catch below 16 000 t (approximate recent Japanese catch levels) for even on the basis of IAC data a stabilizing catch situation would exist (Case B). However, Japan may be prepared to reduce its catch somewhat for renewed access to these SBT stocks in the AFZ if, thereby, it could achieve some cost saving in operating its SBT fleet. At this time, it seems that the brunt of adjustment has fallen on Australia. The Australian response could well have been anticipated by Japan and be why Japan has held out against a global agreement.

The IAC was aware of the strategic nature of the international SBT problem. It points out that a game-theoretic management problem arises when the major fishing nations attempt to formulate strategies designed to maximise their individual returns. The possibility highlights the dilemma faced by Australia in taking unilateral action in respect of the management of the SBT resource, since the exercise of restraint on fishing effort on the part of Australian fishermen for biological reasons could be frustrated by (more than) compensatory increases in the Japanese catch. (IAC, 1984, p. 142)

However, while the IAC thought that the negotiation set or Pareto optimal set of the negotiators might be as described by the curve in Figure 2, it does not go beyond this and suggest the likely negotiated solution.

In this regard, Nash (1953) for example suggests that we might consider the relative threat power of parties (players) to determine the likely negotiated solution. He suggested a solution based on the assumption of transferable utilities and the theoretical solution if a zero-sum two-person threat game happens to be played. The solution to this zero-sum threat game (the optimal threat strategies) provides the status quo for bargaining and the division of co-operative gains (Luce and Raiffa, 1957, note that Nash, 1950 involves a different model).

In relation to the SBT problem it would be necessary to decide the extent to which it is reasonable to assume transferability and comparability of utilities (Shubik, 1983, Ch. 7). One might use for explanatory purposes the discounted net present value obtained from the catch by Australian and Japanese fleets as an indication of relevant utilities. This payoff indicator would be in accordance with the IAC (1984) approach rather than say that of Kennedy and Watkins (1984). Let us suppose net discounted returns are relevant payoffs and let us see what is involved in principle if the Nash extended bargaining model is adopted.

First, the alternative SBT strategies of Japan and of Australia need to be identified and the corresponding payoffs for each fleet in terms of net present returns need to be estimated. The IAC Report (1984) specifies these payoffs for only three alternative sets of strategies but many more are feasible. A payoff matrix like that indicated in (1) needs to be specified. The alternative strategies might be alternative levels of catch (over time) or of effort. A_{ij} represents the payoff to Australia when Australia adopts its ith strategy and Japan adopts its jth strategy. J_{ij} represents the corresponding payoff for Japan.

$$\begin{array}{c} \text{SBT strategies of Japan} \longrightarrow \\[4pt] \begin{array}{cc} \text{SBT} & \\ \text{strategies} & a_1 \\ \text{of Australia} & a_2 \\ \downarrow & \vdots \\ & a_m \end{array} \begin{array}{c} \beta_1 \quad\quad \beta_2 \quad\; \cdots \quad \beta_n \\ \begin{bmatrix} (A_{11}, J_{11}) & (A_{12}, J_{12}) & \cdots & (A_{1n}, J_{1n}) \\ (A_{21}, J_{21}) & (A_{22}, J_{22}) & \cdots & (A_{2n}, J_{2n}) \\ \vdots & \vdots & \vdots & \vdots \\ (A_{m1}, J_{m1}) & (A_{m2}, J_{m2}) & \cdots & (A_{mn}, J_{mn}) \end{bmatrix} \end{array} \end{array} \quad (1)$$

From payoff matrix (1), the zero sum threat matrix (2) can be specified by taking the difference between the payoffs for Australia and Japan. The new matrix is

$$T = \left[A_{ij} - J_{ij} \right] \begin{pmatrix} i = 1, \ldots m \\ j = 1, \ldots n \end{pmatrix} \quad (2)$$

The upper left hand entry in the matrix is, for example $A_{11} - J_{11}$, the extent to which Australia's payoff exceeds that of Japan when both countries adopt strategy one. Matrix (2) represents a zero-sum two-person game and, therefore, has a solution. The threat equilibrium of the game forms the basis for the division of gains in a co-operative game. The same difference in gains is preserved in the negotiated outcome as would prevail in the solution to the threat game.

In relation to the SBT problem, suppose that the shaded set in Figure 3 represents the set of discounted net present values available to Australia and Japan from alternative available strategies. Suppose that the equilibrium threat value corresponds to M, a situation in which both countries make approximately an equal loss with the loss being slightly greater for Japan. The segment KLM represents Pareto optimal points (the negotiation set). Constructing a 45° line through point M, the intersection point of this line with the Pareto optimal set, L, represents the Nash solution. The solution involves almost equal payoff to Australia and Japan and corresponds closely under these assumptions with IAC case B.

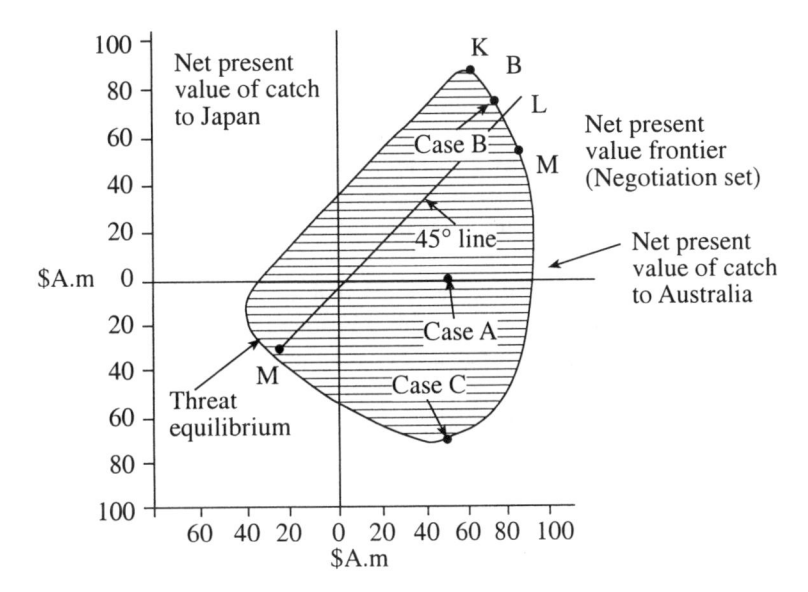

Figure 3 Hypothetical set of alternative payoffs to Japan and Australia from different strategies for SBT exploitation showing a Nash bargaining solution at L assuming the threat equilibrium at M

Is it likely that the negotiated solution will involve about equal payoffs for both fleets? If collapse of the fishery by increased effort happened to be the ultimate threat strategy, the actual operating cost to Australia of bringing about this collapse would most likely be less than for Japan. To some extent, it could be argued, therefore, that Australia should get some advantage on any negotiated settlement. On the other hand, we cannot be sure that net present discounted returns are the only significant variables in this matter. Politically, in the light of the communities and number of individuals involved, the SBT fishery would seem to be a more sensitive problem for Australia than Japan. Australia would be unlikely to carry out a policy leading to its collapse unless it was clear that Japan was intent on suicidal action. Even then, an Australian government might find it politically more acceptable to be able to blame Japan for the collapse rather than to have 'the bone pointed at it' for contributing to the collapse. This complicates the matter. Furthermore, to what extent should consumers' surplus be taken into account? Considering all these factors (and the principle of insufficient reason) it might be concluded that there is no clearcut reason why one fleet should be given an advantage over the other in a negotiated solution. The problem is an interesting one, even though only a few of its dimensions have been considered here.

International migratory species raise many interesting issues as can be seen from the case of the southern bluefin tuna, and some other species that have been discussed elsewhere (Tisdell, 1984). The case of the Atlantic bluefin tuna provides a further example. It has been argued that the international management of this species by the International Commission for Conservation of Atlantic Tuna has been deficient and that the USA should withdraw from it and manage Atlantic bluefin tuna in its 200-mile zone in its own exclusive interest (Hoover, 1983).

Recreational fishing and recreational use of the sea

Recreational use of the sea in Australia is important. It places considerable demands on marine resources and sometimes involves conflict with commercial fishing and with other marine uses or values placed on marine resources conservation. Sometimes also different forms of recreational use of the marine environment are in competition with one another, for example, spearfishing with line fishing or with the goal of species preservation.

An indicator of the importance of recreational fishing comes from a survey undertaken on behalf of the Australian Recreational Fishing Federation. The survey indicates that in 1983–84, 4.5 million people (aged 10 and over) went fishing in Australia and that Australians spent $2200 million on recreational fishing and fishing-related equipment. (For comparative purposes, the value of Australian exports of fish and related products was $412 million in 1983–84.) More than half of Australian households own some type of recreational fishing equipment (*Australian Fisheries*, 1984, Vol. 4, No. 10, p. 23). While the expenditure method has limitations as a measure of the economic value of an activity, it does appear that the gross value of goods and services used in recreational fishing exceeds the gross value of commercial fish and related products caught in Australia. The political power of recreational fishermen appears to have been less than that of commercial fishermen (which would seem to accord with the prediction of the theory of Downs, 1957) but some increase in political influence is expected with the recent formation of the Australian Recreational Fishing Federation.

A number of management issues are raised by recreational fishing. To what extent should commercial fisheries for particular species or in particular areas be restricted so as to benefit amateur fishing, and vice versa? As population increases in an area and as leisure time increases, demands for (and value of) recreational fishing in the area are likely to increase. Would it be desirable to restrict commercial fishing increasingly, or even exclude it from the area? Pressure from amateur fishing is extremely high within Sydney Harbour and there are many other areas in, or close to, the Newcastle–Sydney–Wollongong conurbation where amateur fishing raises urgent management problems (Henry, 1984). Incidentally, a sample survey by the NSW Department of Agriculture found that the most common primary motive (32 per cent of respondents) for a recreational fishing trip in NSW was escapism (to seek privacy or relax and unwind), followed by the attraction of nature (to be outdoors or to enjoy the natural beauty) stated by 23 per cent, and that the harvest (to eat or give away fresh fish or the thrill of catching fish) was only the primary motive in 21 per cent of the cases sampled (Henry and Virgona, 1984). There appears to be as yet no available in-depth economic study of recreational fishing in Australia and its management.

It is not possible to trace here all the consequences that recreational fishing may have for the commercial fisheries and vice versa and for the preservation of species. There are many issues; for example, do spear fishermen change the composition of marine species, and could this be a factor in the reported increase in the prevalence of the crown-of-thorns starfish on the Barrier Reef? To what extent can international use of the AFZ conflict with the interest of recreational fishermen? In this regard, it was mentioned earlier that Japanese longlining off the Queensland coast was restricted partly because of its possible adverse affect on recreational fishing for black marlin.

Management of other recreational activities and their 'harmonization' is also required. These activities include boating, sailing, cruising and in-shore activities such as surfing, scuba diving and skin-diving. Littoral areas are in considerable demand for recreational activities, residential and other uses. To some extent different recreational activities in littoral areas can conflict. For instance, sunbathing, love-making and walking may conflict with the use of dune buggies and off-road vehicles and with hang-gliding. A management device commonly used to deal with such conflict is zoning. Economists have not subjected this management method to the serious consideration and analysis that it deserves (although Mishan (1967) has alluded to its importance and optimality in some cases), but have tended to dismiss it as inefficient because it makes use of fiat or regulation rather than employing a pricing system as some systems of royalties or licences do. While zoning as a management method is not always efficient, it may be optimal when enforcement and other costs of regulation are taken into account and in any case economists should be able to make a positive contribution by making suggestions for optimal zoning.

Other uses of marine areas and marine parks
Sea-bed mineral exploitation is one important use of marine areas and this has proven to be significant in Australia's case for supplies of both oil and natural gas. A particular risk associated with off-shore oil exploitation is the chance of a 'blow-out' or oil-spill with subsequent damage to fisheries, wildlife and tourism. In some areas, there can be adverse visual impacts from oil platforms and obstruction to shipping. Zoning is often used to reduce these risks and conflicts in seabed oil exploitation. For example, oil production is not to be permitted in the Great Barrier Reef Marine Park.

Other minerals are extracted from the sea-bed, for example mineral sands and minerals may be extracted from sea-water itself (e.g. salt). The diversity of uses of the sea are immense: transport lanes for shipping (these also involve pollution risks), space for storage (Japan is reported to be storing coal in inshore marine areas and has built artificial islands for storage purposes), a sink for the disposal of wastes (this ranges from sewage for a city like Sydney, includes various industrial wastes tipped or burnt at sea or discharged directly into waterways, and the controversial dumping of nuclear wastes in international waters), supplies of water for industrial uses where the salt content is of little importance, discharge of waste heat (for example from electricity generating plants such as those on Lake Macquarie) and electricity may be generated from various energy forms in the sea itself (for example, tidal movements).

In the last 15 years, economists have taken much greater interest in the conservational value of species and of nature generally. Some marine species seem to have high existence values. These include mammals such as whales and dugongs and reptiles such as turtles. Australian waters contain (at various times) significant numbers of these species (Tisdell, 1984; Paterson, 1984; Needham and Long, 1984). What is the existence value of these species? To what extent should they be conserved? For example, should there be limited commercial use of green turtles? To what extent should measures be taken to prevent 'incidental take' of species like dugong, the existence of which is claimed to be threatened? A systematic economic study of marine species in Australian waters with existence value would be interesting.

Australia has the largest marine park in the world, namely the Great Barrier Reef Marine Park, and the Great Barrier Marine Park Authority is responsible for its management. The Park management makes extensive use of zoning: some areas are set aside as wilderness preservation areas, others for recreational use and some zones for commercial fishing (see for example, Tisdell, 1983). Mineral exploitation is excluded from the Park. The management of the Park raises many interesting issues that have been barely considered by Australian economists. A second Australian marine park is proposed in the North West Cape area of WA, and it is proposed that zoning should also be extensively used to resolve conflicts and help ensure conservation, particularly of corals. This proposed park, Ningaloo Marine Park, differs from the Great Barrier Marine Park in that it involves both a marine and a terrestrial component. In many instances, terrestrial and littoral land management has important consequences for the productivity of the sea. For example, mangrove areas may provide nurseries for some marine species and the damming of rivers may affect the nutrient and detritus flushes following rainfall and reduce marine productivity near the mouth of such rivers and further afield. Given the interdependence of nature, one can see that at times the management issues raised by marine resources can range widely.

Concluding comments

Australian economists have paid less attention to the economics of natural resources than might have been expected given the overall relative abundance of natural resources in Australia. While marine resources may not be Australia's most important asset, they are important and may become increasingly so. Many economic and related issues with a rich variety of facets are involved in the management of these resources but in-depth interest by Australian economists in these matters is relatively recent.

Possibly Australian economists have given greatest attention to marine resources involved in direct commercial exploitation, such as the fisheries. Nevertheless studies are still in their infancy and there is no strong fisheries economics school in Australia. Australian economic studies of recreational fishing, recreational uses of the sea, conservational values associated with marine resources and various marine environmental issues are rare or non-existent. At best they are in the foetal stages and no significant school of marine resource economics has yet developed in Australia. While no Australian centre of excellence in marine resource economics may be required (and indeed, I go along with the view that centralism in research can be dangerous), there is a strong case for more and broader Australian economic research in marine resources.

Note

1. This article is based on a presentation to a fisheries session of the Annual Conference of the Australian Agricultural Society. I wish to thank Dr Robert Bain, Dr John Kennedy and Dr Geoffrey Waugh for their comments on the article. This article originally appeared in 'Economic problems in managing Australia's marine resources', *Economic Analysis and Policy*, **13**(2), September 1983, pp. 113–41. The content of this article is the same as in the original (except for minor changes) but its presentation has been improved. Consequently, pagination is no longer the same as in the original.

References

Bain, R. (1984), 'Commonwealth fisheries policies – problems priorities and progress', *Australian Fisheries*, **43**(8), 14–18.
Bowerman, M. (1984), 'ITQs for SET – what it all means', *Australian Fisheries*, **43**(9), 2–4.
Branford, J.R. (1984), 'Taiwanese fisheries in the north – monitoring and benefits for Australia', *Australian Fisheries*, **43**(2), 14–16.
Byrne, J.L. (1982), 'The South Australian prawn fishery: a case study in licence limitation', in N.H. Sturgess and T.F. Meany (eds), *Policy and Practice in Fisheries Management*, Canberra: AGPS, pp. 205–24.
Clark, C. (1976), *Mathematical Bioeconomics*, New York: Wiley.
Downs, A. (1957), *An Economic Theory of Democracy*, New York: Harper and Row.
Firth, F.E. (1969), *The Encyclopaedia of Marine Resources*, New York: Van Nostrand Reinhold.
Gopalakrishnan, C. (1980), *Natural Resources and Energy*, Ann Arbor: Ann Arbor Science.
Henry, G.W. (1984), 'Good angling catches from Australia's busiest waterway', *Australian Fisheries*, **43**(3), 2–5.
Henry, G.W. and Virgona, J. (1984), 'Why is angling so popular?', *Australian Fisheries*, **43**(5), 32–3.
Hoover, D.C. (1983), 'A case against international management of highly migratory marine fishery resources: the Atlantic bluefin tuna', *Boston College Environmental Affairs Law Review*, **11**, 11–61.
Industries Assistance Commission (1984), *Report Southern Bluefin Tuna*, Canberra: AGPS.
Kennedy, J.O.S. and Watkins, J.W. (1984), 'Optimal quotas for the southern bluefin tuna fishery', Discussion Paper No. 4/84, School of Economics, La Trobe University, Bundoora.
Luce, R.D. and Raiffa, H. (1957), *Games and Decisions*, New York: Wiley.
Lyle, J.M. (1984), 'Mercury concentrations in four carcharhinid and three hammerhead sharks from coastal waters in the Northern Territory', *Australian Journal of Marine and Freshwater Research*, **35**, 441–51.
Majkowski, J., Hearn, W.S. and Sandland, R.L. (1984), 'An experimental determination of the effect of increasing the minimum age (or size), of fish at capture upon the yield per recruit', *Canadian Journal of Fisheries and Aquatic Science*, **41**, 736–43.
Mishan, E.J. (1967), 'Pareto optimality and the law', *Oxford Economic Papers*, **19**(3), 247–87.
Nash, J.F. (1950), 'The bargaining problem', *Econometrica*, **18**, 155–62.
Nash, J.F. (1953), 'Two-person co-operative games', *Econometrica*, **21**, 128–40.
Needham, D.J. and Long, J.K. (1984), 'South Australians watching for whales?', *Australian Fisheries*, **43**(8), 49–50.
Paterson, R. (1984), 'More good news on whales', *Australian Fisheries*, **43**(5), 32–3.
Rogers, P.P. (1982), 'Boat replacement policy in the West Coast Rock Lobster Fishery – an historical review and a future option', in N.H. Sturgess and T.F. Meany (eds), *Policy and Practice in Fisheries Management*, Canberra: AGPS, pp. 189–204.
Senate Standing Committee on Trade and Commerce, Parliament of the Commonwealth of Australia (1982), *Development of the Australian Fishing Industry*, Canberra: AGPS.
Shubik, M. (1983), *Game Theory in the Social Sciences*, Cambridge, MA: The MIT Press.
Simon, H. (1957), *Administrative Man*, 2nd edn, New York: Macmillan.
Stanistreet, K. (1982), 'Limited entry in the abalone fishery of Victoria', in N.H. Sturgess and T.F. Meany (eds), *Policies and Practices in Fisheries Management*, Canberra: AGPS, pp. 139–52.

Sturgess, N.H., Dow, N. and Belin, P. (1982), 'Management of the Victorian scallop fisheries: retrospect and prospect' in N.H. Sturgess and T.F. Meany (eds), *Policies and Practice in Fisheries Management*, Canberra: AGPS.

Tisdell, C.A. (1983), 'The Great Barrier Reef: a regional case of tourism and natural resources', *Australian Parks and Recreation*, May, 33–42.

Tisdell, C.A. (1984), 'Development and marine-resource conflicts in the southeast Asian/Australasian region: living resources, including turtles, as illustrations', *Occasional Paper No. 107*, Department of Economics, University of Newcastle, 2308.

Optimal Economic Fishery Effort in the Maldivian Tuna Fishery: An Appropriate Model

R. SATHIENDRAKUMAR

University of Newcastle
New South Wales, Australia

C. A. TISDELL

Department of Economics
University of Newcastle
New South Wales, Australia

Abstract The estimation of a production function for Maldivian tuna fishery is a two-step process. First, it is necessary to find the relationship between catch and effort and second, to find the most efficient combination of inputs to produce the various levels of effort and hence output. The paper discusses the selection of an appropriate model to explain the relationship between tuna catch and effort and presents a technique for estimating the effort level required for an optimal allocation of resources which maximize the economic benefit of the fishery to the society. It also considers the extent to which the present pricing policy of the State Trading Organization for tuna has prevented the fishery reaching the open-access equilibrium yield level of effort and dissipating resource rent.

Introduction

The Maldivian tuna fish catch is predominantly obtained by pole-and-line fishing (Christy, et al. 1981; Nichols 1985; Anderson and Hafiz 1985). Tuna accounts for about 80% of the fish catch of the Maldives (Sathiendrakumar and Tisdell 1986). Skipjack tuna (*Katsuwonus pelamis*) is the major type of tuna caught. It made up 77% of the tuna catch in 1980 and 62% in 1983. Yellowfin tuna (*Thunnus albacares*), frigate tuna (*Auxis thazard*), and eastern little tuna (*Euthynnus affinis*) account for the balance of tuna catch.

Pole-and-line fishing is confined to an area up to 25 km from the atoll reef[1] and is conducted in locally built wooden boats, known locally as "masdhoni." Mechanization of masdhonis commenced in 1974–1975 and now virtually all pole-and-line fishing is conducted by mechanized masdhonis. Since these boats do not carry ice they return the same day with their catch and fishing is done only once a day. Thus the number of fishing trips is synonymous with the number of fishing days.

---

[1] The reason for this limited range is given in Christy, et al. (1981), and Sathiendrakumar and Tisdell (1986).

Models Applied to Tuna Catch and Effort Relationship for the Maldives

Models that have been applied to the relationship between catch and effort for the Maldivian tuna fishery treat the fish population as a whole (total biomass) without reference to its structure.[2] In this paper three models of this type are evaluated for their applicability to the Maldivian tuna fishery. These are the Schaefer model (Schaefer 1957), the Fox model (Fox 1970), and a production function threshold-type model suggested by us. For brevity we shall refer to our model as the Sathiendrakumar and Tisdell (S&T) model. The Schaefer model and the Fox model are based on the steady-state relationship between stock size, fishing effort, and yield; sustainable yield is a function of total effort and stock size; stock size is a function of effort. Therefore, given the biological and environmental parameters, sustainable yield is determined by the effort applied. Thus the Schaefer model and the Fox model, which relate catch or sustainable yield to effort, can be interpreted as a production relationship in which effort is the input and catch is the output. Such a relationship is long-run since it reflects complete stock adjustment to changes in the effort level. On the other hand, in the S&T model decreasing returns to fishing effort are attributed to competition among vessels and not to long-term population adjustments. The Schaefer model and the Fox model have been applied to the Maldivian tuna industry by Anderson and Hafiz (1985).

In the Schaefer model, the catch-fishing effort relationship is

$$C_t = af_t - bf_t^2 \tag{1}$$

Where C_t = total landing of fish at time t, f_t = fishing effort at time t, and a,b = parameters. Equation 1 implies that the fishery can be fished to extinction.

In the Fox model the catch-effort relationship is

$$C_t = f_t \cdot Ae^{(-bf_t)} \tag{2}$$

Where A and b are constants. Therefore,

$$Ln(C_t/f_t) = Ln\ A - bf_t \tag{3}$$

Even though the Fox model does not imply that the fishery can be fished to extinction, both this model and the Schaefer model suggest that increased effort will lead to a decline in the stock of breeding tuna.

But in the Maldivian context, where pole-and-line fishermen do not extend their range of fishing beyond 25 km from the atoll reefs, we hypothesize that their increased fishing effort is most unlikely to lead to a decline in the stock of breeding tuna to a level which reduces recruitment.[3] This is because only a part of the range of migratory tuna is being fished. The unfished population outside the Mal-

---

[2] These models differ from those concerned with the growth and mortality rates of individuals (Beverton and Holt 1957).
[3] Surface tuna (skipjack) is highly migratory in nature and is widely distributed in the tropical and subtropical waters of the world's oceans (Comitini and Hardjolukito 1984).

divian range of fishing is sufficient to prevent the population of breeding tuna stock from falling to levels where recruitment declines. As shown in Figure 1, it is assumed that recruitment is a rising function of the population of breeding tuna stock up to a point (the critical population level) and then reaches a plateau and remains constant over a wide range of population size (Cushing 1972; Johannes 1978; Tisdell 1983; Industries Assistance Commission 1984). Recruitment overfishing is unlikely given present practices in the Maldives. It becomes more likely when there is heavy over-fishing in the entire range covered by the migratory tuna species. Even though there is some poaching by foreign vessels in the Maldivian exclusive economic zone (EEZ) (Sathiendrakumar and Tisdell 1986), this illegal effort appears to be insufficient to reduce the total available catch significantly.

The following also suggests that the stock of tuna is not being reduced below the critical level necessary to maintain recruitment of tuna at its plateau. The government of the Republic of Maldives has estimated in its National Development Plan (Ministry of Planning and Development 1985) a potential (target) catch level of 30,000 to 40,000 tons for near-shore tuna and a potential of 9,000 to 21,000 tons from the remaining area in its EEZ. At present the Maldives only exploits near-shore tuna. The average tuna catch of the Maldives was 29,982 tons per annum in the period 1970–1974 (prior to mechanization) and 29,754 tons per annum in 1980–1984 (Sathiendrakumar and Tisdell 1986). Actual catch is therefore 50% below the average target which the government believes is sustainable. Further-

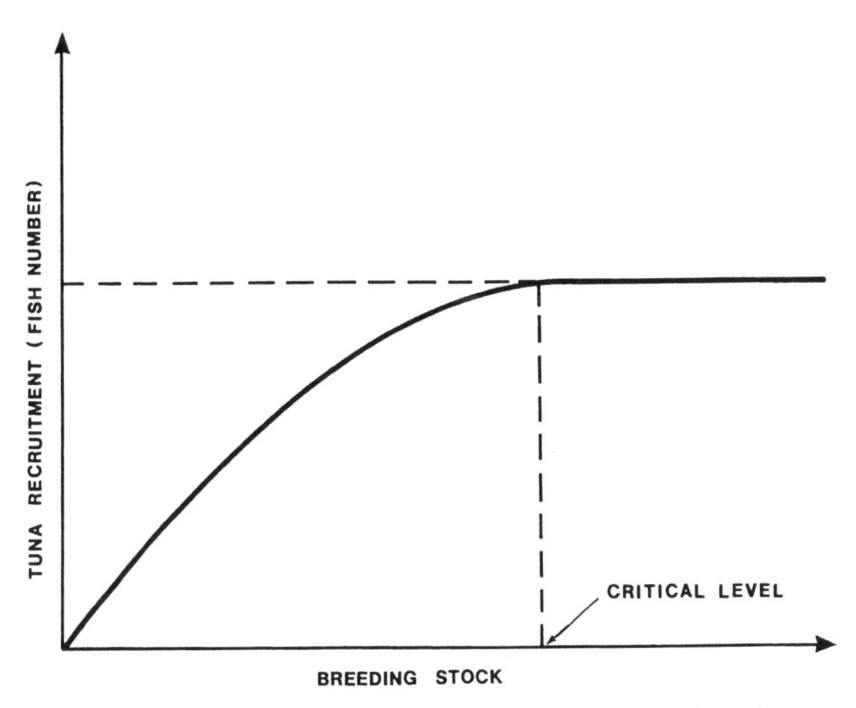

Figure 1. Suggested relationship between parental biomass of tuna and recruitment.

more, surface tuna are highly migratory in nature and tuna are widely distributed laterally and vertically (Comitini and Hardjolukito 1984). Since the present tuna fishery in the Maldives is confined to surface tuna within an area of up to 25 km from the atoll reef, it is considered likely that the population of tuna in the area fished by the Maldives given current levels of effort is unaffected by the fishing effort of the Maldives. It is therefore assumed that a pure catch-effort production model applies to this case (Howe 1979, 259).

In the Maldivian tuna fishery and similar fisheries, it is hypothesized that at low levels of effort, an increase in effort leads to a constant increase in catch until some threshold is reached (e.g., from o to x in Fig. 2). Thereafter due to competition between boats, increased effort in the same area leads to the catch increasing at a decreasing rate up to a point (from x to y in Fig. 2) and the total catch then approaches a maximum value (determined by the fish in the area) asymptotically as shown in Figure 2.

In this model, the catch increases at a decreasing rate with effort and approaches a maximum asymptotically. This relationship can be represented by the mathematical form:

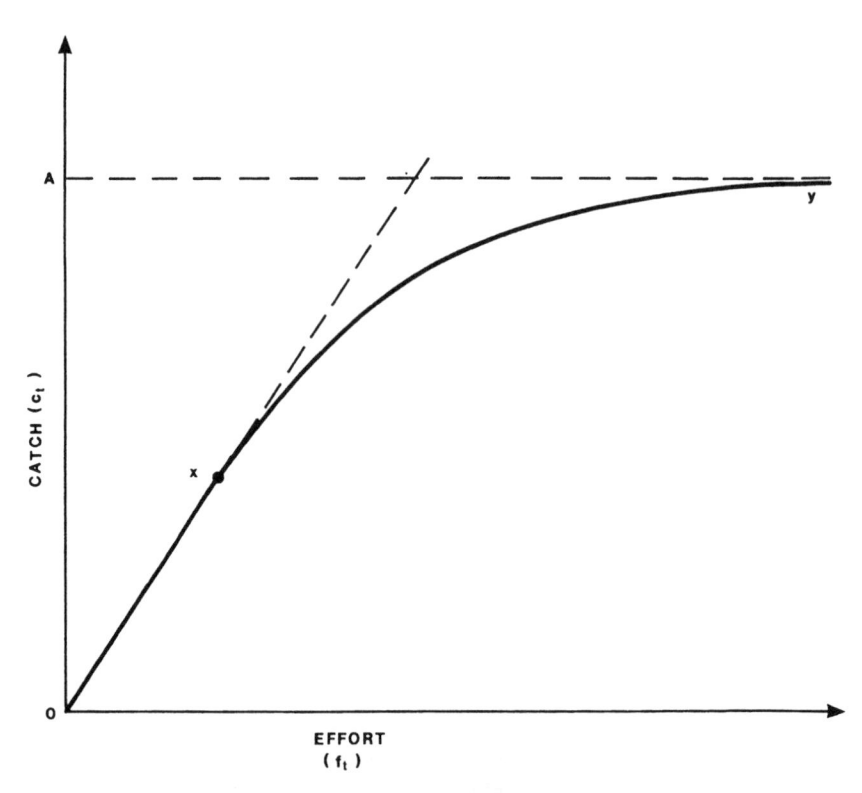

$$C_t = A - Be^{(-kf_t)} \qquad (4)$$

Figure 2. Nature of likely relationship between effort and catch for the Maldivian tuna fishing industry.

Rearranging:

$$A - C_t = Be^{(-kf_t)} \tag{5}$$

Therefore:

$$\text{Ln}(A - C_t) = \text{Ln } B - kf_t \tag{6}$$

A, B, and k are parameters. Because we do not know the value of A (the limiting value of the function corresponding to the maximum catch), an initial value of A is taken which is slightly higher than any of the previous recorded catch levels and then the initial value of A is changed iteratively until the regression of Ln (A − C_t) against f_t gives the best fit in terms of maximizing R^2, the value of the explained variance relative to the total variance (Crowe and Crowe 1969, 111). In accordance with the general recommendation of Crowe and Crowe (1969, 111), the value of A which gives the best linear fit for Equation 6 is used in our empirical analysis. However, in all but one of our cases R^2 was not very sensitive to variations in the value of A. In practice, A may correspond to the abundance of fish in the area being fished and in principle could be estimated by population sampling procedures. In the case of the Maldives however, no suitable census data on the abundance of tuna are available.

Measurement and Standardization of Fishing Effort and the Selection of The Appropriate Model

Measurement and Standardization of Effort

Time series data on tuna catch and effort are used in order to select the model (out of the three models mentioned above) which best represents the relationship between catch and effort. The best available record of fishing effort from the Ministry of Fisheries in Male is the "boat-day." This time series is used for the empirical testing for the period from 1970 to 1984.

Three problems were encountered in the estimation of fishing effort for the Maldivian tuna fishery.[4] The first problem is that during the time period considered two types of technology were employed, that is, sailing masdhonis and the mechanized masdhonis. To overcome this difficulty the number of fishing trips by sailing boats was standardized to the number of fishing trips by mechanized pole-and-line boats. This was done by dividing the number of fishing trips by sailing masdhonis by the ratio of the average catch of tuna per trip of mechanized masdhonis (1980–1984) to the average total catch of tuna per trip of sailing masdhonis before mechanization (1970–1974). See Table 1 for average catches.

The second problem is that between 1975 and 1978, the breakdown of number of fishing trips is not available for each category of boats. The available data is only for the total fishing fleet which comprises both sailing and mechanized masdhonis. This was overcome by first estimating the number of trips by mechanized

---

[4] See Rothschild and Suda (1977) and Treschev (1978) for the number of practical difficulties involved in the measurement of fishing effort.

Table 1
Catch and Effort Data of Tuna Fishing by Sailing and Mechanized Pole-and-Line Boats for Two Periods (1970 to 1974 and 1980 to 1984) for the Republic of Maldives

	Before Mechanization (1970 to 1974)				After Mechanization (1980 to 1984)			
	North	Cent.	South	Rep.	North	Cent.	South	Rep.
Sailing P & L Boats								
Av. No. of boats	887	556	617	2060	305	330	310	945
Av. No of fishing trips	88197	58172	41198	187567	3198	5984	1482	10664
Av. No of trips/boat	99	104	67	91	10	18	5	11
Av. catch of skipjack (mt)	10837	7051	4797	22685	122	164	192	478
Skipjack per trip (kg.)	123	121	116	121	38	27	129	45
Av. catch of yellowfin tuna (mt)	1643	926	215	2784	46	170	3	219
Yellow fin tuna/trip (kg.)	19	16	5	15	14	28	2	20
Av. catch of other tuna (mt)	3250	1104	159	4513	113	122	0	235

Other tuna/trip (kg.)	36	18	4	24	35	20	0	22
Total tuna/trip (kg.)	178	155	125	160	87	75	131	87
Mechanized P & L Boats								
Av. No. of boats	0	0	0	0	412	452	229	1093
Av. No of trips	0	0	0	0	35749	41593	29578	106920
Av. No of trips per boat	0	0	0	0	87	92	129	98
Av. catch of skipjack (mt)	0	0	0	0	7364	6256	7935	21555
Skipjack/trip (kg.)	0	0	0	0	206	150	268	201
Av. catch of yellowfin tuna (mt)	0	0	0	0	2416	1858	513	4787
Yellowfin tuna per trip (kg.)	0	0	0	0	67	44	17	45
Av. catch of other tuna (mt)	0	0	0	0	1266	1161	58	2485
Other tuna/trip (kg.)	0	0	0	0	35	28	2	23
Total tuna/trip (kg.)	0	0	0	0	308	222	287	269

Source: Based on Ministry of Fisheries, *Fishery Statistics* (1970 to 1984), Maldives.

masdhonis during that period[5] and then subtracting this from the total number of trips by combined fleet to obtain the number of fishing trips by the sailing masdhonis.

The third problem is that after 1978 (that is, four years after mechanization), even though sailing boats had withdrawn to a large extent from tuna fishing to reef fishing within the atoll reef (Anderson and Hafiz 1985), the available data are limited to the number of fishing trips by sailing boats and not the number of fishing trips for tuna fishing by sailing boats. Hence, in order to get the number of fishing trips by sailing pole-and-line boats for tuna, these figures had to be adjusted by multiplying the number of fishing trips by sailing boats after 1978 by the ratio of the average catch of tuna per trip by sailing boats after mechanization (1980–1984) to the average catch of tuna per trip by sailing boats before mechanization (1970–1974).

The estimated effort data expressed in terms of mechanized masdhoni days (standardized effort) for the period 1970 to 1984 is given in Table 2. The implicit assumption is that all mechanized boats are similar in their power,[6] and that fishing crews including skippers are equally efficient. The only conceivable relevant variation is the difference in the age of the boat. But it is quite likely that this would be reflected in the number of fishing trips. Thus the only adjustment that had to be made is for the difference in power between sailing masdhonis and mechanized masdhonis.

Selection of Appropriate Model

The time series, standardized effort data given in Table 2 and the tuna catch data given in Table 3 were used to test the three models described above for their applicability. The appropriate functions were fitted to these data by ordinary least squares.

The results tabulated in Table 4 reveal that the regression on catch per unit of effort (CPUE) on effort is not significant for the Schaefer model and for the Fox model for the skipjack tuna, which is the major tuna caught in the Republic of Maldives.[7] However, for the Fox model, the effort variable is significant for yellowfin tuna in the northern atolls and for other tuna in the southern atolls. For the Schaefer model, the effort variable is only significant for yellowfin tuna in the

---

[5] The number of fishing trips by mechanized masdhonis is estimated by multiplying the number of mechanized masdhonis available during that period by the average number of fishing trips by mechanized masdhonis between 1980–1984.

[6] This is a reasonable assumption for the Maldives, since the boats are of indigenous origin with similar length, tonnage, and hull construction. The horsepower of the engines of all mechanized masdhonis is almost uniform.

[7] Results of Anderson and Hafiz (1985) also show that the regression of CPUE (for tuna fishery) on effort was not statistically significant when the Schaefer and Fox models were used. They also state that averaging fishing effort and catch for skipjack over two years (since annual observations were unlikely to be equilibrium observations) (O'Rourke 1971) did not increase the significance of their results. In spite of this non-significance they still use these models and estimate MSY as 19,600 mt for skipjack tuna, with an effort of about 113,000 mechanized masdhoni days. This figure implies that skipjack fishery in the Maldives is already being carried at close to or slightly over the biological optimum level.

Table 2
Data on the Number of Fishing Trips by Pole-and-Line Vessels in Catching Tuna (1970–1984), The Maldives

	1970	1971	1972	1973	1974	1975	1976	1977	1978	1979	1980	1981	1982	1983	1984
1. No. of fishing trips by sailing P & L boats														
In the northern atolls.	95428	81389	76204	102652	85314	74674	52289	38048	29207	3504	2095	1747	1154	1225	1614
In the central atolls.	57007	50418	50576	68452	64411	61702	55665	32893	14188	6147	4604	3664	3125	1772	1197
In the southern atolls.	38986	37430	31764	44174	53637	35923	47927	37487	11636	4657	3009	2653	1170	148	433
In the republic.	191421	169237	158544	218278	203362	172299	155881	108428	55031	14308	9708	8064	5449	3145	3244
2. No. of fishing trips by sailing P & L boats expressed in terms of mechanized boats.														
In the northern atolls.	55161	47046	44049	59336	49314	43164	30225	21993	16883	2025	1211	1010	667	708	933
In the central atolls.	39865	35257	35368	47869	45043	43148	38927	23002	9922	4299	3220	2562	2185	1239	837
In the southern atolls.	16950	16274	13810	19206	23320	15619	20839	16299	5059	2025	1308	1153	509	64	188
In the republic.	111976	98577	93227	126411	117677	101931	89991	61294	31864	8349	5739	4725	3361	2011	1958
3. No. of fishing trips by mechanized P & L boats.														
In the northern atolls.	—	—	—	—	—	2697	14268	20532	20880	28654	29231	27052	27585	44432	50445
In the central atolls.	—	—	—	—	—	1012	4416	13800	17664	27516	31486	32549	37337	43353	63322
In the southern atolls.	—	—	—	—	—	—	774	3483	14964	18734	22417	24130	32169	29387	39788
In the republic.	—	—	—	—	—	3709	19458	37851	53508	74904	83134	83731	97091	117172	153460
4. Total no. of fishing trips (2 + 3) expressed in terms of mechanized P & L boat days.														
In the northern atolls.	55161	47046	44049	59336	49314	45861	44493	42525	37763	30679	30442	28062	28252	45140	51378
In the central atolls	39865	35257	35368	47869	45043	44160	43343	24382	27586	31815	34706	35111	39522	44592	64064
In the southern atolls	16950	16274	13810	19206	23320	15619	21613	19782	20023	20759	23725	25283	32678	29551	39976
In the republic.	111976	98577	93227	126411	117677	105640	109449	86689	85372	83253	88873	84456	100452	119183	155418

Source: Based on *Fisheries Statistics* (1970 to 1984), Ministry of Fisheries Male, The Republic of Maldives.

Note: The average figures used for converting the sailing P & L boats days to mechanized P & L boat days is obtained from Table 1.

Table 3
Data on Tuna Catch by Pole-and-Line Boats in Metric Tons (1970 to 1984), The Maldives

	1970	1971	1972	1973	1974	1975	1976	1977	1978	1979	1980	1981	1982	1983	1984
1. Skipjack Tuna															
In the northern atolls	15450	13216	8591	8785	8146	6530	9099	5543	6228	6008	8518	4950	3655	8869	11444
In the central atolls	8123	8479	5385	6556	6704	4958	4933	3316	2350	5443	6482	5942	3772	5404	10497
In the southern atolls	3495	6505	3658	3420	6911	3112	5571	5173	4971	6346	8075	9306	8266	5218	9773
In the republic	27068	28200	17634	18761	21761	14600	19603	14032	13549	17797	23075	20198	15693	19491	31714
2. Yellowfin Tuna															
In the northern atolls	1169	616	1209	3072	2150	2395	3098	2450	2140	2097	2111	2354	2276	3190	2385
In the central atolls	416	320	629	1844	1422	916	1021	1289	840	1379	1423	2025	1301	2309	3080
In the southern atolls	214	146	100	318	296	201	361	385	234	216	112	361	193	485	1428

Economics and Ecology in Agriculture and Marine Production 287
Optimal Effort in the Maldivian Fishery 25

In the republic	1799	1082	1938	5234	3868	3512	4480	4124	3214	3692	3646	4740	3770	5984	6893
3. *Other Tuna*															
In the northern atolls	2272	2407	1847	5274	4448	3220	1847	2153	1196	1111	910	843	1183	2305	1651
In the central atolls	626	561	1301	1592	1438	777	467	563	263	452	625	814	1385	1808	1784
In the southern atolls	120	100	109	149	316	44	296	213	129	39	51	42	104	69	26
In the republic	3018	3068	3257	7015	6202	4041	2610	2929	1588	1602	1586	1699	2672	4182	3461
4. *Total Tuna*															
In the northern atolls	18891	16239	11647	17131	14744	12145	14044	10146	9564	9216	11539	8147	7114	14364	15480
In the central atolls	9165	9360	7315	9992	9564	6651	6421	5168	3453	7274	8530	8781	6458	9521	15361
In the southern atolls	3829	6751	3867	3887	7523	3357	6228	5771	5334	6601	8238	9709	8563	5772	11227
In the republic	31885	32350	22829	31010	31831	22153	26693	21085	18351	23091	28307	26637	22135	29657	42068

Source: *Fisheries Statistics* (1970 to 1984), Ministry of Fisheries Male, The Republic of Maldives.

Table 4
Values of R^2, and Parameter Values for a, b, and (t) Values for the Three Specified Models for the Three Types of Tuna Caught in the northern, central, and southern Atolls and the Entire Republic of Maldives

Form of Equation	Northern Atolls R^2	a	b	Central Atolls R^2	a	b	Southern Atolls R^2	a	b	Republic of Maldives R^2	a	b
Skipjack tuna												
$C_t/f_t = a - bf_t$ (Schaefer model)	0.0152	0.166	(10^{-5}) 0.066 (.448)	0.0001	0.151	(10^{-5}) −0.0039 (.033)	0.005	0.282	(10^{-5}) −0.069 (−.265)	0.026	0.232 (−.58)	(10^{-5}) −0.036
$\operatorname{Ln}(C_t/f_t) = a - bf_t$ (Fox's model)	0.016	−1.81	0.335 (.452)	0.002	−1.99	0.121 (.147)	0.003	−1.31	−0.184 (−.19)	0.022	−1.49 (−.54)	−0.16
$\operatorname{Ln}(A - C_t) = a - bf_t$, (where $a = \operatorname{Ln} B$, and $b = K$) (S&T model)	0.406	10.24	−1.819 (−2.98)**	0.554	11.901	−9.48 (−4.02)**	0.651	10.502	−10.96 (−4.93)**	0.503	13.125	−3.83 (−3.6)**
Value of A used	21700 tons			10550 tons			10000 tons			32000 tons		

Yellowfin tuna											
$C_t/f_t = a - bf_t$	0.448	0.116	-0.145	0.001	0.032	0.006	0.296	0.0008	0.057	0.001	0.035	0.002
			(-3.2)**			(.124)			(2.33)			(.127)
$\text{Ln}(C_t/f_t) =$	0.304	-1.75	-2.97	0.004	-3.66	0.392	0.113	-4.95	2.39	0.003	-3.5	0.12
$a - bf_t$			(-2.38)*			(.237)			(1.29)			(.189)
$\text{Ln}(A - C_t) =$	0.035	7.61	-2.786	0.548	10.891	-9.38	0.553	8.153	-1.96	0.598	11.830	-3.84
$a - bf_t$			(-.69)			(-3.97)**			(-4.01)**			(-4.4)**
(where $a = \text{Ln } B$, and $b = K$).												
Value of A used	3200 tons			3100 tons			2600 tons			7000 tons		
Other tuna												
$C_t/f_t = a - bf_t$	0.355	-0.003	0.12	0.135	0.008	0.038	0.181	0.012	-0.026	0.137	0.007	0.022
			(2.67)			(1.42)			(1.695)			(1.44)
$\text{Ln}(C_t/f_t) =$	0.375	-4.06	2.27	0.143	-4.56	1.81	0.40	-3.63	-8.03	0.137	-4.28	0.687
$a - bf_t$			(2.79)			(1.47)			(-2.95)**			(1.44)
$\text{Ln}(A - C_t) =$	0.528	9.503	-1.705	0.545	8.978	-5.637	0.039	6.50	0.373	0.332	9.708	-6.33
$a - bf_t$			(-3.81)**			(-3.95)**			(.72)			(-2.54)*
(where $a = \text{Ln } B$, and $b = K$).												
Value of A used	8800 tons			2000 tons			850 tons			11900 tons		

* = significant at 5% with the correct sign.
** = significant at 1% with the correct sign.

northern atolls. Yellowfin tuna and other tuna constitutes only a small fraction of total tuna caught in the Republic of Maldives (Tables 1 and 3). This supports the view that these models do not adequately represent the relationship between catch and effort for the Maldivian tuna fishery. But for the S&T model, the parameter for effort is significant with the correct sign for skipjack tuna (the major tuna catch), yellowfin tuna (except in the northern atolls), and other tuna (except in the southern atolls). Therefore the S&T model is selected as the most appropriate model for estimation of optimum economic yield (OEY) in this situation.

Estimation of Revenue and Cost Functions for Tuna Fishery in the Maldives

Revenue Function

Since fishermen in the Maldives catch a mixture of skipjack tuna, yellowfin tuna, eastern little tuna, and frigate tuna (Christy et al. 1981; Anderson and Hafiz 1985; Sathiendrakumar and Tisdell 1986), it is important to develop a physical production function for total tuna catch and effort, rather than for individual species in order to derive a revenue function. The Republic of Maldives is an archipelago and is divided into three regions (northern atolls, central atolls, and southern atolls) by the Ministry of Fisheries in the Maldives. It is assumed in this paper that tuna catches are regionally independent. The S&T model was used for the total tuna catch and effort relationships. The results given in Table 5 show that the parameters for effort are significant with the correct sign for total tuna catch for the three regions and for the Republic of Maldives. Based on Table 5, the

Table 5

Values of R^2 and Parameter Values for a, b, and (t) Values Standard Error of Estimate (S.E.E.) and Durban Watson Statistics for the Model: $Ln(A - C_t) = Ln B - kf_t$, for the Total Tuna Caught in the northern, central, and southern Atolls and the Republic of Maldives, based on data 1970 to 1984 inclusive

Region	R^2	a = Ln B	b = k	S.E.E.	D.W.S.
Northern atolls (Value of A = 26600 tons)	0.77305	10.5329	−0.000024031 (−6.65)**	0.131851	1.243*2
Central atolls (Value of A = 21050 tons)*1	0.69385	10.3178	−0.000022352 (−4.619)**	0.146959	1.501*3
Southern atolls (Value of A = 12250 tons)*1	0.70171	10.0872	−0.000068016 (−5.385)**	0.322557	2.109*3
The republic (Value of A = 47500 tons)*1	0.72166	11.7032	−0.000017697 (−5.157)**	0.223120	1.487*3

**. (t) values significant at 1% level.
*1. Value of A is determined in the manner explained in the text.
*2. No serial correlation at 2.5% level.
*3. No serial correlation at 5% level.

tuna catch functions for the northern, central, and southern atolls and the republic are:

Northern atolls:

$$C_t = 26600 - 37528.65e^{(-.0000240031 f_t)} \tag{7}$$

The value for parameter A of 26,600 was determined in this case by selecting a figure just above the highest recorded total tuna catch in the region (18,891) (Table 3) and increasing it iteratively until the improvement in R^2 value was only 1%. Thus in this case, it is not the value of A for a maximum R^2 value. Increasing the value of A beyond 26,600 only leads to a very small increase in R^2 value. In order to obtain a maximum R^2 value, A has to be increased in this case to a value greater than the one used for the whole republic which is not logically sustainable as an actual situation.[8] Hence, our decision not to use the maximum R^2 criterion in this case. However, in all other cases, Equations 8, 9, and 10, the value of A is determined iteratively so as to maximize R^2. The results are:

Central atolls:

$$C_t = 21050 - 30300.21e^{(-.000022592 f_t)} \tag{8}$$

Southern atolls:

$$C_t = 12250 - 25372.21e^{(-.000072045 f_t)} \tag{9}$$

Republic of Maldives:

$$C_t = 47500 - 120952.11e^{(-.000017697 f_t)} \tag{10}$$

The revenue function is given by:

$$TR = C_t \cdot P \tag{11}$$

Where TR = total revenue, C_t = catch at time t, and P = average price of tuna.

Estimation of Average Price for Tuna

The price paid to tuna fishermen for their catch is generally well below the export price of tuna realized by the State Trading Organization (STO) in the international market even after accounting for the costs of fish collection and export. The STO is owned by the Maldivian government and has a monopoly on the export of tuna. The lower price paid to the tuna fishermen by the STO implies that the government extracts substantial resource rents from the tuna fishermen. This rent is used by the STO to purchase flour, rice, sugar, and baby food which is then sold to the consumer at the administrative rate (or at half the cost) (Christy et al. 1981). Furthermore, "estimates indicate that the 50% resource rent rate in Maldives is not necessarily too high, even though it may seem so" (Christy et al. 1981, 31). Our study takes for granted the present resource rent rate of 50% extracted by the government and considers economic optima (social) on this basis.

The average price paid to fishermen by STO per kg of tuna varies depending

---

[8] Our data in this case seems inadequate for estimating the "true" value of A.

on the size of tuna caught. In 1984, Maldivian fishermen received a price of Rf. 1.50 per kg for tuna weighing over 2 kg and Rf. 1.10 per kg for tuna weighing below 2 kg (Ministry of Fisheries 1984).[9] In order to estimate the average price per ton received by the tuna fishermen in the three regions and the republic, skipjack tuna, yellowfin tuna, and the other tuna was expressed as a percentage of the total tuna catch for each region. Skipjack tuna and yellowfin tuna expressed as a proportion of total tuna landed were added together and multiplied by Rf. 1500 (since the average weight of skipjack and yellowfin tuna is over 2 kg) and the other tuna expressed as a proportion of total tuna landed was multiplied by Rf. 1100 (since the average weight of other tuna is below 1 kg). The above two figures were then added together to obtain the average price per ton paid to tuna fishermen in each region. This was multiplied by 2 to give the world price (social value as per standard international trade texts) for Maldivian tuna since it is assumed (and available evidence indicates that on average this is the case) that the present resource rent rate is 50%.[10] Table 6 provides the details.

The estimated social revenue function for the three regions and the republic, based on the tuna catch functions (Equations 7 to 10) and average world price (Table 6), are:

Northern atolls:

$$TR = 2866.4[26600 - 37528.65e^{(-.00002403 1f_t)}] \qquad (12)$$

Central atolls:

$$TR = 2907.2[21050 - 30300.21e^{(-.000022592f_t)}] \qquad (13)$$

Southern atolls:

$$TR = 2983.2[12250 - 25372.21e^{(-.000072045f_t)}] \qquad (14)$$

Republic of Maldives:

$$TR = 2904.0[47500 - 120952.11e^{(-.000017697f_t)}] \qquad (15)$$

Cost Function

In the short-run the fishery cost function is of the form:

$$TC = AFC \cdot N + AVC \cdot f_t \qquad (16)$$

---

[9] The average weight of skipjack and yellowfin tuna was 2.12 kg and of little tuna and frigate tuna was 0.95 kg (Ministry of Fisheries 1984).
[10] Different world prices for the three regions and for the republic are due to different catch composition in each region.

Table 6
Percentage Catch of Skipjack tuna, Yellowfin Tuna, and Other Tuna and the Average Price per Ton of Tuna (Rf.) Paid to Tuna Fishermen and the Average World Price[1] (Social) per Ton of Tuna (Rf.) in the northern, central, and southern Atolls and the Republic of Maldives

	Northern Atolls	Central Atolls	Southern Atolls	Republic of Maldives
Skipjack tuna[2] catch as a % of total tuna catch	64.6	71.6	92.8	73.6
Yellowfin tuna[2] catch as a % of total tuna catch	18.7	16.8	05.1	14.4
Other tuna[2] catch as a % of total tuna catch	16.7	11.6	02.1	12.0
Av. price per ton paid to fishermen (Rf.)[3]	1433.20	1453.60	1491.60	1452.00
Av. world price per ton of domestic tuna[4] (Rf.)	2866.40	2907.20	2983.20	2904.00

1. Average world price for Maldivian tuna is estimated on the assumption that the present resource rent rate is 50%.
2. Based on average catch between 1970–1984 from Table 3.
3. Skipjack tuna and yellowfin tuna are priced at Rf. 1500/ton and other tuna are priced at Rf. 1100/ton.
4. Different prices in each region are due to different catch composition.

Where TC = total cost of catching tuna, AFC = average fixed cost per tuna boat, N = number of tuna boats operating (Note: N = $f_t/180$ and this should be an integer.), AVC = average variable cost per fishing trip, and f_t = total number of fishing trips. The above form of the cost function implies that the total variable cost as a function of effort as measured by the number of boat trips is a straight line (variable cost per unit of effort [per boat trip] is constant). This implies that the total variable cost per fishing trip is given by input prices and is not related to the size of the biological population, that is, cost of effort is independent of catch.

In the long-run, all costs are variable or escapable. This is also true for planning of future resource use possibilities, that is; for the consideration of alternatives for resource use. Given the conditions assumed below, long-run average cost per unit of effort in the tuna fishery in the Maldives as a whole is constant. It consists of a constant cost for operating each boat annually for a constant number of fishing trips per year (180 trips) plus an annual constant allowance for the capital cost of each boat. Hence, long-run marginal cost and long-run average cost coincide and by finding either we can estimate the level of effort needed for optimum economic yield. It is appropriate that long-run curves be used to determine the long-term economic efficiency of resource use bearing in mind that resources have alternative uses. This is the approach adopted here. It means, among other things, that allowance must be made for the cost of capital used in production.

Capital Cost

The average cost and lifespan of hull, engine, and equipment was estimated from the survey[11] conducted by one of the authors, R. Sathiendrakumar. Since actual expenditure does not measure actual social cost in many LDCs (Little and Mirrlees 1974), it is important to consider the opportunity cost of capital. The opportunity cost of capital in the Maldives was estimated at 10% and 15%. These two estimates were taken because, "using too low a rate of interest to discount social profits would lead the economy to attempt to invest too much, with inflationary effect, and too high a rate of interest could leave savings unutilized and cause excessive unemployment . . ." (Little and Mirrlees 1974, 291). Furthermore, accounting rates of interest (ARIs) between 10% and 15% are used as cut-off points for project appraisal in many LDCs. Based on these two estimates, capital is depreciated using the annuity method (Owler and Brown 1975, 208). The formula for an annuity is:

$$PR^n = A(R^n - 1)/(R - 1) \qquad (17)$$

Rearranging Equation 17:

$$A = PR^n(R - 1)/(R^n - 1) \qquad (18)$$

Where A = annual sum to be provided for replacement of the asset, P = the value of the asset, $R = 1 + (r/100)$, r = the rate of interest, and n = the number of years (life span). Table 7 gives the annual sum needed for replacement of the asset at 10% ARI and 15% ARI.

Operating Cost

The operating cost consists of fuel cost, repair and maintenance cost, labor cost, and cost of catching bait. Bait fish are first caught in the morning using dip nets in the nearby reef. This operation takes about 15 to 30 minutes (Sathiendrakumar and Tisdell 1986). The cost of catching bait is not taken into account separately but is assumed to be included in the labor cost associated with the tuna fishing boat. In the Maldives, the tuna catch is shared equally between the boat owner and the crew, and as such the labor involved in fishing does not receive a fixed wage rate, the remuneration of labor is a share of the catch. The opportunity cost of labor, the reward forgone by not undertaking the most rewarding alternative job (Tisdell 1972) is taken as the cost of labor. Three alternative levels of opportunity cost are considered.

Most LDCs have a low degree of mobility of labor and capital (Todaro 1981; Gunatilake 1973) and considerable social heterogeneity (Tisdell 1985). Furthermore, in many island economies, especially coral based islands such as the Republic of Maldives, opportunities for alternative employment within the economy are quite limited (Tisdell and Fairbairn 1983; Sathiendrakumar and Tisdell 1985).

---

[11] The survey was conducted between November 1985 and January 1986 for mechanized pole-and-line boat owners in two fishing islands (Velidhoo and Holudhoo in Noonu atoll).

Table 7
Cost of Operating a Mechanized Pole-and-Line Boat for Three Alternative Levels of Opportunity Cost for Labor and for Two Alternative Rates of Interest (ARI) for Capital

1) *Capital Cost* Item	Av. Cost[1] in Rf.	Life Span[2] in Years	Annual Sum to be Provided[3] for Replacement at 10% ARI	15% ARI
Hull	50937.50	25	5618.53	7889.67
Engine	20750.00	10	4719.60	5778.31
Equipment	5500.00	05	3169.05	3383.14
Total capital[4] cost			13510.00	17050.00
2. Operating cost (for 180 trips)				
2a) With zero opportunity cost of labor				
Fuel cost			18720.00	18720.00
Repair and Maintenance cost			2730.00	2730.00
2a(i) Total operating cost			21450.00	21450.00
2b) With Rf. 10 as opportunity cost per[5] unit of labor				
Labour cost (10*10*180)			18000.00	18000.00
Other cost [2a(i)]			21450.00	21450.00
2b(i) Total operating cost			39450.00	39450.00
2c) With Rf. 15 as opportunity cost[6] per unit of labor				
Labour cost (15*10*180)			27000.00	27000.00
Other cost [2a(i)]			21450.00	21450.00
2c(i) Total operating cost			48450.00	48450.00
3. Total cost				
3a. Total cost (with zero opp. cost of labor) [1 + 2a(i)]			34960.00	38500.00
Av. cost = Marginal cost			194.20	213.90
3b. Total cost (with Rf. 10 as opp. cost per unit of labor), [1 + 2b(i)]			52960.00	56500.00
Av. cost = Marginal cost			294.20	313.90
3c) Total cost (with Rf. 15 as opp. cost per unit of labor) [1 + 2c(i)]			61960.00	65500.00
Av. cost = Marginal cost			344.20	363.90

1. Based on the survey conducted by R. Sathiendrakumar in Velidhoo and Holudhoo in Noonu atoll, between Nov. 1985 and Jan. 1986.
2. Source: Ministry of Fisheries, Male.
3. Only capital cost is depreciated at 10% and 15% ARI.
4. Error due to rounding up.
5. See Note 1.
6. 50% higher than the survey figure.

Alternative employment is usually unavailable in the vicinity of the fishing community (Munasinghe et al. 1980) and even when present, the educational levels and skills of fishermen often do not meet the requirement of such employment.

One possibility is to assume that if fishermen are displaced from fishing they will have no other alternative job to go to. The opportunity cost of labor may then be assumed to be zero. The other option is to take the casual wage rate per day (Rf. 10) as the opportunity cost of labor.[12] For comparative analytical purposes, a third opportunity cost equal to 1.5 times the casual wage rate is also considered, even though it is quite likely that this figure is an excessive value for the opportunity cost for the fishermen.[13] Since ten fishermen on an average operate each pole-and-line boat for 180 days per year, the opportunity cost per unit of labor in operating a pole-and-line boat was taken as Rf. 0, Rf. 1800, and Rf. 2700 for the three alternatives just mentioned.

Table 7 gives alternative cost estimates, namely the total cost per year and the average cost per fishing trip incurred in operating a mechanized pole-and-line boat for zero opportunity cost of labor, for Rf. 1800 as annual opportunity cost of labor, for Rf. 2700 as annual opportunity cost of labor, and for the two ARIs for capital mentioned before. These three variable cost levels enable us to estimate the sensitivity of optimum economic yield level of effort for changes in cost levels.

Estimation of Optimum (Social) Economic Yield Level of Effort (OEY) and Optimum Price to be Paid to Tuna Fishermen and Open Access-Equilibrium Yield Level of Effort (OAEY)

The theory of OEY has been covered in many fisheries economics and management books (Christy and Scott 1965; Gulland 1974; Anderson 1977; Waugh 1984) and thus this paper will not go into the details of this. As is well known and as explained graphically by Anderson (1977, Ch. 3) the open-access equilibrium will usually be economically inefficient. An excessive number of boats from an economic point of view can be expected to operate in an open-access equilibrium.

OEY Level of Effort

The OEY level of effort (from a social economic efficiency viewpoint) corresponds to the level for which marginal social revenue product of effort (MSR) is equal to social marginal cost (MC). From the total revenue function given in Equations 12 to 15 as based on world tuna prices rather than the prices paid by STO, the

---

[12] The survey conducted by R. Sathiendrakumar in late 1985 and early 1986 reveals that the casual wage rate in the northern atolls of the Republic of Maldives is about Rf. 10. This figure is used for all three regions and the republic.

[13] Even though limited job opportunities are available due to the expanding tourist industry and building and construction industries in the republic, these sectors are unlikely to provide substantial alternative job opportunities for fishermen because they have very low levels of education. Nevertheless, one referee has suggested that a higher opportunity figure than the casual wage rate of Rf. 10 may be relevant.

[14] $R = e^{(-af)}$. Therefore, dR/df = marginal function = $-ae^{(-af)}$, (Chiang 1967, 293).

MSR productivity functions[14] for the tuna effort in the northern, central, southern atolls and for the Republic of Maldives are given below:
Northern atolls:

$$dTR/df_t = 2866.4[37528.65(.000024031)e^{(-.000024031f_t)}] \quad (19)$$

Central atolls:

$$dTR/df_t = 2907.2[30300.21(.000022592)e^{(-.000022592f_t)}] \quad (20)$$

Southern atolls:

$$dTR/df_t = 2983.2[25372.21(.000072045)e^{(-.000072045f_t)}] \quad (21)$$

Republic of Maldives:

$$dTR/df_t = 2904.0[120952.11(.000017697)e^{(-.000017697f_t)}] \quad (22)$$

The MSRs given in Equations 19 to 22 were each equated to the three levels of MC (which each equal average total costs in this case and which are assumed to correspond with social MC). Because dTR/df_t is declining throughout and MC is constant for all of our cases, the second-order conditions for a maximum are satisfied when the first-order conditions are met. The results for the three regions and for the Republic of Maldives are given in Table 8.

Table 8 also shows the sensitivity of the OEY level of effort to changes in MC levels brought about as a result of variations in the opportunity cost of labor and the rate of interest. Sensitivity is indicated by variations in the optimal number of boats and the optimum number of pole-and-line boat trips. Even though MC was changed for trial purposes by 87% for the northern atolls, central atolls, southern atolls and for the Republic of Maldives, the variation required in number of mechanized pole-and-line boats to achieve OEY level of effort was only 24% for the northern atolls, 27% for the central atolls, 19% for the southern atolls, and 18% for the republic.

Optimum Price to be Paid to the Tuna Fishermen by STO to Ensure Maximum Economic Efficiency

In order to maximize economic efficiency of resource use, the STO should pay a price to the tuna fishermen such that the optimum social economic yield level of effort (that is at world prices) is equal to the open access equilibrium level of effort at the STO price. This will enable a social optimal allocation of effort in the Maldives. Thus the STO optimum price condition will be of the functional form:

$$P^*[A - Be^{(-kf_t^*)}] = AC \cdot f_t^* \quad (23)$$

Rearranging:

$$P^* = AC \cdot f_t^*/[A - Be^{(-kf_t^*)}] \quad (24)$$

Table 8
Optimum Effort Levels at World Price[1] (Social) in Number of Mechanized Pole-and-Line Fishing Trips and in Number of Mechanized Pole-and-Line Boats and the Optimum Total Profit (Social) (in Rf.) and the Optimum Average Profit per Boat[2] (in Rf.) for the northern, central, and southern Atolls and for the Republic of Maldives

Region	Op. Cost of Labor per Day in Rf.	ARI for Capital (%)	Av. Cost = M.C in Rf.	No. of P&L Boat Trips	No. of Boats[3]	Total (Rf. 000)	Av.[2] (Rf.)
Northern atolls	0.0	10	194.20	107720	598	47245	79000
	0.0	15	213.90	103699	576	45164	78410
	10.0	10	294.20	90435	502	37397	74500
	10.0	15	313.90	87738	487	35643	73190
	15.0	10	344.20	83903	466	33043	70910
	15.0	15	363.90	81587	453	31413	69340
Central atolls	0.0	10	194.20	103003	572	32597	56990
	0.0	15	213.90	98726	548	30611	55860
	10.0	10	294.20	84617	470	23279	49530
	10.0	15	313.90	81748	454	21641	47670
	15.0	10	344.20	77669	431	19277	44610
	15.0	10	363.90	75206	417	17721	42500
Southern atolls	0.0	10	194.20	46291	257	24859	96730
	0.0	15	213.90	44950	249	23960	96220
	10.0	10	294.20	40525	225	20538	91280
	10.0	15	313.90	39626	220	19748	89760
	15.0	10	344.20	38347	213	18567	87170
	15.0	15	363.90	37574	208	17819	85670
Republic	0.0	10	194.20	195852	1088	88932	81740
	0.0	15	213.90	190392	1057	85128	80540
	10.0	10	294.20	172380	957	70601	73770
	10.0	15	313.90	168718	937	67242	71760
	15.0	10	344.20	163511	908	62209	68510
	15.0	15	363.90	160366	890	59020	66310

1. World price is taken as twice the present STO price.
2. Optimum average profit per boat is obtained by dividing optimum total profit by optimum number of boats.
3. The average number of trips per boat is taken as 180 per year.

Where f_t^* = optimum social economic yield level of effort (with world price), AC = average cost per unit of effort, and P* = optimum STO price which will make the optimum social economic yield level of effort equal to the private open access equilibrium yield level of effort.

Figure 3 can be used to illustrate the procedure so as to be representative of

Optimal Effort in the Maldivian Fishery

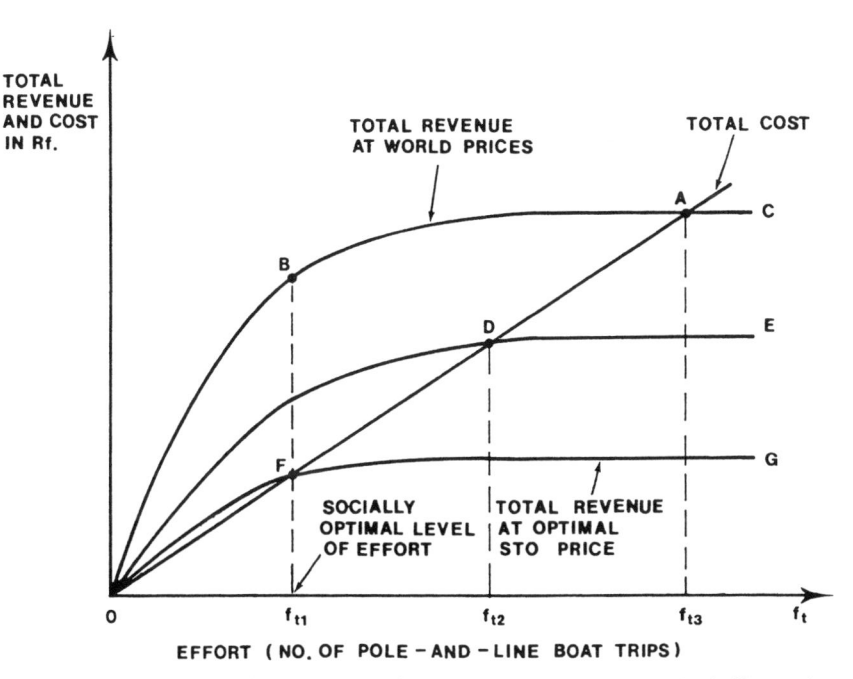

Figure 3. Determining the optimal STO price so as to give optimal level of effort under conditions of common-access.

the Maldivian situation.[15] Line ODA represents the total cost of fishing effort and curve OBC represents the total revenue to be earned at world prices as a result of fishing effort. In the absence of intervention by the STO, the tuna fishing industry would reach long-term open-access equilibrium at point A with an effort of f_{t3}. In order to maximize the social profit (i.e., the difference between curve OBAC and line ODA) for the Maldives, however, effort needs to be only of the order of f_{t1}. In a long-term open-access situation, this may be achieved if the price paid by STO to fishermen for tuna is reduced sufficiently below the world price to cause the total revenue curve applying to the fishermen to intersect ODA for f_{t1}. This would be the case if curve OFG applies. The problem is to find (solve mathematically for) the optimum STO price, P*. The data below indicate that the STO price is in excess of P* but lower than the world price so that in practice total revenue curves like ODE are faced by fishermen as a whole. A curve of this type leads to long-term open-access equilibrium at D, corresponding to f_{t2} of effort. The level of effort is higher than socially optimal but less than would occur in the absence of STO intervention and closer to the social optimum.

The above therefore establishes a procedure for finding the socially optimal STO price. Looked at differently, the above provides a procedure for determining the optimal tax to be imposed through the STO on tuna traded at world prices.

---

[15] The curves can be specified empirically for the various situations given our data. The actual P* values are specified in Table 9.

Table 9
Optimum STO Price (in Rf. per ton) to be Paid to Tuna to Make the Optimum Economic Yield Level of Effort with World Price (Social) the Open Access Equilibrium Level of Effort (Private) for the northern, central, and southern Atolls and for the Republic of Maldives

Region	Op. Cost of Labor per Day in Rf.	ARI for (%)	Av. Cost = M.C in Rf.	OEY Level of Effort	Optimum Price for Tuna (Rf. per ton)	Present STO Price[1] for Tuna (Rf. per ton)	Required Effective Tax on World Price for Optimum Resource Allocation (%)
Northern	0.0	10	194.20	107720	879.67	1433.20	69.3
atolls	0.0	15	213.90	103699	944.09		67.1
	10.0	10	294.20	90435	1191.55		58.4
	10.0	15	313.90	87738	1249.42		56.4
	15.0	10	344.20	83903	1336.82		53.4
	15.0	15	363.90	81587	1392.76		51.4
Central	0.0	10	194.20	103003	1105.56	1453.60	62.0
atolls	0.0	15	213.90	98726	1186.83		59.2
	10.0	10	294.20	84617	1502.32		48.3
	10.0	15	313.90	81748	1577.12		45.6
	15.0	10	344.20	77669	1691.01		41.8
	15.0	10	363.90	75206	1764.57		39.3
Southern	0.0	10	194.20	46291	792.30	1491.60	73.4
atolls	0.0	15	213.90	44950	854.29		71.4
	10.0	10	294.20	40525	1095.71		63.3
	10.0	15	313.90	39626	1152.85		61.4
	15.0	10	344.20	38347	1239.52		58.4
	15.0	15	363.90	37574	1295.20		56.6
Republic	0.0	10	194.20	195852	869.93	1452.00	70.0
	0.0	15	213.90	190392	939.71		67.6
	10.0	10	294.20	172380	1213.97		58.2
	10.0	15	313.90	168718	1279.49		55.9
	15.0	10	344.20	163511	1379.34		52.5
	15.0	15	363.90	160366	1443.80		50.3

1. Actual effective tax is assumed as 50%. Thus the present world price for Maldivian tuna is double the present STO price.

The above analysis can of course be also illustrated in terms of per-unit curves, that is, average total revenue (and marginal revenue) per unit of effort and average and marginal cost per unit of effort. Actual f_t levels for the situation illustrated in Figure 3 are specified in Tables 10 and 11.

Table 9 gives the estimated optimum STO price that needs to be paid to the tuna fishermen to make the optimum social economic yield level of effort equal to the private open-access equilibrium yield level of effort. According to Table

Table 10

The Open Access Equilibrium Effort Level in Number of Mechanized Pole-and-Line Fishing Trips and in Number of Mechanized Boats (for the six average costs given in Table 7) for the northern, central, and southern Atolls and for the Republic of Maldives for the Present STO Price and World Price[1]

Region	Op. Cost of Labor per Day in Rf.	ARI for Capital (%)	Av. Cost = M.C in Rf.	No. of Mechanized P&L Boat Days STO Price	World Price	No. of Boats[2] STO Price	World Price
Northern	0.0	10	194.20	193671	392573	1076	2181
atolls	0.0	15	213.90	174426	356361	969	1979
	10.0	10	294.20	119145	258430	662	1435
	10.0	15	313.90	108953	241875	605	1343
	15.0	10	344.20	94711	219934	526	1221
	15.0	15	363.90	86090	207506	478	1152
Central	0.0	10	194.20	149886	314751	832	1748
atolls[3]	0.0	15	213.90	132800	285447	737	1585
	10.0	10	294.20	78717	205100	437	1139
	10.0	15	313.90	65607	191223	364	1062
	15.0	10	344.20	—	172611	—	959
	15.0	10	363.90	—	161929	—	899
Southern	0.0	10	194.20	93863	188177	521	1045
atolls	0.0	15	213.90	85037	170845	472	949
	10.0	10	294.20	60456	124182	335	690
	10.0	15	313.90	56090	116364	311	646
	15.0	10	344.20	50112	106065	278	589
	15.0	15	363.90	46586	100272	258	557
Republic	0.0	10	194.20	353411	710292	1963	3946
	0.0	15	213.90	319567	644862	1775	3582
	10.0	10	294.20	222870	468565	1238	2603
	10.0	15	313.90	204800	438966	1137	2438
	15.0	10	344.20	178835	399893	993	2221
	15.0	15	363.90	162159	377856	900	2099

1. World price is taken as twice the present STO price.
2. The average number of trips per boat is taken as 180 per year.
3. In the central atolls with Rf. 15 as opportunity cost of labor the total cost was above total revenue with the STO price.

9, except in the central atolls where the present price is below the optimum price when the opportunity cost of labor is Rf. 10 per day, in all the other atolls and in the Republic the present price paid to the fishermen by the STO is higher than the optimum price. This means that a greater level of effort than is socially optimal is encouraged in all but the central atolls, and this is confirmed by the evidence that follows.

Table 11
Present Number of Mechanized Pole-and-Line Boats (in 1984), Number Required for Open Access Yield (OEY) with Present STO Price (Private) and Number Required for Optimum Economic Yield (OEY) with World Price[1] (Social) (for the six average costs given in Table 7) for the northern, central, and southern Atolls and for the Republic of Maldives

Region	Op. Cost of Labor per Day in Rf.	ARI for Capital (%)	Av. Cost = M.C in Rf.	No. of Boats[2] Present in 1984	OAEY Level of Effort with Present STO Price	OEY Level of Effort with Present World Price
Northern	0.0	10	194.20	453	1076	598
atolls	0.0	15	213.90		969	576
	10.0	10	294.20		662	502
	10.0	15	313.90		605	487
	15.0	10	344.20		526	466
	15.0	15	363.90		478	453
Central	0.0	10	194.20	567	832	572
atolls[3]	0.0	15	213.90		737	548
	10.0	10	294.20		437	470
	10.0	15	313.90		364	454
	15.0	10	344.20		—	431
	15.0	10	363.90		—	417
Southern	0.0	10	194.20	276	521	257
atolls	0.0	15	213.90		472	249
	10.0	10	294.20		335	225
	10.0	15	313.90		311	220
	15.0	10	344.20		278	213
	15.0	15	363.90		258	208
Republic	0.0	10	194.20	1296	1963	1088
	0.0	15	213.90		1775	1057
	10.0	10	294.20		1238	957
	10.0	15	313.90		1137	937
	15.0	10	344.20		993	908
	15.0	15	363.90		900	890

1. World price is taken as twice the present STO price.
2. The average number of trips per boat is taken as 180 per year.
3. In the central atolls with Rf. 15 as opportunity cost of labor the total cost was above total revenue with the STO price.

OAEY Level of Effort

OAEY is obtained when total revenue is equal to total cost, that is,

$$TR = P(A - Be^{(-kf_t)}) = TC \tag{25}$$

Where TR = total revenue function and TC = total cost function given in Equation 16.

Since Equation 25 is a transcedental equation, the solution for f_t (effort) cannot be obtained in a closed form, so it was solved numerically by using Newton's method (Gerald 1978). The results are given in Table 10. This table specifies open-access equilibrium-yield levels of effort in terms of number of boats for the northern, central, and southern atolls and for the Republic of Maldives for the present STO price (present private OAEY) and present world price (present social OAEY).

Table 11 compares the present stock of mechanized pole-and-line boats of the Maldives with the number required for private open access equilibrium yield (OAEY) at the present STO price and for optimum social economic yield (OEY). Any number of pole-and-line boats fishing for tuna above the optimal social level is defined as over-exploitation.

Conclusion

Since the local Maldivian tuna fishery is confined to 25 km from the atoll reef, the Maldivian tuna fishery fishes only a fraction of the total geographic area occupied by the species in their migration. Therefore, in the Maldivian context, it seems inappropriate to assume that changes in CPUE reflect actual changes in abundance. But it might be safer to assume that changes in CPUE reflect changes in migration or distributional behavior of the fish.

Thus the Schaefer model and the Fox model, used by Anderson and Hafiz (1985) for the Maldivian tuna fishery, are not appropriate models for representing the relationship between tuna catch and effort in the Maldivian context. When applied both show very low R^2 values and parameter values that are not statistically significant on the whole. However, these models could become applicable if Maldivian fishing effort becomes more widespread so as to encompass a significant section of ocean covered by the tuna stock in their migration. In the future, this might be achieved by the Maldivian tuna fishing fleet or by the combined effort of other fishing fleets operating in that region. But in the present context a production function type of model (e.g., the S&T model) seems more appropriate to represent the relationship between tuna catch and effort in the Maldives. The R^2 values for the S&T model are much higher and estimates of parameter values are statistically more significant than for either the Schaefer model or the Fox model. It might be mentioned once again that, except for the northern atolls, values of R^2 in the S&T model were relatively insensitive to the chosen value of A so the results in the S&T model are not an artifact of the procedure for estimating parameter A.

Having selected the S&T model as the appropriate model to represent the relationship between catch and effort for the Maldivian tuna fishery, the economic evaluation carried out shows that the number of boats present in the northern, central, and southern atolls and in the Republic of Maldives is below the present private open access equilibrium level when the opportunity cost of labor is zero for the two ARIs of 10% and 15% (Table 11). As can be seen from the second to last column of Table 11, in the central atolls and in the Republic of the Maldives as a whole, the existing number of boats exceeds the estimated equilibrium open access number at the present STO price if the opportunity cost of labor is Rf. 10 per day or more, and ARI is 10% or more. This suggests that in the central atolls and in the Republic of Maldives the opportunity cost of labor is below Rf. 10 or that some of the boats present in these areas are being used for other activities like cargo transport in spite of their being registered as fishing boats. If the latter is the case, then this affects the results since only a part of the cost can be allocated to fishing.

The present price paid to fishermen by STO for tuna is, given the present world price of tuna, slightly higher than the price required to maximize social profit, except in the central atolls for opportunity cost of labor of Rf. 10 and above. But in spite of this, the present pricing policy by the STO has prevented the fisheries reaching the open access equilibrium yield level of effort (Table 10) where profits would be completely dissipated. Thus; although the main purpose for which the pricing policy was adopted by the Maldives was to subsidize imported staple commodities to consumers from the resource rent extracted from the fisheries, it has prevented to a considerable extent socially wasteful employment and investment in the fisheries sector within the 25 km range, even though it has not completely prevented some excess employment and investment.

There is also the need in assessing the adequacy of the policies of the STO to take account of the fact that politically economic efficiency is only one possible objective. In addition, world prices of tuna are subject to fluctuations and uncertainty. In the light of this, it might be argued that the policies of the STO have been reasonably effective in preventing dissipation of the resource rent from the Maldivian tuna fishery even though this appears to have been a side-benefit rather than avowed political purpose for imposing an effective tax on the export of tuna.

Acknowledgments

The authors wish to thank the Ministry of Fisheries, Maldives, and especially Mr. Jameel for assistance with data and for facilitating R. Sathiendrakumar's recent field work in the Maldives, and two anonymous referees of this journal for suggestions that were useful in revising an earlier draft of this paper. The usual *caveat* applies.

References

Anderson, L. G. 1977. *The Economics of Fisheries Management*. Baltimore: The John Hopkins University Press.

Anderson, R C., and A. Hafiz. 1985. *The State of Maldivian Tuna Stock: Analysis of Catch and Effort Data and Estimation of Maximum Sustainable Yield*. Male, The Republic of Maldives: Ministry of Fisheries.

Beverton, R. J. H., and S. J. Holt. 1957. On the Dynamics of Exploited Fish Population. *Fisheries Investigation*, London, Ser. 2 (19).

Chiang, A. 1967. *Fundamental Methods of Mathematical Economics*. New York: McGraw Hill.

Christy, F. T. Jr., and A. Scott. 1965. *The Common Wealth in Ocean Fisheries*. Baltimore: The Johns Hopkins University Press.

Christy, F. T. Jr., L. C. Christy, P. W. Allen, and R. Nair. 1981. *Management of Fisheries in EEZ, Maldives*. Rome: FAO, F1:GCP/INP/334/NOR.

Comitini, S., and S. Hardjolukito. 1984. Economic Benefits and Costs of Alternative Arrangements for Tuna Fisheries Development in the Exclusive Economic Zone: The

Case of Indonesia. Paper presented at the seminar on Marine Resource Economics: Southeast Asian Seas in Transition, April 24-28, 1984, Environment and Policy Institute, East-West Center, Hawaii.
Crowe, A., and A. Crowe. 1969. *Mathematics for Biologist*. London: Academic Press.
Cushing, D. H. 1972. A History of Some of the International Fisheries Commissions. *Proc. Roy. Soc. Edinburgh*, Sect. B., 73:361-390.
Fox, W. W. Jr. 1970. An Exponential Surplus Yield Model for Optimizing Exploited Fish Population. *Am. Fisheries Soc. Trans.* 100:80-88.
Gerald, C. F. 1980. *Applied Numerical Analysis*. 2d ed. Cambridge, Mass.: Addison-Wesley Publishing Company.
Gulland, J. A. 1974. *Management of Marine Fisheries*. Seattle: The University of Washington Press.
Gulland, J. A., ed. 1977. *Fish Population Dynamics*. London: John Wiley and Sons.
Gunatilleke, G. 1973. The Urban-Rural Balance and Development: The Experience of Sri Lanka. *Marga Quarterly Journal* 2(1);35-68.
Howe, C. W. 1979. *Natural Resource Economics: Issues, Analysis, and Policy*. New York: John Wiley and Sons.
Industries Assistance Commission. 1984. *Report Southern Bluefin Tuna*. Canberra: Australian Government Publication Service.
Johannes, R. E. 1978. Traditional Marine Conservation in Oceania and Their Demise. *Annual Review of Ecology and Systematics* 9:349-364.
Little, I. M. D., and J. A. Mirrlees. 1974. *Project Appraisal and Planning for Developing Countries*. London: Heinemann Educational Books.
Ministry of Fisheries. 1970-1984. *Fishery Statistics, 1979 to 1984*. Male, Maldives: The Republic of Maldives.
Ministry of Planning and Development. 1985. *Republic of Maldives: National Development Plan, 1985-1987*. Male, The Republic of Maldives.
Munasinghe, H., K. P. N. S. Karunagoda, W. Gamage, and S. Fernando. 1980. Socio-Economic Conditions of Small-Scale Fisheries in Sri Lanka. Marga Institute Doc., SEM/100/80(3), Oct. 1980.
Nichols, E. H. 1985. *International Arrangement and Manpower Resources in Relation to Fisheries Management in the EEZ (Maldives)*. Rome: FAO, F1:GCP/INT/398(NOR).
O'Rourke, D. 1971. Economic Potential of the California Trawl Fishery. *American Journal of Agricultural Economics* 53:583-592.
Owler, L. W. J., and J. L. Brown. 1975. *Wheldon's Cost Accounting and Costing Methods*. 13th ed. London: Macdonald and Evans Ltd.
Rothschild, B. J., and A. Suda. 1977. Population Dynamics of Tuna. In *Fish Population Dynamics. See* Gulland 1977.
Sathiendrakumar, R., and C. A. Tisdell. 1985. Tourism and the Development of the Maldives. *Massey Journal of Asian and Pacific Business* 1(1):27-34.
Sathiendrakumar, R., and C. A. Tisdell. 1986. Fishery Resources and Policies in the Maldives: Trends and Issues for an Island Developing Country. *Marine Policy* 10(4):279-293.
Schaefer, M. B. 1954. Some Aspects of the Dynamics of Populations Important to the Management of the Commercial Marine Fisheries. *Inter American Tropical Tuna Commission Bulletin* 11(6).
Tisdell, C. A. 1972. *Microeconomics: The Theory of Economic Allocation*. Sydney, Australia: John Wiley and Sons.
Tisdell, C. A. 1983. Economic Problems in Managing Australian Marine Resources. *Economic Analysis and Policy* 13(2):113-141.
Tisdell, C. A. 1985. The World Conservation Strategy, Economic Policies and Sustainable Resource Use in Developing Countries. *Environmental Professional* 7:102-107.
Tisdell, C. A., and T. I. Fairbairn. 1983. Development Problem and Planning in a Resource

Poor Pacific Country: The Case of Tuvalu. *Public Administration and Development* 3:341–359.

Todaro, M. P. 1981. *Economic Development in Third World.* 2d ed. London: Longmans.

Treschev, A. I. 1978. Application of the Fished Volume Method for Measuring Fishing Effort. *Co-operative Research Report No. 79.* Copenhagen: ICES.

Waugh, G. 1984. *Fisheries Management: Theoretical Developments and Contemporary Applications.* Boulder, Colorado: Westview Press.

DEVELOPMENT OF AQUACULTURE AND THE ENVIRONMENT: COASTAL CONFLICTS, AND GIANT CLAM FARMING AS A CASE

CLEM TISDELL

Department of Economics, University of Queensland, Brisbane 4072 (Australia)

(Received in final form: January 25, 1991)

The paper considers coastal land-use conflicts and the development of mariculture in Australia, compatibility or otherwise of global aquacultural development with conservation and environmental goals, and environmental consequences of giant clam farming as a case. The Brundtland Report favours the expansion of aquaculture, but takes insufficient account of the environmental consequences of such development. While the development of farming, including seafarming, can help to save farmed species from extinction, it is not without serious consequences for the natural environment and may well reduce overall biodiversity.

KEY WORDS: Aquaculture, coastal conflicts, seafarming, environment.

INTRODUCTION

Economic development and conservation of the *coastal zone* appear rarely to be completely compatible, despite the fact that the Brundtland Report claims that the development of aquaculture is likely to be environmentally benign on the whole. Use of the coastal area for the development of aquaculture can have a number of adverse environmental effects which are outlined in this paper. In Australia, aquaculture is relatively underdeveloped and a recent government report, the McKinnon Report, sees scope for its expansion. The Brundtland Report prepared for the United Nations is also optimistic about the global prospects of expanding supplies of fishery products by means of aquaculture. But environmental constraints on and production sustainability problems for mariculture are largely unrecognised in these reports. Giant clam farming is considered to be an especially "environmentally friendly" form of aquaculture but even in this case some adverse environmental impacts of economic development occur. Giant clam farming is a new mariculture possibility, developed in response to dwindling wild stocks of giant clams. Most species of giant clams are now listed in the Convention on International Trade in Endangered Species (CITES). Giant clam farming has a number of desirable environmental features and self-sustaining properties compared to other forms of mariculture, but that is not to say that it is without some adverse environmental consequences. A decision, therefore, has to be made about whether the economic benefit from farming giant clams can be expected to outweigh any adverse environmental consequences. Account must also be taken of the fact that the environmental impact of mariculture of giant clams, like many other forms of aquaculture, varies with the techniques adopted for their cultivation. Giant clam farming is unusual in that its environmental and conservation benefits appear to be large.

The use of the coastal zone involves considerable social conflict and this conflict has intensified with economic growth. Costs in terms of opportunities forgone of one development rather than other possible developments or of no development at all have become greater. This has happened throughout the whole world as levels of population and income have risen, as new technology has expanded the number of possible uses for the coastal zone and provided improved access to it, e.g. via better transport. In addition, utilization of the coast for recreational purposes has become increasingly popular (an historical change of taste) and individuals now have more leisure and income to take advantage of its recreational opportunities. In Australia, as in many other countries, increasing demand to employ the coastal zone for outdoor recreation (and as a venue for indoor recreation too) comes not only from local inhabitants but from foreign tourists as well. Proposals for the use of the coastal zone for aquaculture including new forms of aquaculture such as Pacific Giant Clam mariculture need to be assessed against this background.

The Queensland Government has recognised Coastal Management as an area requiring greater and more efficient government attention and has suggested in its Green Paper, *Review of estuarine and coastal management procedures in Queensland*,[1] new administrative procedures for coastal and estuarine management. It suggests that in the past demands to use the coastal area could be settled independently. But this is no longer workable: "Increasing population, growing economic activity and greater recreational opportunities now mean that in some coastal areas, close to major population centres, there is a high level of competition for available resources".[2]

Views differ about the extent to which mariculture (aquaculture in marine or brackish waters) is compatible with the conservation of nature, with the protection of the environment, and with alternative uses of the coastal area. Virtually no economic development is compatible with leaving Nature undisturbed. To that extent economic development and conservation of the natural environment are incompatible, and it is nonsense (or at the very best wishful thinking) to suggest otherwise. But some forms of economic development including some forms of mariculture are more conducive to conservation and environmental goals than others, as will become apparent from the discussion below. Also it should be noted that conservation as a goal is not a straightforward goal. For example, some environmental changes may favour some species but be detrimental to others. Then one has to evaluate whether there is a satisfactory change overall in nature preservation or not, taking into account the relative value of species favoured and of those disadvantaged. The answer may not be clear-cut because of the difficulties of valuation and differences in valuation by different members of the community.

DEVELOPMENT OF MARICULTURE IN AUSTRALIA AND IN QUEENSLAND

Aquaculture is relatively underdeveloped in Australia compared to other regions of the world. By far the greatest producer of aquaculture products in the world is Asia, followed by Europe and then North America.[3] It is estimated that approximately 40% of world aquaculture production by weight is accounted for by marine products.[4] But in Australia possibly more than 90 per cent of all aquaculture production is from mariculture. In part, this is a reflection of the relative lack of freshwater in Australia compared to those continents in which aquaculture production is concentrated.

In order of estimated value of production at the farm gate, the estimated relative importance of mariculture industries in Australia in 1987–88 were (1) Pearl oysters ($53 M), (2) Table oysters ($39.6 M), (3) Salmon and trout ($6.7 M), (4) Microalgae ($2.8 M), (5) Prawns (shrimp) ($1.0 M), (6) Barramundi ($0.8 M), (7) Mussels ($0.8 M), (8) Crabs ($0.4 M), (9) Brine shrimp ($0.1 M), (10) Macroalgae (seaweed) (negligible) and (11) Other molluscs (negligible).[5] Considerable expansion in the short term has been predicted for the value of Australian mariculture production, with expansion being especially rapid in production of prawns, barramundi, salmon and trout. One study[6] predicted a doubling of the value of mariculture production in Australia in a period of two years compared to production in 1987–88, with approximately a ten-fold expansion in prawn production, an eight-fold rise in salmon and trout output and five-fold growth in the value of barramundi production. Such an expansion will undoubtedly mean that those in the Australian mariculture industry will want to make greater use of coastal and estuarine areas for aquaculture. While expansion in salmon and trout farming will on the whole be confined to Tasmania, much of the expansion in prawn production, barramundi farming and production of other molluscs, such as giant clams, is likely to be concentrated in Queensland and require land and water space. In particular, Queensland will need to give continuing attention to the environmental impact and economic consequences of mariculture of crustaceans in the State since Queensland accounts for about 80 per cent of Australian commercial activity concerned with aquacultured crustaceans (principally penaeid prawns)[7] and an extremely large expansion in activity is predicted.

While the recent McKinnon Report[8] on wealth from Australia's ocean resources was enthusiastic about the prospects for expanding mariculture in Australia, it also pointed to a number of difficulties:

"overseas competitors have many years of experience, which is lacking in Australia;

a relatively small database of basic biological data about species which could be cultured [especially native species];

use conflicts, when operators and/or government attempt to alienate space for aquaculture operations the over-complex regulatory environment in Australian government agencies, not yet adjusted to the needs of aquaculture; and

the low level of awareness of international market factors."[9]

Although the Report brings attention to the conflict problem over coastal space and suggests the need for action by the States to ease complexity of regulations adverse to aquaculture, as well as the need for governments to take positive action to promote the development of aquaculture, it says little about the possible environmental impact of mariculture and appropriate methods for resolving conflict about use of coastal space.

AQUACULTURE AND GLOBAL CONSERVATION GOALS

Given worldwide concern about the environment, especially the environmental future and the sustainability of economic development, the General Assembly of the United Nations established the World Commission on Environment and Development to, amongst other things, propose long-term environmental strategies for achieving sustainable development and to consider ways and means by which the international community can deal more effectively with environmental concerns. The recommendations of this Commission were published in *Our Common Future*.[10] One of its recommendations is that in order to improve food security, "the expansion of aquaculture should be given high priority in developing and developed countries."[11] It argues that the capture fisheries will be unable to meet increasing demand. Supply from marine fisheries by early next century is likely to be well short of expected demand and most of the naturally available freshwater fish stocks are already fully exploited or damaged by pollution. Therefore, there is a gap to be filled by aquaculture.

At the present time about 10% of all the world's fisheries products are supplied by aquaculture. Approximately 7% of fresh fish supplies, 75% of mollusc production and 75% of seaweed supplies are obtained from aquaculture.[12] Asia is the main producer of fish products by aquaculture and within Asia, China and Japan stand out as being major producers, but other Asian countries are also substantial producers.

While there may be scope for substantially expanding the world's supply of living aquatic products through aquaculture, especially mariculture, it is surprising that *Our Common Future* does not bring attention to the environmental difficulties of doing this and, in certain circumstances, the adverse consequences of aquaculture development for food supplies, conservation and the sustainability of production as well as the distribution of income. The consequences depend upon the type of product grown by aquaculture, the techniques used to grow it and the location in which it is grown. Different techniques are liable to have different environmental spillovers. Economic or conservation opportunities forgone as a result of the development can be expected to vary according to the location in which aquaculture is placed. Some types of mariculture developments, such as those associated with particular types of prawn (shrimp) production, involve low labour-intensities and high capital plus land ratios, and tend to increase inequality of income, at least initially, in communities in less developed countries. Furthermore, they can have serious adverse effects on conservation, the environment and on the sustainability of production.

SOME ENVIRONMENTAL AND SUSTAINABILITY EFFECTS OF MARICULTURE

Pullin[13] divides environmental impact of Third-World aquaculture systems into three sets:

1. *Extensive*. Those requiring no feed or fertilizer inputs.
2. *Semi-intensive*. Having some feed and/or fertilizer inputs but not mainly reliant on these.
3. *Intensive*. Mainly reliant on external food inputs.

Most of Australia's mariculture production is based upon extensive techniques (e.g. pearl and table oysters) but more intensive forms of mariculture such as salmon, prawn and barramundi production are developing.

Coastal and/or brackish water ponds for shrimp and prawns and for fish production, e.g. mullet, sea bass or barramundi, can have the following adverse environmental impacts: first, destruction of natural ecosystems, especially mangroves; second, salinization/acidification of soils/aquifers and third, the release of effluents/drainage high in biological oxygen demand (BOD) and suspended solids. Mariculture using sticks, rafts, pens, cages, etc., e.g. oysters, seaweed, some fish species, may (a) present a navigational hazard, (b) be incompatible with use of the area for recreational purposes and for fishing, (c) have an adverse visual impact, and (d) exclude wild species or lead to destruction of wild species because of habitat change, e.g. seaweed culture which occupies pristine coral reefs.

Brackish water ponds in estuarine areas generally result in destruction of native habitat and a loss of breeding and hatchery areas for natural fish stocks, and may result in the destruction of mangroves, which are valuable timber resources in many developing countries, and loss of detritus and natural food sources for wild fish stocks. In some developing countries the "seed" for mariculture is captured from the wild, but as mariculture expands it can reduce the availability of wild seed. This for example has happened in Ecuador. The construction of new ponds for shrimp not only demands more postlarvae but reduces their availability from the wild. Despite an approximate doubling of effort to catch postlarvae in the wild to satisfy the increased demand for seed as a result of expansion of the area of cultivation of shrimps, the number of semilla (postlarvae) caught has barely risen.[14]

Furthermore, inequality in incomes may rise as a result of pond culture in estuarine or coastal areas. Areas which were previously common property become alienated and villagers are likely to be denied free access to them. Furthermore, areas alienated in some countries may be distributed according to privilege as appears to be the case in Ecuador.[15] Again, those reliant on fishing resources downstream of estuaries and along the coast near the estuarine outlets may find that their available wild supplies of fish species are reduced both due to lower availability of food for these species and less recruitment of stock. So the income of coastal fishermen and gatherers can be reduced severely. This has, for example, occurred in parts of Los Negros island in the Philippines due to prawn cultivation.[16]

There can be no hard and fast rule about the optimal form of property ownership from an economic viewpoint. We cannot, for instance, show from a conservation and sustainable development point of view that private ownership is always superior to common property. Where there are few spillovers from mariculture, as in the case of oyster production, private property may be more conducive to sustainability of production and the preservation of oysters than common access. But in other cases, such as pond culture in estuarine areas, common property may be more supportive of *overall* production because of its favourable conservation impact. This *may* be so for common property which involves common access (*res nullius*) or for that which involves communal management (*res communis*).

Sometimes mariculture is seen as a means of preserving a particular species. This, for example, has been claimed to be a benefit of turtle farming[17] and also of giant clam farming. However, some conservationists argue that farming leads to an expansion in demand for the product and increases harvesting pressure on wild stocks,[18] so reducing their numbers. Nevertheless, economically viable farming provides a strong incentive mechanism for the preservation of the farmed stock of the species. Still, farming does tend to result in a reduction in genetic diversity, that is in the number of species and varieties.

This may occur for any of the following reasons:

(1) Farmers concentrate on those varieties and species which are commercially most attractive at the time or are expected to be so in the not too distant future. Those varieties and species which are less profitable to farm are therefore neglected or discarded and tend to disappear. This has happened with many agricultural crops.[19]

(2) Farming systems tend to create uniform managed environments. By reducing diversity of environments, they give rise to reduced variety of habitats and thereby tend to lower genetic diversity.

(3) Farming by continually disturbing natural habitats leads to displacement of wild species and varieties.

(4) Farming by creating artificial managed environments of a relatively uniform nature may reduce diversity arising from evolutionary processes which rely on diversity of natural conditions and which, under natural conditions, involve co-evolution.[20]

(5) Where a species is farmed the areas most suitable for farming it are also likely to be those areas which are most suitable for wild populations of the species. Thus wild populations, which often show considerable diversity, tend to be replaced by farmed populations, because the areas in which they occur are converted to farming.

(6) Farming may result in an expanded demand for the product farmed and this expansion of demand *may* result in increased harvesting of wild stocks and result in the loss of some varieties or species in the wild.[21]

(7) Farming often results in the artificial introduction of species and varieties to new areas where they frequently displace indigenous varieties which have evolved in relative isolation. For example, the introduction of new terrestrial mammals to Australia appears to have resulted in the loss of some marsupials via competition so that globally the number of species has been reduced. Giant clams only exist naturally in the Pacific and Indian Oceans but recently they have been introduced to the Caribbean. What effect will they have on species diversity in this region? In addition, giant clams are being translocated in their natural range as farming develops. Whether or not this genetic mixing will lead to reduced diversity is at present unknown.

Thus, although farming may result in a species being saved from extinction, at the same time it is likely to lead to reduced genetic diversity unless controlled. The situation is certainly complicated.

Another environmental aspect of aquaculture which needs to be considered is its susceptibility to water pollution. Filter feeders in particular are very susceptible to water pollution. This makes it difficult to sustain viable production from molluscs in areas where industrial pollution or pollution from other sources is increasing with economic growth. Indeed, some molluscs, such as oysters, are even susceptible to releases from anti-fouling paints used on the hulls of some boats. Let us consider environmental aspects of the farming of giant clams, molluscs for which mariculture techniques have recently been developed.

GIANT CLAM FARMING AS A NEW MARICULTURE DEVELOPMENT

Pacific giant clams, sometimes called "killer" clams, belong to the family *Tridacnidae* and differ from several other marine molluscs which are also called clams. They are confined in their natural distribution to tropical and semi-tropical waters of the Indian Ocean and the western side of the Pacific Ocean but not all species occur throughout that range. Indeed most species are only naturally present in a limited band fanning out from southeast Asia to the Solomon Islands, Papua New Guinea and northern Australia. All currently recognised species, except *Hippopus porcellanus*, occur in northern Australia. The habitats of northern Queensland in particular seem to be very suitable for them.[22] The family is especially well adapted to tropical coral environments, requires warm water, good penetration of light of the water and an absence of disturbance by freshwater. They therefore tend to thrive on tropical coral atolls.

Populations of giant clams have been greatly depleted throughout their natural range (with the possible exception of Australia) as a result of harvesting by man. This is a consequence of greater demand as a result of human population increase and of new technologies which have made it easier to harvest the species and transport products from it over longer distances for consumers. In particular, Taiwanese operators have taken advantage of these technological developments but so too have some Australian operators in the Pacific islands. Technological developments include improved diving equipment, better ocean vessels, refrigeration and in some cases freeze-drying techniques. But apart from increased pressure on giant clams for commercial markets, greater harvesting pressure has also come from gathering of clams for subsistence food by indigenous peoples. As a result most species of clams are considered to be endangered and many have become locally extinct and the area in which they have become extinct is increasing. Hence, giant clams have been listed under CITES, the Convention on International Trade in Endangered Species.[23] The aim of such a listing is to reduce the possibilities for international trade which draws on natural stocks, lower the demand for their harvest and so help to maintain natural stock. However, not all countries are signatories to CITES, some (such as Japan) excluded a number of items listed under CITES and there are a number of loopholes and "dishonest" practices that may be used to circumvent CITES restrictions on international trade in products from endangered species.

Possibly a more effective way for the conservation of giant clams has been discovered. Methods to breed them in captivity and to farm them have been developed in recent years. All species of giant clams have now been bred in captivity. The most recent species to be bred in captivity is *Hippopus porcellanus*, the China Clam, which is the rarest clam of all and very much sought after for the shell trade.

Techniques for the mariculture of giant clams have been developed mainly at two centres: (1) The Micronesian Mariculture Demonstration Center (MMDC) at Palau with a considerable amount of financial assistance from the United States and (2) James Cook University (Department of Zoology) with the financial support of the Australian Centre for International Agricultural Research (ACIAR). But techniques are also being developed elsewhere, e.g. in the Philippines, Fiji, Solomon Islands (ICLARM) Japan and Papua New Guinea.

Basically three phases are involved in the mariculture of giant clams: (1) the hatchery phase, (2) ocean nursery phase and (3) grow-out phase. In the hatchery, which is typically located on the ocean foreshore, clams are bred and their progeny reared in saltwater tanks. At about nine months of age, the seed clams are then transferred to a position in the ocean where they are protected by some type of covering (e.g. plastic mesh) from predators. This is the ocean nursery phase. At about three years of age (at this time *Tridacna gigas* are about 20 cm across) the clams can be moved to an unprotected ocean situation to commence their grow-out phase.

It is not possible here to assess all the environmental and conservation impacts of giant clam culture. However, the culture appears in many respects to be less environmentally damaging than many other forms of mariculture and it has appealing self-sustainability properties. Giant clams do not require to be fed or fertilized (indeed, much of the clam's food is manufactured by an algae which lives in its mantle and which provides the animal with its brilliant colours). Using the classification suggested by Pullin,[24] the cultivation of giant clams is extensive rather than intensive. Closed breeding cycles have been established for the species which has been reared in captivity for the longest period, namely *Tridacna derasa*, and it appears likely that closed breeding cycles will be established for all species. Thus clam farming does not require continuing capture of broodstock from the wild or the taking of seed from the wild.

However, hatcheries do compete for scarce space when located on the foreshores of coastal areas, and ocean nurseries and grow-outs located on intertidal areas such as rock shelves can have unfavourable visual impacts and crowd out coral and other species. While ocean nurseries and grow-out areas located in the intertidal zone can be expected to be *more economic than those located in subtidal benthic areas*, subtidal benthic locations have the advantage of resulting in less adverse visual impact and may have a smaller adverse impact on other species.

Rack culture is also a possibility. Clams are raised off the sea floor by the racks. The suspension of clams from floating rafts is a fourth possibility. Both of the last mentioned forms of cultivation may interfere with navigation. All types of clam cultivation may interfere with the use of the ocean for recreation, e.g. swimming and boating, except subtidal benthic cultivation.

In protected areas intertidal ocean-nursery culture appears to be more economic than subtidal benthic. At least at Orpheus Island near Townsville, available evidence is that intertidal culture involves easier construction of covers, easier maintenance, less fouling of covers and better growth and survival rates for giant clams than alternative methods.[25] But in exposed situations, damage may occur to protective covers and growth and survival rates of clams may be lower. One can therefore envisage circumstances in which the more economic methods for giant clam cultivation are those which are not so satisfactory from an environmental point of view. The question then arises of the extent to which environmental values should be forgone for economic gain.

Economists have argued that environmental protection or conservation is not an absolute good in itself.[26] They suggest that the economic benefits of a development have to be compared to its environmental costs. A development is, according to this point of view, justified if the net economic benefits from it outweigh its estimated environmental costs. Furthermore, the socially optimal technique to use is that which gives the highest returns after an allowance for environmental costs are subtracted from private returns. Thus, even if intertidal culture of giant clams involves greater environmental damage than the alternative of subtidal, intertidal culture may still be justified in economic terms in some localities, if it is sufficiently more profitable. This is not to deny that there may be difficulties in measuring environmental costs and we also need to be aware that partial evaluation can lead us to overlook global problems.[27] In the case of giant clam mariculture, much more research will be needed before economic benefits and environmental costs can be quantified with accuracy.

CONCLUDING COMMENTS

For many, the oceans are our last frontier for development. Man has conquered the land and has subdued Nature upon it, and is cultivating and husbanding a major portion of it. Man's domestication of sea creatures and plants and man's control over ocean resources lags far behind his control over land. While some international bodies[28] and national bodies[29] see great potential for the economic development of ocean resources to support economic growth, such development will not be without environmental costs[30] and ill-considered ocean development may well be unsustainable.

On the other hand, there is scope for beneficial development of ocean resources including expansion of mariculture, both in the world as a whole and in Australia in particular. Unnecessary government impediments to such development need to be pruned away since they will restrict business enterprise *some* of which may be in the general interest. However, the problem is complex. Some government control must remain since the environmental impacts of ocean developments have to be taken into account and the competing demands of others to use the coast must be considered in dedicating any part to a particular use.[31]

Acknowledgements

I wish to thank the Australian Centre for International Agricultural Research (ACIAR) for its financial support, Carunia Firdausy and René Wittenberg for research assistance and Dr Darrel Doessel for his comments on an earlier draft of this paper. Research for this paper has been partially funded by an ACIAR Research Grant, Economics of giant clam mariculture (Project 8823).

References

1. Premier's Department, *Review of Estuarine and Coastal Management Procedures in Queensland*, Green Paper (Author, Brisbane, 1989).
2. *Ibid.* p. 4.
3. C. E. Nash, "A global overview of aquaculture production" *J. World Aquaculture Society* **19**(2), 51–60 (1988).
4. UN Food and Agriculture Organization, *Yearbook of Fishing Statistics, 60.* (FAO, Rome 1987).
5. Department of Science, Industry and Technology, *Oceans of Wealth?* A Report by the Review Committee of Marine Industries, Science and Technology (R. M. McKinnon, Chairman) (Australian Government Publishing Service, Canberra 1989).
6. C. Garland, "The structure and financial value of the Australian mariculture industry" In: *Proceedings of Maritime Australia 88*, Occasional Papers in Maritime Affairs No. 5 (Australian Centre for Maritime Studies, Canberra, 1989) pp. 105–114.
7. Department of Science, Industry and Technology, *op. cit.* (1989) p. 27.
8. Department of Science, Industry and Technology, *op. cit.* (1989).
9. *Ibid.* p. 29.
10. World Commission on Environment and Development *Our Common Future* (Oxford University Press, Oxford, 1987).
11. *Ibid.* p. 138.
12. C. E. Nash, "Aquaculture attracts increasing share of development aid" *Fishing Farming International* **14**(6), 22–23 (1987).
13. R. S. V. Pullin, "Third-world aquaculture and the environment" *NAGA: The ICLARM Quart.* **12**(1), 10–13 (1989).
14. S. K. Meltzoff and E. LiPuma, "The social and political economy of coastal zone management: Shrimp mariculture in Ecuador" *Coastal Zone Management J.* **14**(4), 349–380 (1986).
15. *Ibid.* pp. 349–380.
16. Personal observations (1987).
17. C. A. Tisdell, "Conflicts about living marine resources in southeast Asia and Australian waters: Turtles and dugong as cases" *Marine Resource Economics*, **3**(11), 89–108 (1986).
18. See Tisdell, *op. cit.* (1986).
19. M. L. Oldfield, *The Value of Conserving Genetic Resources* (Sinauer Associates, Sunderland, Mass., 1989); C. A. Tisdell and M. Alauddin, "New crop varieties: impact on diversification and stability of yields" *Australian Economic Papers* **18**(52), 123–140 (1989).
20. M. L. Oldfield, *op. cit.* (1989).
21. C. A. Tisdell, *op. cit.* (1986).
22. J. S. Lucas, "Giant clams: description, distribution and life history" In: (J. W. Copland and J. S. Lucas, Eds.) *Giant Clams in Asia and the Pacific* (Australian Centre for International Agricultural Research, Canberra, 1988) pp. 31–32.
23. S. Lyster, *International Wildlife Law* (Grotius, Cambridge, 1985).
24. R. S. V. Pullin, *op. cit.* (1989).
25. J. S. Lucas, R. D. Braley, C. M. Crawford and W. J. Nash, "Selecting optimum conditions for ocean-nursery culture of *Tridacna gigas*" In: (J. W. Copland and J. S. Lucas, Eds.) *Giant Clams in Asia and the Pacific* (Australian Centre for International Agricultural Research, Canberra, 1988) pp. 129–144.

26. C. A. Tisdell, *Natural Resources, Growth and Development: Economics, Ecology and Resource-Scarcity* (Praeger, New York, 1990).
27. C. A. Tisdell and J. M. Broadus, "Policy issues related to the establishment and management of marine resources" *Coastal Management* **17**, 37–53 (1989).
28. World Commission on the Environment and Development, *op. cit.* (1987).
29. Department of Science, Industry and Technology, *op. cit.* (1989).
30. R. D. Braley, W. J. Nash, J. S. Lucas and C. M. Crawford, "Comparison of different hatchery and nursery culture methods for the Giant Clam *Tridacna gigas*" In: (J. W. Copland and J. S. Lucas, Eds.) *Giant Clams in Asia and the Pacific* (Australian Centre for Agricultural Research, Canberra, 1988) pp. 110–114.
31. Since the above was first written, a new species of giant clam, *Tridacna trevoro*, the devil clam, has been identified. It occurs in relatively deep water in the Fiji/Tonga area. See J. S. Lucas, E. Leava and R. D. Braley, "A new species of giant clam (*Tridacnidae*) from Fiji and Tonga" *ACIAR Working Paper No. 33* (Australian Centre for International Agricultural Research, 1990).

The Cost of Production of Giant Clam Seed *Tridacna gigas*

CLEMENT A. TISDELL,[1] WILLIAM R. THOMAS AND
LUCA TACCONI

Department of Economics, University of Queensland 4072 Australia

JOHN S. LUCAS

Department of Zoology, James Cook University, Townsville, Queensland 4811 Australia

Abstract

The costs of providing giant clam seed in Australia are examined for alternative annual volumes of production. Considerable economies of scale in production are available, both in relation to labor costs and non-labor costs (mostly capital costs). The fall in per-unit cost of producing giant clam seed is considerable when annual production is expanded from 100,000 to 500,000 seed clams per year. At 10% rate of interest, the unit cost per clam seed falls from $1.43-$2.04 at a production level of 100,000 to $0.41-$0.55 at a production level of 500,000. Per-unit operating costs also fall. They decline from $1.01-$1.22 to $0.29-$0.35. This suggests that there are likely to be cost economies in having large centralized hatcheries.

Considerable effort has been expended on research directed towards commercialization of giant clam *Tridacna* mariculture (Braley 1989). Already several commercial or semi-commercial giant clam farms exist, e.g., in northern Australia, Palau, the Marshall Islands and American Samoa. However, the pioneering industry is still a newly emerging one struggling to establish an economic niche for itself.

Seven species of giant clam, all but the recently discovered new species *Tridacna tevoroa*, have been successfully raised from eggs to juveniles (Pernetta 1987) and beyond. Most effort, however, has been directed towards the mariculture of *Tridacna derasa* and *Tridacna gigas*. A closed breeding cycle has been established for *Tridacna derasa* at the Micronesian Mariculture Demonstration Centre (MMDC) at Palau (Heslinga and Fitt 1987). Because work on *Tridacna gigas* started later than that for *Tridacna derasa* and because *Tridacna gigas* takes several years longer to sexually mature than *Tridacna derasa*, a closed breeding cycle for *Tridacna gigas* has not been completed. It seems likely, however,

the James Cook University Research Station at Orpheus Island (JCUOIRS) will have a closed breeding cycle established for *Tridacna gigas* by the mid-1990s when their first batch of cultured giant clams mature. Production commenced by relying on captured sexually mature wild clams which formed captive broodstock population. However, it is to be expected, as experience at MMDC indicates, that the industry will come to rely on cultured broodstock (Heslinga and Fitt 1987). This reliance should add little to nursery costs given the fecundity of clams.

The work on culturing giant clams has been concentrated on *Tridacna gigas* and *Tridacna derasa* for two reasons. Firstly, these two species are considered to be in danger of extinction in many parts of their range (Munro 1989). Secondly, these two species are the largest and have the fastest growth rates; therefore, they are considered to have the greatest potential for mariculture by biologists.

This paper examines the costs of producing *Tridacna gigas* seed (i.e., juveniles with a shell length of 2–3 cm) based on Australian experience and cost structures using the current techniques most commonly employed in Australia. To some extent the re-

---
[1] Corresponding author.

sults are Australian specific but not entirely so. All prices and costs are in Australian dollars.

Lucas (1988) considered that *Tridacna gigas* reached a shell length of 2–3 cm in about one year. At this size the clams are ready for transferral to ocean based nurseries. There are a number of reasons why seed-size of 2–3 cm should be used as a demarcation point in determining production costs for giant clams. At this seed-size the clams can be transferred to ocean nurseries, although continued growth in shore-based facilities is also a possibility (Munro 1989). If clam seed is to be sold as a product, as occurs in Palau for ocean-staged growth in other areas (Heslinga and Fitt 1987), then transportation costs must be considered. Clams at the next stage of culture when they are ready for unprotected ocean growout are much bigger (approximately 125 times heavier). Transportation of clams as seed stock will generally occur in the range of 2–3 cm because of the much lower transport costs involved. In the case of international sales, quarantine restrictions might make the shipment of the larger clams for regrowth impractical.

Note that it is also possible to transport veligers of giant clams by air transport prior to their settlement. This has been successfully done in some cases, e.g., by Coastal Aquaculture Centre, Solomon Islands, but it is not yet an established practice. At present the normal practice is to transport seed of 2–3 cm in size.

There are two other important reasons why seed-size (2–3 cm) is a good point for a demarcation in production costs. First the conditions optimal for both shore-based hatchery facilities and ocean-based growth facilities may not occur together. This is because a hatchery requires easy access to clean, clear sea water, whereas for an ocean-based ocean nursery a wide intertidal area is necessary if the intertidal method of growout is adopted. Clearly the wider the intertidal area the harder it will be to pump sea water to the hatchery. Crawford et al. (1988) considered intertidal ocean nursery growout to be superior to other methods they have tried. Secondly, factors such as availability of technical staff, power supplies and economies of scale may mean that importation of giant clam seed from centralized hatcheries is a sounder economic proposition for clam growers than growing their own seed. As well as possibly being more economical, centralized hatcheries may be advantageous from the point of view of conserving natural giant clam stocks. A proliferation of small hatcheries could lead to increased competition for already limited natural broodstock.

Materials and Methods

The successful economic production of a giant clam seed will depend heavily on the appropriate selection of a site for a giant clam hatchery. Hatcheries using extensive techniques, which rely heavily on favorable environmental conditions, will be much more vulnerable to the vagaries of nature than those hatcheries using intensive techniques where culture conditions are, to a greater extent, controlled. A venture is, however, more likely to be successful if environmental conditions are favorable.

Besides environmental considerations, other factors such as the availability of labor, power and supplies must be taken into consideration. A hatchery on a mainland would usually be less expensive to run in terms of labor and transportation costs, and the costs of constructing hatchery facilities are likely to be less. However, mainland hatcheries may not be viable due to factors such as poor water quality. Clearly a trade off between all the various environmental and logistical factors must be made. This of course would not be easy and would depend very heavily on the expertise of those involved in the project. If production of mature clams at the same site was being considered, then further environmental conditions such as suitability of the intertidal zone would need to be taken into account if intertidal growout is to be used.

Since the culture of giant clams becomes more extensive as the clams increase in size, the environmental factors must be more fully respected if ocean growth phases are envisioned.

Ideally the area chosen for a site will have a stable water temperature of around 23–30 C and a salinity of 33–35 ppt (Heslinga et al. 1986). Such conditions prevail around equatorial islands in the Pacific; however, access to power, supplies and skilled labor could be a problem. In order to minimize salt water pumping costs, the site should have easy access to clear unpolluted water. The wider the intertidal zone the greater the distance the salt water will have to be pumped. The increased pipe length will add to pipe-laying expenses, as well as increase the frictional forces on the water being pumped and lead to greater fuel usage. The capital and operating costs of the salt water system will depend on both the volume of water to be pumped and the circumstances under which it must be pumped.

Once a site is chosen all the necessary permits required by the various government agencies (depending on the country) must be obtained. For example, to establish a hatchery on an island in the Great Barrier Reef, Australia, a permit has to be obtained from the Great Barrier Reef Marine Park Authority at a cost of a few hundred dollars per year.

The cost of construction of the hatchery will depend heavily on the location of the site. The construction costs of a hatchery on a remote uninhabited island will be much greater than the construction on a mainland close to urban amenities.

The costs of employee accommodation are even more dependent on location than those relating to the hatchery. If the hatchery is located in an isolated area where daily commuting is impractical then full living quarters and amenities will have to be provided for the workers. In the case of a hatchery located within commuting range for the employees such construction would not be necessary. Freshwater needs of humans will usually have to be met by rainfall. Therefore, large rainwater tanks will be required in most instances to store collected water.

The location of a hatchery is an important factor in determining labor costs. For instance, it is usually necessary to pay higher wages to employees of an isolated hatchery than one closer to a city.

It is particularly vital to the operation of the hatchery that a continuous supply of good quality sea water be maintained. In terms of equipment this is best achieved by designing excess capacity into the system. This allows the equipment to be run at speeds that are best suited to the equipment's smooth running. It is also imperative that backup equipment be installed ready for immediate use should the normal equipment fail. Although having surplus capacity and backup equipment increases running costs, it considerably reduces the risk of complete failure in the production process.

The cost of the equipment needed to supply sea water to the hatchery facility will depend on the quantity of water required, and the distance and height that the water must be pumped. The cost will also depend on whether new or secondhand pumps and generators are used. For the purpose of this paper, however, only prices of new equipment will be considered. The location is also important because the cost of laying the pipes will depend on the terrain.

At present fiberglass tanks, concrete tanks and tanks with galvanized iron walls and plastic liners are used for culturing giant clam seed. The relative merits of each type of tank depend on the circumstances peculiar to a given application at a particular location. For example, concrete tanks cost much the same as fiberglass tanks but last much longer and are the cheapest to operate over a long period of time (20–50 yr), but they are difficult to reposition once in place. Fiberglass tanks, by contrast, have considerable mobility. A 10,000 L fiberglass tank costs around $1,400 which makes it a much more expensive capital purchase than the cheap "splasher pools" used at JCUOIRS. Splasher pools are produced commercially in Australia as inexpensive backyard swimming pools. Their sides consist of galvanized iron, usually with small corrugates for added strength. These are set in a surrounding light steel frame and are normally of round or oval shape. A plastic (polythene) liner is placed on the inside of this construction to hold the water. Their height may vary but a common height is about one meter or slightly higher and their diameter can be easily varied. However the short lifespan (1–2 yr) of a 10,000 L splasher pool and liner costing around $270 makes them a less attractive long-term proposition than fiberglass tanks.

The commercial giant clam growers near Cairns have favored fiberglass, while workers at JCUOIRS appear to favor low cost splasher pools. A big advantage of the splasher pools is that they are very convenient for shipping, e.g., to Pacific Islands, since the galvanized iron sides roll up and the plastic liner folds up. A big disadvantage of the galvanized tanks is that they are quite delicate once set up and do not react favorably to careless handling. Fiberglass tanks are also vulnerable to careless handling. Although concrete tanks would appear to be the best long-term proposition, until such time as methods have been refined and the giant clam industry is proven to be economically viable, it is likely that commercial growers will continue to use fiberglass tanks because such tanks are easily repositioned and do not have to be replaced often. The cheap splasher pools do, however, seem well suited to short-term pilot studies where capital costs must be kept low. They allow for flexible adjustment for uncertainty.

It is necessary to facilitate drainage of the tanks by either raising the tanks above the surface level or by excavating drainage passages beside the tanks. JCUOIRS has found raising the tanks above the surface by about 0.5 m to be convenient. The tanks can be raised by numerous methods, however, whichever method is used allowance must be made for the cost of materials and labor used. Estimates given in Table 1 are for the costs of using besser (concrete) bricks and concrete to retain the fill or line the drainage passages.

Large nursery tanks and the large algal culture tanks are kept in the open and generally require some limited protection from the elements. Generally shade cloth supported by galvanized tubing is used for this purpose. The shade cloth does not adversely affect the growth of young clams as might be expected because of the reduction of light available to the symbiotic zooxanthellae contained within the clam's tissues. "Zooxanthellae in small (1–2 cm long) juvenile *Tridacna gigas*, for example, photosynthesize at near maximal rates even at ambient light intensities as low as 500 μ/E/m^2/s, about one-fourth that of natural above-water noontime intensity on a clear day...." (Heslinga and Fitt 1987, p. 335).

The cost of materials alone necessary to provide weather protection to sufficient tanks to culture 500,000 seed clam (i.e., 30 tanks of 4 m diameter) would be about $11,000. The lifespan of the shade cloth would be about 5 yr and tubing would have a life expectancy of about 10 yr.

Algal overgrowth and nutrient enrichment of water are considered together because nutrient enrichment adds to the problem of algal overgrowth. Algal overgrowth has been one of the hardest problems for giant clam growers to solve. It has been found both at the JCUOIRS and "Reefarm" (a commercial farm in North Queensland, Australia), that the addition of nitrogenous fertilizers, such as calcium nitrate, to the water greatly enhances the growth rate of the clams. The cost of nitrogenous supplements for 500,000 seed clams would be around $5,000–$10,000 depending on the fertilizer used.

Nutrient enrichment, however, exacerbates algal overgrowth which adversely affects the juvenile clams by shading them out. Algal grazing snails *Trochus niloticus* have been used to combat the problem of

TABLE 1. *Estimated capital costs associated with giant clam seed production (Australian dollars).*

Production target	100,000 seed	500,000 seed	1,000,000 seed
Hatchery[a]	45,000–100,000	55,000–120,000	60,000–120,000
Workers accommodation and amenities[a]	3 workers 20,000–120,000	4 workers 25,000–120,000	5 workers 30,000–130,000
Seawater supply system	50,000–80,000	50,000–80,000	50,000–80,000
Fiberglass nursery tanks[a] 10,000 L ($1,400 each)	7,000	35,000	70,000
Beser Bricks and cement for tank bases	2,000	10,000	20,000
Hatchery fiberglass tanks 500 L ($300 each)	900	3,000	4,500
Broodstock tanks 10,000 L ($1,400 each)	1,400	2,800	4,200
Tractor	20,000	20,000	20,000
Utility truck	15,000	15,000	15,000
Boats[a]	5,000–50,000	5,000–60,000	5,000–65,000
Shade cloth	625	2,640	5,000
Galvanized piping and fittings for shade cloth	2,000	8,300	16,000
Laboratory equipment	4,000	6,000	8,000
Microscopes (2)	2,500	2,500	2,500
Glassware	1,000	2,000	3,000
Plastic piping and fittings	2,000	5,000	8,000
Diving equipment	3,000	3,000	3,000
Air blowers	3,000	6,000	9,000
Miscellaneous items	10,000	15,000	25,000
Total	194,425–424,425	271,240–516,240	356,800–606,800

[a] Depends very heavily on local circumstances of the hatchery. See text for a more complete explanation.

algal overgrowth (Heslinga and Fitt 1987). These snails, however, do not always manage to keep the algae under control, and also pollute the water with fecal waste. Philippe Dor and Bruce Stevens of the commercial hatchery, Reefarm, believe that they have found a solution to the problem of filamentous algal overgrowth. They have found that by introducing colonies of amphipods *Ampithoe* spp. into the tanks, the growth of filamentous algae is very much restricted. The amphipods are detrital feeders and apparently dislodge the filamentous algae in their efforts to obtain the detritus on which they feed. The filamentous algae then floats off and passes through the tank outlet which is positioned at the water surface.

Most of the other major expenses depend heavily on the particular circumstances of a hatchery. Expenses incurred on the acquisition and maintenance of boats could be very considerable if the hatchery was very isolated and required its own boat for transporting staff and supplies. On the other hand, a hatchery situated on the mainland, which had good road access, could keep boat expenses to a minimum and get by with just small dinghies and occasionally hiring larger vessels when required for collecting broodstock. One item of major expense that would be required in the majority of hatcheries would be a small agricultural tractor. A tractor could be useful to help clear areas to be used for tanks. It would also be needed for moving the broodstock from the sea to the brood tanks when spawning was to be included and for harvesting clams in the intertidal area. The price of the tractor quoted in Table 1 refers to a new one; however, a second-hand tractor would generally be quite satisfactory and would cost considerably less. Another major expense would be the cost of a utility truck for transporting supplies and equipment.

Other minor expenses include items such as microscopes, glassware and antibiotics. The cost of lease permits are included under Miscellaneous Items which also accounts for items which had not been originally considered. Cost underestimates may result from uncounted components and services or increased requirements (Huguenin and Colt 1989).

After obtaining the necessary permits, foundation *gigas* broodstock are collected from reefs and transferred to holding areas in close proximity to the hatcheries. The broodstock are generally kept subtidally as this restricts unauthorized access to them. During the period when clams become ready for spawning (October–March in Australia) some of the broodstock are transferred to shore facilities and kept in a tank. Spawning can be artificially induced; however, it is not clear whether or not such induction leads to the release of less viable eggs (Fitt et al. 1984). Eggs are collected in plastic bags as they are expelled from the spawning clam. The eggs are then placed in small (50 L) containers and fertilized with sperm obtained from another clam. The developing eggs sink to the bottom and are siphoned up and transferred to larger tanks (500 L). Within 24 h at 28–30 C the fertilized eggs hatch into ciliated trochophore larvae. It has been found that trochophore survival is greatest at densities of less than five per milliliter (Fitt et al. 1984). This means that a 500 L container can accommodate 2,500,000 trochophores comfortably. Sufficient 500 L containers would be required to accept as many eggs as was required since a large *Tridacna gigas* can release hundreds of millions of eggs in one day (Heslinga and Fitt 1987).

The trochophores develop into bivalved veligers within 2 d after fertilization at 28–30 C and can now be placed into large nursery tanks. At 3 d post-fertilization, the veligers are fed cultured unicellular algae. They are fed until metamorphosis occurs at about 9 d post-fertilization. Zooxanthellae are introduced at 7 d post-fertilization. The metamorphosed clams appear to obtain all their energy requirements from their symbionts once the symbiosis is well established. The metamorphosed giant clams can be kept in the large outdoor tanks until they are ready for sale to other growers or for transfer into an ocean nursery. Depending on the survival and growth rate, the young giant clams may need to be thinned out into other tanks prior to sale or transfer. At JCUOIRS up to 50,000 six month old clams have been produced from one 10,000 L tank. Generally (barring total failure) a range of 10,000 to 50,000 can be expected. It should be possible to keep at least 20,000 one-year old seed clams in the larger 10,000 L tanks (surface area 16 m^2); therefore, 25 tanks would be required to produce 500,000 one-year old seed clams.

Results

Tables 1 and 2 present cost estimates based on the methods used by JCUOIRS and Reefarm. The estimates for the cost of building hatchery facilities are based largely on a quote obtained by Richard Braley for the Quantity Surveying Section, Ministry of Works, Cook Islands. Apart from the hatchery the cost estimates shown for capital expenditure have been calculated by the authors by estimating material requirements for the various production targets; prices for these materials were obtained from local suppliers. With regard to operating costs, labor and fuel costs are based largely on extrapolations of information obtained from JCUOIRS and Reefarm. The cost estimates for repairs and maintenance are derived by adopting the usual accounting rates.

These cost estimates are summarized in Table 3. Figs. 1 and 2 illustrate the relationship between the unit cost of production of giant clam seed and the volume of production when the rate of interest is 10%. Fig. 1 was constructed on the assumption that the minimum cost estimates prevail and Fig. 2 on the basis that maximum cost estimates prevail. Cost curves of the nature suggested by these estimates have been

TABLE 2. *Estimated annual operating costs associated with giant clam seed production (Australian dollars).*

Production target	100,000 seed	500,000 seed	1,000,000 seed
Labor ($30,000 per employee)	(3) 90,000	(4) 120,000	(5) 150,000
Fuel	5,000–10,000	15,000–20,000	20,000–35,000
Repairs and maintenance			
pumps and generators	1,000–4,000	2,000–8,000	3,000–10,000
boats	2,000–15,000	3,000–20,000	4,000–25,000
glassware & laboratory consumables	1,000	2,000	3,000
tractor	2,000	2,500	3,000
(a) Total operating cost	101,000–122,000	144,500–172,500	183,000–226,000
Operating cost per clam	1.01–1.22	0.29–0.35	0.183–0.226
(b) Depreciation of worker accommodation and amenities at 2.5% pa	500–3,000	625–3,000	750–3,250
(c) Depreciation at 10% pa allowed for all capital items	17,443–30,443	24,624–39,624	32,680–47,680
(d) Interest charges[a] assuming a real rate interest of 10%	24,493–48,543	34,349–60,249	44,830–71,980
Total (a + b + c + d)	142,936–203,986	204,098–275,373	261,260–348,910
Total cost per clam[b]	1.43–2.04	0.41–0.55	0.26–0.35
If real rate of interest of 5% is assumed: Interest charges	12,246–24,271	17,175–30,125	22,415–35,990
Total	131,389–179,714	186,924–245,249	238,845–312,920
Total cost per clam	1.31–1.79	0.37–0.49	0.24–0.31

[a] Assumption is made that all funds used for capital and operating costs are borrowed, and that operating costs expended progressively through the year.

[b] Lower and upper limits of these costs are graphed in Figs. 1 and 2 respectively.

drawn in freehand. They show significant economies of scale in production of seed. For other rates of interest, e.g., 5%, similar curves showing economies of scale can be derived.

Table 3 indicates that at an output of 100,000 clam seed per year, at the very minimum a price of $1.00 per clam seed needs to be obtained just to recover operating cost and possibly a price of up to $1.22 per clam is needed. Recovery of total costs, assuming a 10% real rate of interest, would require a price of $1.43–$2.04 per seed depending upon whether the most optimistic or pessimistic assumption about the cost level prevails. At a 5% real rate of interest the comparable required prices for breaking even are $1.31–$1.77. A commercial hatch-

TABLE 3. *Per-unit cost of producing giant clam seed as a function of volume of annual output (Australian dollars).*

	Number of seed clams		
Type of cost	100,000	500,000	1,000,000
Operating cost	1.01–1.22	0.29–0.35	0.18–0.23
Total cost (5% interest)	1.31–1.79	0.37–0.49	0.24–0.31
Total cost (10% interest)[a]	1.43–2.04	0.41–0.55	0.26–0.35

[a] Lower and upper limits of those costs are graphed in Figs. 1 and 2 respectively.

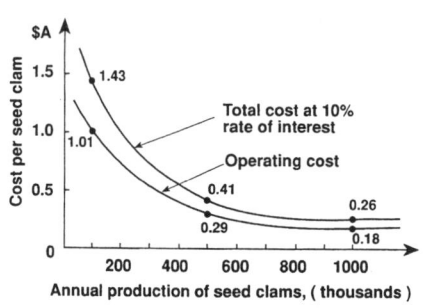

FIGURE 1. *Operating costs and total costs of production for clam seed given lower bounds of predictions and a 10% rate of interest.*

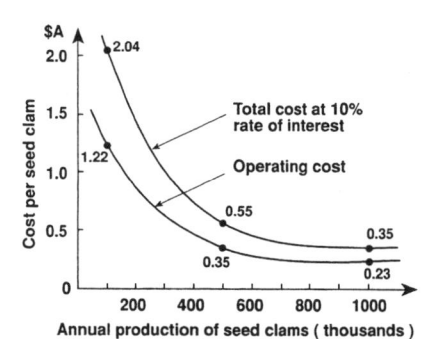

FIGURE 2. *Operating costs and total costs of production for clam seed given upper bounds of prediction and a 10% rate of interest.*

ery on this scale is unlikely to be feasible given that $1.00 per clam seed is likely to be about the maximum price that could be attained. However at a higher volume the economic prospects are considerably improved because of economies of scale. Although MMDC has sold giant clam seed of approximately 1 yr of age for US$1.00 each (Heslinga and Fitt 1987), it is likely to be difficult to make a profit from ocean growout of giant clams at this price (Tisdell et al., in press).

At an output of 500,000 clam seeds per year operating costs per clam seed lie in the range $0.29–$0.35. Total costs per seed produced are in the range of $0.41–$0.55 per clam seed given a 10% rate of interest, and at a 5% rate of interest total costs per seed are in the range of $0.37–$0.49 per seed clam. At such volumes, much lower prices are required to break even or to make a profit compared with an output of 100,000 seed clams per year. Even lower per-unit costs occur for an output of 1 million seed clams per year. Although cost per seed clam produced appears to decline as the volume of output rises, it does so at a decreasing rate.

Note also that no allowance has been made for possible marketing costs and in larger enterprises possibly a separate cost item should be included for administration. However, allowance for these items is unlikely to alter the general conclusion that substantial economies of scale in production exist.

Discussion

An examination of the estimated cost of producing giant clam seed (Tables 1, 2, 3) indicates that there are considerable economies of scale in producing giant clam seeds. Although the figures quoted are estimates, the figures do suggest that large hatcheries would produce giant clam seeds more cheaply than small hatcheries. While these results are based on Australian experience there is evidence that strong economies of scale in hatchery and operations also apply elsewhere (Hambrey 1991). Thus, large centralized hatcheries which supply seed to a network of growers could be more profitable than a proliferation of small hatcheries operated independently by individual growers. Whether a system of centralized hatcheries which supply small growers would be economically more viable than small localized hatcheries will depend on costs associated with the transportation of the giant clam seed. Indications are, however, that the transport of giant clam seeds would be a viable proposition (Solis and Heslinga 1989).

If giant clam farming does not prove to be commercially profitable, it is still highly likely that giant clam seed will be required for restocking of reefs. Restocking of reefs is required for two reasons. Firstly, the larger of the giant clams species (particularly *Tridacna gigas*) have been made locally extinct or brought to the point of extinction on many reefs. Restocking reefs where local populations have become extinct should be advantageous to the survival of the species, although the decrease in genetic variability brought about by restocking should be considered carefully. Secondly, giant clams have been used as a traditional food source by Pacific Islanders for many centuries and giant clam farming may allow Pacific people to maintain their custom and also, if clam farming is profitable, to increase their income by marketing clam products.

Acknowledgments

The authors are grateful to Richard Braley, Geoff Charles, Philippe Dor, Ken Frankish, Steven Lindsay, Bruce Marcum, Bruce Stevens, Rene Wittenberg and others for providing useful information. This study was funded by the Australian Centre for International Agricultural Research as part of ACIAR Research Project No. 8823, The Economics of Giant Clam Mariculture, and Project No. 8733, The Culture of the Giant Clam (Tridacnidae) for Food and Restocking of Tropical Reefs.

Literature Cited

Braley, R. D. 1989. Farming the giant clam. World Aquaculture 20(1):7–17.

Crawford, C. M., J. S. Lucas and W. J. Nash. 1988. Growth and survival during the ocean-nursery rearing of giant clams, *Tridacna gigas*—1. Assessment of four culture methods. Aquaculture 68: 103–113.

Fitt, W. K., C. R. Fisher and R. K. Trench. 1984. Larval biology of tridacnid clams. Aquaculture 39: 181–195.

Hambrey, J. 1991. Economies of village based giant clam farming in the Solomons: financial analysis. Working Paper No. 2. ICLARM, Honiara, Solomon Islands.

Heslinga, G. A. and W. K. Fitt. 1987. The domestication of reef-dwelling clams. BioScience 75(5): 332–339.

Heslinga, G. A., T. C. Watson and T. Siamu. 1986. Cultivation of giant clams: Beyond the hatchery. Pages 53–57 in J. L. Maclean, L. B. Digon and L. V. Hosillos, editors. The first Asian fisheries forum. Asian Fisheries Society, Manila, Philippines.

Huguenin, J. E. and J. Colt. 1989. Design and operating guide for aquaculture seawater systems. Elsevier, Amsterdam, Holland.

Lucas, J. S. 1988. Giant clams: description, distribution and life history. Pages 21–32 in J. W. Copland and J. S. Lucas, editors. Giant clams in Asia and the Pacific. Australian Centre for International Agricultural Research, Canberra, Australia.

Munro, J. L. 1989. Fisheries for giant clams (*Tridacnidae* Bivalvia) and prospects for stock enhancement. Pages 541–558 in J. F. Caddy, editor. Marine invertebrate fisheries: Their assessment and management. John Wiley and Sons, Chichester, England.

Pernetta, J. 1987. Giant clams: A new potential food source in tropical small island states or another source of biological contamination? Science in New Guinea 13(2):92–96.

Solis, E. P. and G. A. Heslinga. 1989. Effect of desiccation on *Tridacna derasa* seed: Pure oxygen improves survival during transport. Aquaculture 76: 169–172.

Tisdell, C. A., L. Tacconi, J. R. Barker and J. S. Lucas. In press. Economics of ocean culture of giant clams *Tridacna gigas*: Internal rate of return analysis. Aquaculture.

AQUA 20036

Economics of ocean culture of giant clams, *Tridacna gigas*: internal rate of return analysis

C.A. Tisdell[a], L. Tacconi[a], J.R. Barker[b] and J.S. Lucas[c]

[a]*Department of Economics, University of Queensland, Brisbane, Qld., Australia*
[b]*Reefarm Hatcheries, Abbott Street, Cairns, Qld., Australia*
[c]*Department of Zoology, James Cook University, Townsville, Qld., Australia*

(Accepted 14 July 1992)

ABSTRACT

Tisdell, C.A., Tacconi, L., Barker, J.R. and Lucas, J.S., 1993. Economics of ocean culture of giant clams, *Tridacna gigas*: internal rate of return analysis. *Aquaculture*, 110: 13–26.

The paper estimates the internal rate of return of investment in a giant clam farm involved in the ocean phase of mariculture as a function of the period of ocean growout of *Tridacna gigas*. The ocean farm is assumed each year to place 100 000 seed clams of approximately 1 year of age. The optimal length of time to hold them depends on the farm-gate price of clam meat and is estimated to be 11 years when they are sold at $A5 per kg. This yields an estimated internal rate of return of 18.0% and maximises the net present value or capitalised value of the farm. If 40% drip loss occurs in the meat the internal rate of return is 11.25% and the optimal period to hold batches of clams is 14 years.

INTRODUCTION

Despite technical advances in the mariculture of giant clams, Tridacnidae, its economics remains uncertain and data for evaluating it are limited compared to established aquaculture industries. Nevertheless, even though data are limited, it is sensible to use these rather than wait for many years for accumulation of data before attempting an economic evaluation. This is reasonable even though the results are likely to be indicative rather than definitive and use of econometric techniques may be impossible. Economic uncertainty is likely to be high in the early stages of development of an industry. Any economic evaluation which reduces that uncertainty and/or identifies economic factors which have an important bearing on profitability is significant for entrepreneurs and for government planning in relation to the emerging industry. Furthermore, even primitive models provide a basis for refinement

---

Correspondence to: C.A. Tisdell, Department of Economics, University of Queensland, Brisbane, Qld. 4072, Australia.

0044-8486/93/$06.00 © 1993 Elsevier Science Publishers B.V. All rights reserved.

as an industry develops. This statement provides the rationale for this paper which considers the likely profitability of farming (of ocean growout) of the giant clam, *Tridacna gigas*, for meat, taking the length of time for which giant clams are grown out in the ocean before harvest as the main controlled variable. The sensitivity of economic returns to variations in factors such as the price of clam meat, drip weight loss in meat, discount or interest rates and mortality rates of clams are discussed.

The main objective of farming giant clams is assumed to be the maximisation of profit. This implies that a firm specialising in the ocean growout of giant clams should manage this so as to maximise the net present value or capitalised value of net earnings from its operations over time (cf. Hicks, 1946). This approach looks at the profitability of the enterprise's operations as a whole. It is superior to an approach tried earlier by Tisdell et al. (1991) which allocated costs to individual batches of clams assumed to be grown out serially and estimated the profit from and optimal period to grow out the first batch of clams. It is superior because it avoids the somewhat arbitrary allocation of joint costs involved in overall operations to different sets of clams. Note that while the net present value of each set of clams grown out serially (not just the first set) might be assessed by the earlier method, this would not eliminate the arbitrary allocation of joint costs incurred in the production of all batches of clams.

MATERIAL STUDIED AND METHOD

To estimate profitability of giant clam farming, biological production data, production cost data and market information are needed. Information on production and costs was obtained from research work at Orpheus Island Research Station and from the practical experience of one of the authors on a commercial giant clam farm on Fitzroy Island, North Queensland, Australia. Market surveys conducted by Shang et al. (1991) and Tisdell and Wittenberg (1990a,b) provided a guide to realistic market prices for clam meat and the likely size of the market.

The hypothetical farm operating in North Queensland considered here is supposed to acquire each year 100 000 seed clams of approximately 1 year of age for intertidal ocean culture. In the first year of the ocean phase of culture the clams are kept in lines (using steel mesh 'boxes' for protection, see Fig. 1) and moved to exclosures in the second year (Fig. 2) (Barker et al., 1988). The exclosures form a suspended netting fence to exclude predators. In the third year the clams are ready to be transferred to the open ocean (Crawford et al., 1988) without any protection. They are assumed to be placed in an intertidal area as is done at Orpheus Island Research Station.

For an Australian farm assumed to operate according to the above method, long-term capital costs, short-term capital costs and operating costs have been

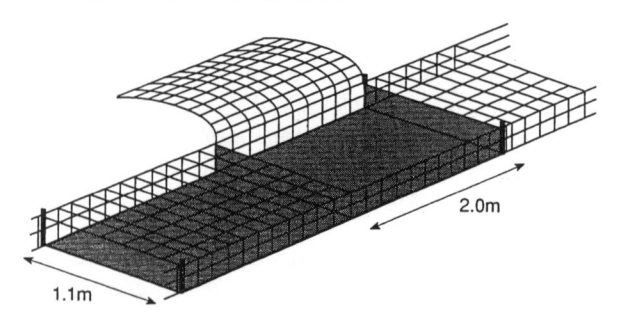

Fig. 1. Protective line showing individual cell — 15 cells/line.

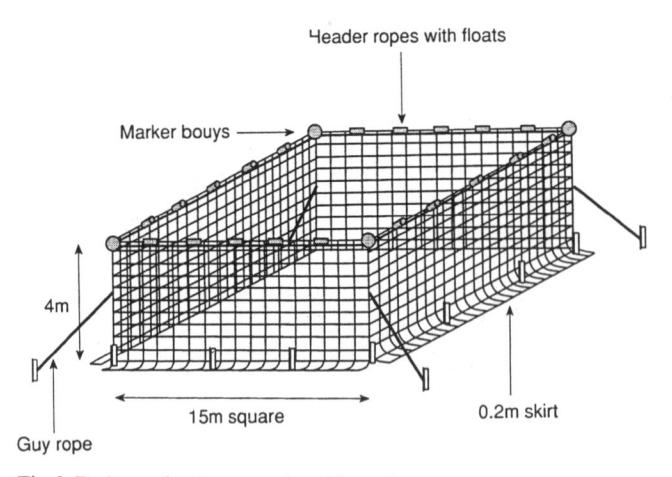

Fig. 2. Exclosure for the protection of juvenile clams.

estimated in Australian dollars (Table 1 and Table 2). Note that costs would include the annual cost of a lease of about 10 ha on the Great Barrier Reef. This would cost about $A800 per year. In addition about $A2000 would have to be spent on advertising (6 weeks in a national newspaper) so that any person can acquire knowledge of the proposed lease and object to it with the Great Barrier Marine Park Authority. The Marine Park Authority also requires a bank guarantee of $A6000 in order to cover any environmental cleanup needed. However, this bank guarantee does not imply any additional cost for the lessee. Long-term capital costs include worker accommodation, a tractor, two boats and an electricity generator. As it is assumed that the depreciation rate for the tractor, the boats and the generator is 10% per annum, they have to be replaced in year 11 of operation. Short-term capital costs comprise lines and exclosures. Lines have to be replaced every 3 years and exclosures every second year. Operating costs include miscellaneous expenditures (e.g.

TABLE 1

Capital costs ($A) for ocean phase of giant clam farm

Year	Capital costs						Total
	Long term				Short term		
	Worker house	Tractor	Boats	Generator	Lines	Exclosures	
1	80 000	20 000	25 000	3000	15 000	—	143 000
2					—	27 000	27 000
3					15 000	—	15 000
4					—	27 000	27 000
5					—	—	—
6					15 000	27 000	42 000
7					—	—	—
8					—	27 000	27 000
9					15 000	—	15 000
10					—	27 000	27 000
11		20 000	25 000	3000	—	—	48 000
12					15 000	27 000	42 000
13					—	—	—
14					—	27 000	27 000
15					15 000	—	15 000
16					—	27 000	27 000
17					—	—	—
18					15 000	27 000	42 000

office costs), fuel, wages (one full-time worker and 60 man/days of casual work per year), and the expenses to purchase 100 000 seed clams per year (Tisdell et al., 1990, 1991). The insurance premium for this operation has been estimated to be $A9000 per year on the basis of experience at Reefarm.

Given that clam mariculture is still in an early stage of commercial development, several assumptions have to be made in order to assess its economic viability. Tisdell et al. (1991) assumed that the output (about 180 t/year or 108 t/year if post-harvest drip loss of 40% occurs) of a farm that acquires 100 000 seed clams a year could be absorbed by the market. This is a reasonable assumption given that market studies indicate an annual market potential at several hundred tonnes of meat (e.g. Tisdell and Wittenberg, 1990a,b; Shang et al., 1991). Furthermore, it is likely to be more profitable for the farm to supply the market on a continuous basis rather than at widely spaced intervals. For example, an extreme alternative to growing out 100 000 clams annually and holding each batch for 10 years might be to grow out 1 million initially and sell after 10 years of growout. But this would mean that purchasers would not have a regular supply and the market could be flooded by the sales when these occur. The farm would also have little scope for gradual development of the market for its produce. This pattern of production would

TABLE 2

Annual operating costs ($A) for ocean phase of giant clam farm*

Clam seeds (100 000 @ 75c)		75 000
Wages:		
1 full-time worker	30 000	
Casual work:		
40 days/year × $80 day	5 000	
		35 000
Insurance		9 000
Fuel		1 800
Repair and maintainance:		
house	800	
tractor	200	
boats	250	
generator	50	
		1 300
Lease fees		800
Miscellaneous expenditures (e.g. office costs)		6 100
Total		129 000

*In addition to the costs shown in Tables 1 and 2, an initial advertising cost of $2000 is needed. This is allowed for in calculations of IRR.

in addition add greatly to the initial liquidity problem of the firm (cf. De Silva and Tisdell, 1986). This discontinuous pattern of operations would also mean that workers are discontinuously employed, and expertise built up in clam farming may be lost. Also a person may have to be specifically employed to guard clam stocks, whereas with continuous operations personnel can incidentally provide surveillance while engaged in other activities. It is for these reasons that the economics of continuous operation is explored.

Further factors favouring production on a rotating basis involving one set of clams being grown out each year and eventually one batch being harvested per year are:
- the progressive establishment of the farm (i.e. adding one set of clams in each year) allows for a process of trial-and-error learning: e.g. size of lines and exclosures, positioning of the clams in the ocean, how to deal with pests and possible diseases.
- skilled labour could be difficult to find for the handling and packaging of the clams if the production is on a synchronised basis. This factor depends to a certain extent on the handling and packaging technology. Also, if specific machinery is required for the packaging, its availability could be a constraint.
- the lines and the exclosures can be used for more than one year; therefore, if the farm is set up all at once, capital investment in lines and exclosures would be higher than in the progressive setting up case.

A scale of growing on 100 000 clam seed per year was considered (rather than one, say, of growing on twice as many or half as many clam seed per year) because that level of operations would, with some casual help, keep one person fully employed. Operations at half this level would involve part-time employment of the staff member or his or her time would not be fully utilised if employed full-time. The latter would not be economic and in the former case staff may not be available part-time and/or housing costs are likely to be the same whether the staff member is employed full-time or part-time. Operations on a larger scale are possible but more staff would be required, but this is not specifically analysed. A permanent staff member would be required approximately for each 100 000 clam seed or part thereof grown out annually. While some economies of scale might be achieved in housing and equipment (capital costs), few economies in operating costs seem likely. In other words, the ocean phase of clam farming is likely to show only weak economies of scale. Size of operations, however, can be expected to be constrained by the risk factor and by the availability of space suitable for growout. For the size of operations assumed in this study, 12–13 ha of intertidal reef is needed to accommodate the ocean growout of the clams when the farm is fully operational. Operations on a larger scale will involve a larger area and this may not be easily found in the *same* locality. For these reasons, the size of operations considered here appears to be a reasonable economic size to consider to gauge the likely economics of giant clam farming.

As mentioned earlier, we proceed on the assumption of Hicks (1946) that the objective of a firm is to maximise its net present value (its net capitalised value) of earning possibilities. This is equivalent to profit maximisation. In order to ascertain the net present value or net capitalised value of a farm, the stream over time of all its expected costs and benefits has to be taken into account. Thus arbitrary allocation of joint costs is avoided. This method has for example been used by Firdausy and Tisdell (1991) to evaluate the economics of a seaweed farm in Bali, Indonesia.

The net present value (NPV) of an investment project or management plan is given by:

$$\text{NPV} = \sum_{t=0}^{n} \frac{(B_t - C_t)}{(1+r)^t} \tag{1}$$

where B_t = benefits at time t; C_t = costs at time t; and r = discount rate

A measure closely related to the NPV is the internal rate of return (IRR). The IRR of return is the discount rate, r, that makes NPV equal to zero:

$$\text{NPV} = \sum_{t=0}^{n} \frac{(B_t - C_t)}{(1+r)^t} = 0 \tag{2}$$

In other words, the discounted benefits and discounted costs are equated. The IRR indicates the maximum rate of interest that can be paid for funds used in an investment if the investment is to break even (see for example Shang, 1981; Meade, 1989, Ch. 11).

The economically optimal length of the rotation cycle (of clams in this case) for the farm can be found by maximising IRR. The controlled variable in this case is the length of the rotation cycle. The economically optimal length of the rotation cycle is shorter than the biologically optimal cycle (cf. Bowes and Krutilla, 1985; Hartwick and Olewiler, 1986, Ch. 11; Tisdell, 1991). The economics of giant clam farming is analogous to that for forestry in many ways. Mainly this is because there is a long lag (gestation period) involved in clam and forest production, that is between establishment of a forest or a clam farm and the harvest. Thus some of the forestry economic models may be applied to clam farming (Yamaguchi, 1977; Tisdell, 1986; Watson and Heslinga, 1988).

Even if funds for financing the clam farm are not borrowed (or if only some funds are) but supplied as equity in the business, the opportunity cost of the capital or funds used in the business needs to be taken into account. The opportunity cost of funds used in the business is the return on the funds which could be earned by investing them elsewhere. This is at least equal to the rate of interest on government bonds. If investment in clam farming is to be economical, the internal rate of return from it must at least be equal to the rate of interest which could be earned by investing the same funds in government bonds.

In determining the optimal length of the clam rotation cycle, account should be taken not only of reinvestment opportunities or returns available in the market (as indicated by the prevailing rate of interest) but also the returns available from reinvestment in the farming of clams (cf. Mishan, 1971, Ch. 28). It will not, as a rule, pay to hold a batch of clams until the marginal increase in its net value equals the rate of interest because one can dispose of a batch of older clams and replace these by a younger set which will have a greater growth efficiency. It is usually optimal to harvest a batch earlier and take advantage of the rapid growth in weight of clams in the earlier part of their life-cycle to earn a higher rate of return than the prevailing rate of interest. This seems to be the situation potentially in relation to the ocean culture of giant clams.

In Fig. 3 a hypothetical relationship is shown for illustrative purposes between the internal rate of return and the number of years clams are held in their ocean phase (equals the number of batches of various ages in place). In the case shown, the maximum internal rate of return is shown as OB and this suggests that the optimal policy is to establish a rotation cycle of x_1 years. If the rate of interest is less than OA and net present value is maximised *without* account being taken of the possibility for returns from reinvestment in clam

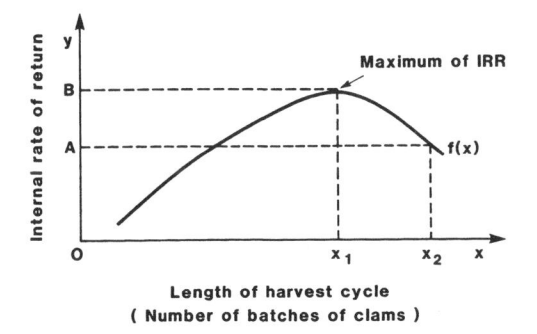

Fig. 3. Internal rate of return as a function of the length of the harvest cycle — theoretical relationship.

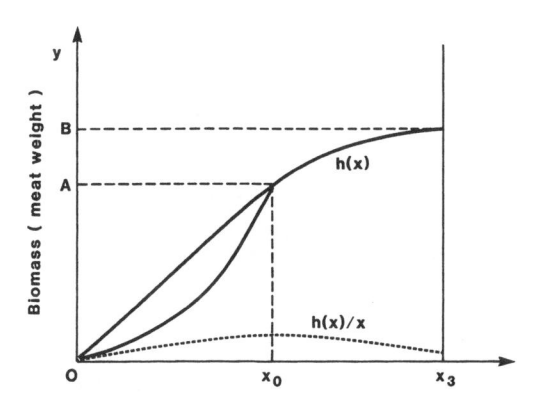

Fig. 4. Time in ocean phase needed to result in two types of maxima for biological mass (meat weight in this case).

farming, the computed rotation cycle will exceed x_1. For example, if the rate of interest is OA, IRR equals the rate of interest for a cycle of x_2 years. A cycle of this length maximises the net present value of the project when account is not taken of reinvestment opportunities in clam farming. But this cycle will be too long in comparison with the most economic cycle which in this case is approximately x_1 years in length (cf. Tisdell, 1991). Observe that if the rate of interest exceeds OB, clam farming would be uneconomic — it would be better to invest any available funds at the going rate of interest.

Note that both maximisation of the internal rate of return taking full account of reinvestment opportunities, and net present value maximisation, not taking reinvestment opportunities in clam farming into account, result in cycles of shorter length than (a) the length of cycles required to maximise biomass (meat mass in this case) on average per unit of time taking into account replacement cycles, and (b) the length of cycles needed to maximise biomass (meat weight in this case) taking a single cycle in isolation into account. Watson and Heslinga (1988) have estimated (b) for *T. derasa*.

The latter two concepts are illustrated in Fig. 4. The function $h(x)$ shows biomass (meat mass) hypothetically as a function of time, also the number of batches put down. In this case the production function (biomass) may only reach a maximum when constrained by the available space because the greater the number of batches put down the greater will be the biomass usually. Space may, however, constrain the farm to x_3 batches, for example, and maximum production would then be achieved by using all the space and putting down x_3 batches. But in the case shown this will not maximise production per unit of time. This occurs for a cycle of length x_0 which corresponds to the maximum of the average function which has been dotted in.

RESULTS AND DISCUSSION

The process of maximising IRR is shown in Fig. 3. In this case the maximum IRR occurs for x_1 number of batches of clams and is equal to OB. Given that one set of clams is put down every year, this is also the optimal length of the rotation cycle (x years). We shall show that at a clam meat price of A$7 per kg, the farm examined here maximises its present discounted value by holding nine sets of clams of ages for each batch (at the end of each year) ranging from 2 to 10 years (Table 4 and Fig. 5).

The first step is to estime the IRR function for this case. As there is not a direct way of determining r from equation (2), the IRR has to be determined in a recursive way by trial and error (Gittinger, 1982). In the present case, to determine the maximum IRR for the project, the recursive procedure was repeated for several different years, that is, values of x, and for different output prices. The gross value of the farm is given by the market value of the stock of clams held at the end of the year considered (e.g. 10) plus the value of all the assets that have a realisable market value, i.e. worker accommodation, tractor, boats and generator. Their market value is assumed to be the purchasing cost less depreciation. Annual depreciation rates used employing the straight-line method of depreciation are: house 2.5%; tractor, boats and generator 10%.

When discounting, it is assumed that expenditures occur at the beginning of the year. This implies that costs for year 1 are not discounted; costs in year 2 are discounted by a time factor of one. Benefits are instead assumed to accrue at the end of the year; therefore, in year 10, the time factor is 10.

The market value of the clams was calculated using gate-prices for clam meat of $A7 per kg, $A5 per kg and $A3 per kg. The case of a post-harvest drip loss of 40% is also considered for each output price. As reported by Hambrey (1991) and from experience at the commercial farm on Fitzroy Island, clam meat can loose up to 50% of its weight in the first 24 h post-harvest. This 'drip loss' can, however, be reduced by freezing the meat.

The above price range was adopted on the basis of evidence arising from marketing studies covering several Asian and Pacific countries (Tisdell and Wittenberg, 1990a,b; Shang et al., 1991). It appears that giant clam meat could attract a market retail price of up to $A10 per kg or more. The farm-gate price of $A5 per kg allows at least a 100% mark up at retail level. Retail prices for giant clam adductor muscle were found to be in the range of $US8–30 per kg in Taiwan.

Sales of shells and clams as species for aquaria were not considered. Also, the price of clam meat is not assumed to vary with the age of clams (i.e. a higher-priced sushi and sashimi market for young clams is not considered). The biological data relating to the first 20 years of life of a set of 100 000 giant clams are presented in Table 1. The growth rates are the average rates shown in Munro (1988). The flesh weight is proportional to total weight and for the first 5 years data are based on work undertaken at Orpheus Island Research Station (OIRS); for clams older than 5 years the figures are those reported by Munro (1989). The survival rates of clams shown in Table 3 are assumed to be realistic on the basis of OIRS experience.

TABLE 3

Biological data

Age BP	Length (cm) at BP	Weight (kg) per clam Total(a)	Weight (kg) per clam Flesh	Survival rate	Number of clams BP	Total flesh weight (kg) EP
1	6.869	0.039	0.005	0.58	100 000	6.203
2	14.158	0.509	0.107	0.82	58 000	22.088
3	20.721	1.786	0.464	0.95	47 560	37.906
4	26.630	3.995	0.839	0.95	45 182	50.359
5	31.950	7.111	1.173	0.95	42 923	74.175
6	36.739	11.025	1.819	0.95	40 777	99.650
7	41.051	15.590	2.572	0.95	38 738	125.390
8	44.933	20.650	3.407	0.95	36 801	150.273
9	48.429	26.050	4.298	0.95	34 961	173.463
10	51.575	31.653	5.223	0.95	33 213	194.388
11	54.409	37.338	6.161	0.95	31 552	212.699
12	56.960	43.006	7.096	0.95	29 975	228.231
13	59.256	48.575	8.015	0.95	28 476	240.959
14	61.324	53.983	8.907	0.95	27 052	250.961
15	63.185	59.183	9.765	0.95	25 700	258.387
16	64.861	64.141	10.583	0.95	24 415	263.432
17	66.370	68.836	11.358	0.95	23 194	266.321
18	67.729	73.253	12.087	0.95	22 034	267.288

Legend: (a), including shell; BP, beginning of period; EP, end of period.

TABLE 4

Internal rate of return for ocean farming of *Tridacna gigas*

Year of growout	Internal rate of return						
	$3/kg	$3/kg −40%DP	$5/kg	$5/kg −40%DP	$7/kg	$7/kg −40%DP
8	6.10	0.00	16.30	6.10	23.10	12.80
9	8.40	0.00	17.50	8.40	23.50	14.40
10	9.60	1.30	17.90	9.60	23.30	15.00
11	10.50	3.00	18.00	10.50	22.90	15.40
12	11.00	4.00	17.80	11.00	22.20	15.50
13	11.20	4.80	17.50	11.20	21.60	15.30
14	11.25	5.30	17.00	11.25	20.80	15.05
15	11.21	5.70	16.57	11.21	20.10	14.75
16	11.05	5.87	16.05	11.05	19.30	14.35
17	10.86	6.03	15.57	10.86	18.60	13.96
18	10.60	6.00	15.05	10.60	17.90	13.55

Legend: DP, post-harvest drip loss.

The estimated IRR function for this problem is given in Table 4 and Fig. 4. The maximum IRR occurs in year 9 of the ocean phase when the gate-price is $A7 per kg and is 23.5%. In this case, for the investor who wants to maximise returns on the investment it is optimal to hold the clams for 9 years in the ocean phase.

As is evident from Fig. 5, a lower farm-gate price and a post-harvest drip loss result in an increase in the optimal length of time to harvest. For a lower farm-gate price, it pays to keep the clams longer or, conversely, for a higher price a shorter time. The latter occurs because an increase in the price of clam meat increases the relative meat value per unit of time of smaller clams more than that of older clams because the growth rate of smaller clams is faster. So the opportunity cost of not replacing older clams by smaller ones rises.

A lower farm-gate price also implies a lower IRR. At a price of $A3 per kg and $A5 per kg with drip loss, the IRR is just above 10% which can be regarded as a threshold rate for the investment to be considered profitable given that current real interest rates in Australia are around 10%. At a price of $A3 per kg with drip loss, the IRR is consistently below that threshold.

The sensitivity of the IRR to changes in the price of clam seeds was also analysed. In the foregoing analysis, a price of $A0.75 per clam seed has been used. The sensitivity analysis also considered a price per seed of $A1 and $A0.50. Table 5 presents the results for the case of clam meat price of $A5 per kg.

A clam seed price of $A1 implies a decrease in the maximum IRR (in year 11) of about two percentage points. At a clam seed price of $A0.50, the opti-

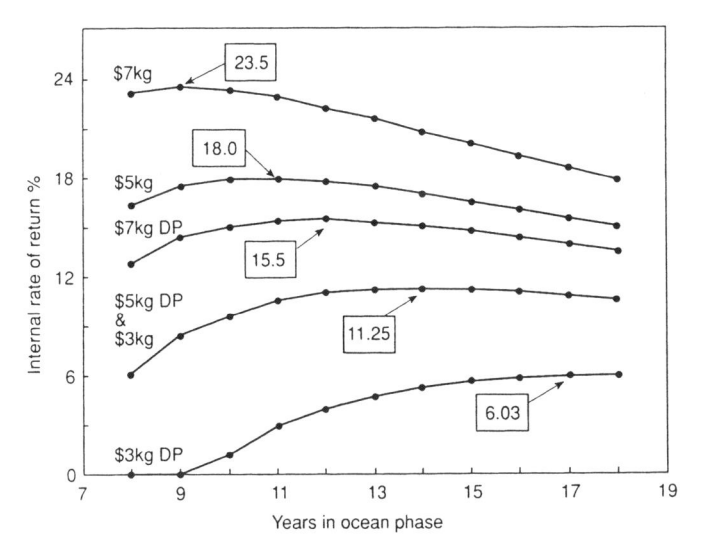

Fig. 5. Internal rate of return curves for ocean farming of *Tridacna gigas*. DP = 40% post-harvest drip loss.

TABLE 5

Internal rate of return: sensivity analysis to the price of clam seeds (clam meat farm-gate price $5/kg)

Year of growout	Clam seed price		
	$1.00	$0.75	$0.50
	Internal rate of return		
8	13.50	16.30	19.40
9	15.00	17.50	20.30
10	15.60	17.90	20.40
11	15.95	18.00	20.30
12	15.92	17.80	19.90
13	15.75	17.50	19.40

mal harvesting time is decreased by 1 year and the maximum IRR is 20.4%, an increase of over two percentage points compared to a seed price of $A0.75 per clam.

CONCLUSION

The present analysis shows that a giant clam farm facing the environmental and economic conditions assumed in the paper can be profitable. Only in the event of the farm-gate price for clam meat being $A3 per kg with a 40% drip loss is the IRR lower than 10% for the cases illustrated. To increase economic profitability, it is evident that reduction of the drip loss is important to clam farming activity.

One should exercise some caution in applying these results to other countries because they are based on Australian conditions and would need to be adjusted for application elsewhere. Nevertheless, this paper, apart from identifying those factors to which economic returns (that is, returns considerably in excess of the prevailing rate of interest) from clam farming can be expected to be sensitive, suggests that very high returns from ocean culture of giant clams are unlikely. On the other hand, improved techniques of clam farming could lower costs, and additional marketing opportunities, other than those assumed, may arise, making farming more economic than estimated here. Considerable uncertainties remain but this is to be expected in an embryonic industry which is in the process of development.

ACKNOWLEDGEMENTS

We wish to thank Professor Masashi Yamaguchi, Department of Marine Sciences, University of Ryukyus, Japan, the participants to the Second Giant Clam Project Leaders' Meeting, University of the Philippines, February, 1991, and two anonymous referees for helpful comments on earlier drafts of this paper. The usual caveats apply. This research has been partially funded by the Australian Centre for International Agricultural Research (ACIAR) Project No. 8823, Economics of Giant Clam Mariculture.

REFERENCES

Barker, J.R., Crawford, C.M., Shelley, C.C., Braley, R.D., Lucas, J.S., Nash, W.J. and Lindsay S., 1988. Ocean-nursery technology and production data for the giant clam *Tridacna gigas*. In: J.W. Copland and J.S. Lucas (Editors), Giant Clams in Asia and the Pacific. ACIAR, Canberra, pp. 225–228.
Bowes, M.D. and Krutilla, J.V., 1985. Multiple use management of public forestlands. In: A.V. Kneese and J.L. Sweeney (Editors), Handbook of Natural Resource and Energy Economics, II. Elsevier Science Publishers, Amsterdam, pp. 531–543.
Crawford, C.M., Lucas, J.S. and Nash, W.J., 1988. Growth and survival during the ocean-nursery rearing of giant clams, *Tridacna gigas*. 1. Assessment of four culture methods. Aquaculture, 68: 103–13.
De Silva, N.T.M.H. and Tisdell, C.A., 1986. Supply-maximising and variation-minimising replacement cycles of perennial crops and similar assets: theory illustrated by coconut cultivation. J. Agric. Econ., 38(3): 243–251.
Firdausy, C. and Tisdell, C.A., 1991. Economic returns from seaweed (*Eucheuma cottonii*) farming in Bali, Indonesia. Asian Fish. Sci., 4: 61–73.
Gittinger, J.P., 1982. Economic Analysis of Agricultural Projects, 2nd Edition. Johns Hopkins University Press, Baltimore, MD.

Hambrey, J., 1991. Estimation of farm-gate value of giant clam (*Tridacna gigas*): products of source in the Solomon Islands. Economics of Village Based Giant Clam Farming in the Solomons, Working Paper No.1, ICLARM South Pacific Office, Honiara, Solomon Islands.

Hartwick, J.M. and Olewiler, N.D., 1986. The Economics of Natural Resource Use. Harper and Rowe, New York, NY.

Hicks, J.R., 1946. Value and Capital, 2nd Edition. Clarendon Press, Oxford.

Meade, J.W., 1989. Aquaculture Management. Van Nostrand Reinhold, New York, NY.

Mishan, E.J., 1971. Cost-Benefit Analysis. Allen and Unwin, London.

Munro, J.L., 1988. Growth, mortality and potential aquaculture production of *Tridacna gigas* and *Tridacna derasa*. In: J.W. Copland and J.S. Lucas (Editors), Giant Clams in Asia and the Pacific. ACIAR, Canberra, pp. 218-220.

Munro, J.L., 1989. Fisheries for giant clams (Tridacnidae: Bivalvia) and prospects for stock enhancement. In: J.F. Caddy (Editor), Marine Invertebrate Fisheries: Their Assessment and Management. John Wiley and Sons, Chichester, UK, pp. 541-558.

Shang, Y.C., 1981. Aquaculture Economics: Basic Concepts and Methods of Analysis. Westview Press, Boulder, CO.

Shang, Y.C., Tisdell, C. and Leung, P.S., 1991. Report on a Market Survey of Giant Clam Products in Selected Countries. Centre for Tropical and Subtropical Aquaculture, Publication No. 107.

Tisdell, C.A., 1986. The Economic and Socio-Economic Potential of Giant Clam (Tridacnid) Culture: a Review. Department of Economics, University of Newcastle, NSW 2308, Research Report or Occasional Paper No. 128.

Tisdell, C.A., 1991. Economics of Environmental Conservation. Elsevier Science Publishers, Amsterdam.

Tisdell, C.A. and Wittenberg, R., 1990a. Report on Possible Demand for Giant Clam Meat by Tongan Descendants in Australia: Inferences from Interviews Conducted in the Brisbane Area. Department of Economics, University of Queensland, Research Reports and Papers in Economics of Giant Clam Mariculture, No. 8.

Tisdell, C.A. and Wittenberg, R., 1990b. The Potential Market for Giant Clam Meat in New Zealand: Results of Interviews with Pacific Island Immigrants. Department of Economics, University of Queensland, Research Reports and Papers in Economics of Giant Clam Mariculture, No. 15.

Tisdell, C.A., Lucas, J.S. and Thomas, W.R., 1990. An Analysis of the Cost of Producing Giant Clam (*Tridacna gigas*) Seed in Australia. Department of Economics, University of Queensland, Research Reports and Papers in Economics of Giant Clam Mariculture, No. 11.

Tisdell, C.A., Barker, J.R., Lucas, J.S., Tacconi, L. and Thomas, W.R., 1991. Ocean Culture of Giant Clams (*Tridacna gigas*): an Economic Analysis. Department of Economics, University of Queensland, Research Reports and Papers in Economics of Giant Clam Mariculture, No. 18.

Watson, T.C. and Heslinga, G.A., 1988. Optimal harvest age for *Tridacna derasa*: maximising biological production. In: J.W. Copland and J.S. Lucas (Editors), Giant Clams in Asia and the Pacific. ACIAR, Canberra, pp. 221-224.

Yamaguchi, M., 1977. Conservation and cultivation of giant clams in the tropical Pacific. Biol. Conserv., 11: 13-20.

Economic Returns from Seaweed (*Eucheuma cottonii*) Farming in Bali, Indonesia

CARUNIA FIRDAUSY and CLEM TISDELL

Department of Economics
The University of Queensland
Queensland 4072, Australia

Abstract

The farming of seaweed is becoming more common in Indonesia. The species known as *Eucheuma cottonii* is the most commonly cultured. There appears, however, to be no estimates of the economic returns from this activity for Indonesia. After providing a brief background on culture technique and marketing aspects, we analyze the investment, cost and revenue data from a 1-ha *E. cottonii* farm in Jungut Batu, Bali. It is estimated that the payback period for this activity on this farm is 7.8 months and that seaweed farming gives an accounting rate of return of 123% and an economic rate of return (IRR) of 153%. Thus *E. cottonii* farming is a potentially attractive investment in Indonesia and more so since it is relatively labor-intensive and does not require significant quantities of processed or imported inputs such as fertilizers, chemicals, fuel and food.

Introduction

Seaweed is the major non-food fishery item exported from Indonesia. Total export production of seaweed in 1984 was estimated to be 3,061.1 t with a value of US$658,842; by 1988 this had increased to 8,366 t with a value of $2,880,510 (Anon. 1989). In terms of volume, seaweed ranks fourth among Indonesia's fisheries exports following shrimp, tuna and other fish (Table 1).

Due to increasing export demand for seaweed, the Indonesian Government has encouraged seaweed farming in coastal areas (Directorate General of Fisheries 1988). Many coastal rural dwellers in Bali have adopted seaweed culture, but in other coastal rural areas the rate of adoption of this culture is slow. This is partly because Bali has some advantages such as natural seed availability, suitable

Table 1. Indonesia's export of fisheries products by type of commodities, 1984-1987 (in t).

Commodities	1984	1985	1986	1987	Average growth/year (%)	Percentage of total exports
Food items	66,392	72,629	92,579	122,270	13.4	87.1
Prawn	28,025	30,980	36,101	44,267	14.0	31.1
Tuna/skipjack	14,702	17,889	24,236	33,995	17.5	24.2
Other fishes	8,623	9,158	10,611	18,902	15.8	13.5
Frog thigh	2,200	2,802	3,752	3,078	-2.5	2.2
Sea cucumber	1,318	3,123	2,362	2,517	30.6	1.8
Jelly fish	2,556	1,875	4,762	3,372	-15.1	2.4
Crabs	2,143	1,749	1,944	2,049	1.3	1.5
Others	6,823	5,053	8,811	13,730	25.4	9.8
Nonfood items	9,303	11,868	14,866	18,108	16.7	12.9
Ornamental fish	204	235	859	530	61.5	0.4
Seaweeds	3,061	5,446	7,111	9,882	34.4	7.0
Sea shell	2,603	2,832	2,389	2,740	5.2	2.0
Others	3,435	3,355	4,507	4,956	6.0	3.5
Total	75,695	84,497	107,443	140,378	13.8	100
Total value (FOB US$000)	248,063	259,444	374,117	475,524	18.1	

Source: Central Bureau of Statistics, Jakarta, 1987 cited in Directorate General of Fisheries, Jakarta, 1988.

coastal sites, developed transportation and market networks. These attractive conditions are not satisfied in other coastal areas of Indonesia. For example, seaweed farmers in Sibolga, North Sumatra; Seribu Islands, Jakarta and in Kabupaten Barru, South Sulawesi, depend on seed from Bali and market their seaweed through middlemen mostly from Bali.

As yet no estimates appear to have been made of the potential economic return from seaweed farming in Indonesia. Therefore, this paper aims to provide estimates of potential economic returns from *E. cottonii* farming using data obtained from a farm in Bali and analyze the available cost data. Although the data and the results are specific to Bali, the techniques used to evaluate and analyze the data are of general applicability to seaweed farming. Before analyzing the data, background information is given about the species maricultured, the culture technique used and the marketing aspects.

Species, Culture Technique and Marketing Aspects

Five species of seaweed are of economic importance in Indonesia: *Gracilaria* sp., *Gelidium* sp., *Sargassum* sp., *Eucheuma cottonii* and *Hypnea* sp. *E. cottonii* is the most extensively cultured and the market price of this species is higher than for other seaweed species.

For example, in December 1988, at the farm gate in Bali, the market price of dried *E. cottonii* was Rp. 450/kg (Indonesian currency unit: Rp. 1,715 = US$1), but the price of dried *E. spinosum, Gracilaria* sp. and *Gelidium* sp. was only Rp. 250/kg as one author found from interviews in Bali.

The culture technique most commonly used is the off-bottom monoline system employed by *E. cottonii*-seaweed farmers in Bali because of the suitable nature of coastal sites. Materials required for this technique include wood, bamboo, nylon lines or twine and plastic rafia. Plants are tied along the nylon lines. Propugales are spaced along the nylon line at intervals of 0.2 m or at a planting density of 5 plants·m^{-1}. The lines are tied to stakes driven into the bottom of the seafloor. The stakes are spaced at distances of 10 m along the rows which are 0.5 m apart.

The market structure of seaweed in this coastal area (Bali) can be described as one of oligopsony. At the village level, seaweed farmers sell their products to few buyers (collectors and/or middlemen) in dried form. These village collectors collect the seaweeds on the site, and therefore, seaweed farmers do not need to pay transportation costs. Collectors at the village level then sell the seaweed to large traders in Bali and or outside Bali (Jakarta and Ujungpandang). Finally, these large traders of seaweed sell to exporters or directly to overseas buyers in countries such as Denmark, USA and Singapore.

It is worth noting that buyers at the village level are not necessarily agents of any parent company, but have established trading relationships with large traders as well as with other interested international buyers. Trading relationships between village/local traders and seaweed farmers extend beyond the mere sale of seaweed. Village seaweed traders provide a multiplicity of economic services to seaweed farmers such as finance for investment purposes (loans).

In Bali the price of *E. cottonii* at the village level is much the same everywhere and, in 1988, was Rp. 450/kg. Seaweed collectors are able to make a profit because of the higher international market price. In 1988 they were paid $0.35/kg (Rp. 600/kg) for seaweed at export ports in Bali.

In Kabupaten Barru, South Sulawesi and Sibolga, North Sumatra, seaweed farmers are paid low prices compared to Bali ranging between Rp. 250 and 350/kg. There are greater variations in price than in Bali due in part to less competition between seaweed buyers. Furthermore, the lower price may reflect monopsonistic elements, higher transport and transaction costs than on Bali as well as the availability of less information by farmers. As a result, coastal rural dwellers in these areas are not enthusiastic about adopting seaweed farming as an alternative economic activity.

Some possible solutions to these problems could include government intervention and the establishment of a cooperative marketing system among farmers. The establishment of farmers' cooperatives may result in some positive influence on the marketing system, e.g., by establishing minimum prices paid to the farmers for their production and providing market information. Seaweed-farmer cooperatives have only been established in Bali. Perhaps, this is one reason why farmers in Bali obtain a higher price for seaweed.

The risks involved in seaweed farming can be categorized into two kinds, namely, natural risks and risks from non-natural causes. Natural risks arise from variations in biological and environmental conditions. Unusual strong wave action in sandy areas can slow the growth rate of the plants. Animal 'pests' such as fish (*Siganus* sp.) and turtles can damage seaweed. Tropical cyclones or typhoons can destroy the crop and the capital investment. Rain can affect the harvesting schedule, handling and processing of seaweed. To reduce the latter risks, most seaweed farmers in Bali have their own storage facilities. Many also fence their seaweed areas using netting to keep out animal pests. On the economic side, risks may arise from variations in the price of seaweed or costs.

Analysis and Data Limitations

Many methods can be used to evaluate the economic desirability of business projects. Project evaluation methods include the payback period method, the average rate of return approach and discounting methods. The last set of methods include the net present value (NPV) or discounted cash flows (DCF), the internal rate of return (IRR), and benefit-cost ratio methods (Tisdell 1972; Shang 1981; Gittinger 1982).

The payback period and the average rate of return methods, however, fail to make any allowance for the timing of benefits and costs. For instance, the payback period method simply estimates the speed with which the project repays the original investment. Projects which repay the original investment or outlay in the shortest period of time are preferred. The limitations of this method are that it ignores the flow of returns beyond the payback period and does not take into account the receipt pattern within the payback period. No account is taken of the possibility that some projects involve capital outlays in other than the initial period. Thus this measure ignores much of the time pattern and, indeed, some of the net benefits from projects (Tisdell 1972).

Whilst there is wide acceptance by economists of the use of the NPV criterion for evaluating projects, there are differing views on what rate or rates of discount should be used for calculating NPV (Tisdell 1972; Bradford 1975; Mendelsohn 1981; Pearce and Nash 1981; Gittinger 1982; Mishan 1982). Some argue that because capital should be invested where returns are highest, the appropriate rate of discount is the opportunity cost of capital. Although this is appealing theoretically, it is difficult to apply in practice since the opportunity cost of capital is imperfectly known.

Data presented in this paper are based upon the actual costs and production figures collected through interviews with a 'model' seaweed farmer in Jungut Batu village, Nusa Penida, Bali, in December 1988. He had a 1-ha farm, a size larger than the average farm and was able to provide one of the authors with suitable data. He had also received an award from the Indonesian President for enterprise and he may be, therefore, a more efficient farmer than the average seaweed farmer. The amount of seaweed produced by this farmer was 48 $t \cdot ha^{-1} \cdot year^{-1}$ (dried form). His yield was higher than the average product of 30-40 $t \cdot ha^{-1} \cdot year^{-1}$ suggested in the literature (Chapman and Chapman 1980; Abdul Malik and Rahardjo 1988). Thus, the returns calculated in this paper indicate what can be achieved under above-average management. In addition, the analysis presented in this paper assumes costs and revenue for a farm for 10 years under *static* conditions of farm size and technology.

Taxes are not included in the calculations as the Indonesian tax system is not well organized in rural villages. It seems that products from agriculture, aquaculture or fisheries sold directly by farmers to the buyers are not subject to tax. Furthermore, seaweed farmers do not lease their holding because coastal dwellers in Bali believe that the coastal areas adjoining their village belong to them. This implies that village land regulations (customs) will be a constraint to the entry of non-villagers to this business in Bali. However, there is a coastal rental fee paid annually by seaweed farmers for village contribution.

Since there are no comparative data from small farmers in the area, the results of economic analysis may not be representative of many small farmers in the area. Thus, some caveats apply for this analysis. Nevertheless, the returns presented in this paper are indicative of the potential returns possible from seaweed farming in Indonesia at least for 1-ha seaweed farmers.

Economic Returns

Several studies on economic returns from seaweed farming have shown that yields from investment in this activity can be high. Padilla and Lampe (1989), for instance, from their study in the Philippines, found that seaweed farming is a high-yielding investment. They calculated the return on investment in the Philippines to be 78% which is way above the opportunity cost of capital. The accounting return is higher for noncorporate farmers since these farmers do not impute costs for their labor contribution to the farm or for their entrepreneurial skills. However, it is not clear how Padilla and Lampe (1989) calculate this return, that is, whether it is the internal rate of economic return or an accounting-type return calculated along the lines used by Shang (1976).

Shang (1976) estimated the rate of return on *Gracilaria* seaweed farming cultivation in Taiwan to be 56%. He claimed that the cost of farming seaweed per unit area is less than the cost of other types of aquaculture. This culture involves labor-intensive production and requires few facilities and little equipment. In addition, seaweed can be harvested in six weeks, whereas milkfish need six to nine months to achieve market size.

In Bali, Indonesia, although seaweed farming has become well established since the early 1980s, little information is available on economic returns from this activity. Fishermen and coastal rural dwellers in Bali began to grow seaweed because of the low yields from fishing operations and because of government prohibition on the collection of corals. Entry into seaweed farming in Bali was encouraged by financial assistance from seaweed purchasers (collectors/middlemen) and readily available markets. As a consequence, seaweed farming is now, for instance, the main economic activity in the coastal village of Jungut Batu, Nusa Penida in Bali.

We calculate the return to the seaweed operations on two bases here, namely, the accounting-type rate of return estimated by Shang (1976) and the IRR. From an economic point of view, the IRR is the more accurate indicator of returns.

In the accounting-type of approach outlined in Table 2, the straight line method of depreciation is used and assets are assumed to have no residual value at the end of their useful life. The application of this method results in a rate of return of 123%. Using Table 2, we can also see that the payback period is extremely short, namely, 0.6 x 13 months equals 7.8 months.

Assuming a planning period of 10 years (an economic horizon at 10 years with assets having no residual value beyond their expected useful life), cash outflows (capital costs and operating costs) and cash inflows (revenue) are indicated in Table 3 for the planning period. A 10-year economic horizon is assumed for the sake of simplicity to accommodate the duration of useful life of some materials (bull hammers and iron bars) used in the activity.

The initial seedling stock is assumed to last the whole 10 years because seaweed can be reproduced vegetatively and farmers can collect new planting material from their harvest. No boat or raft is used on this 'model' farm. Because the farming area is close to the shore, this farmer can plant seaweed and harvest it when the tide is low. Real prices and costs are assumed not to alter throughout the planning period. On this basis, the IRR from seaweed farming as indicated in Table 4 is 153%.

Tables 2, 3 and 4 show that seaweed farming has the potential to give high returns. The yield provides an income of Rp. 19,200,000 in the first year, is more than twice that of annual operating costs and the initial investment can be paid back in less than a year. The rate of return of 123%/ha/year using the method employed by Shang (1976) is way above the opportunity cost of capital in Indonesia (Table 2). But the method used in Table 2 to calculate returns is deficient from an economic viewpoint since returns and costs are not considered as a stream over the life of the project.

By assuming an economic life for a seaweed farming project of 10 years, it is found that the IRR of this activity is 153%. Therefore, the maximum rate of interest which could be paid for funds to invest in this activity and still break even is 153% (Table 4).

It is worth noting that labor is the major operating cost in seaweed farming. It accounts for 60% of total annual expenses. This cost includes labor for seeding, weeding, harvesting and drying. Thus

Table 2. Cost and return analysis for a selected 1-ha seaweed farm *(Eucheuma cottonii)* in Bali, 1988.

	Cost (Rupiah)	Life (years)
A. Initial investment (Cash outflows)		
- 20,000 kg seed stock at Rp. 50/kg	1,000,000	
- 1,000 kg nylon plastic (4 mm)	4,000,000	2
- 100 kg nylon plastic (8 mm)	400,000	2
- 8,000 pcs bamboo at Rp. 200 each	1,600,000	2
- 300 kg rolls plastic at Rp. 1,000/kg	300,000	1
- 2 bull hammers at Rp. 5,000 each	10,000	10
- 1 iron bar at Rp. 3,000	3,000	10
- 1 knife at Rp. 500	500	5
- 15 pairs of gum boots at Rp. 5,000 a pair	75,000	1
- 2 pcs mask at Rp. 25,000	50,000	1
- 15 baskets at Rp. 1,000	15,000	6 months
- 2 scoop nets at Rp. 3,000	6,000	1
- 100 gunny sacks at Rp. 400	40,000	6 months
- 1 axe at Rp. 4,000	4,000	5
- 1 wood saw at Rp. 5,000	5,000	5
- 50-m net at Rp. 2,000/m	10,000	2
- initial set up labor cost, e.g., setting up the bamboo posts	750,000	
Total initial cash outflows	8,268,500	
B. Operating costs		
- 15 laborers at Rp. 30,000 for a year	5,400,000	
- license (including coastal rental fee)	50,000	
- Depreciation (derived from initial investment)	3,549,200	
Total production cost	8,999,200	
C. Cash inflows (there are 6 harvests in a year, 48,000 kg/year at Rp. 400/kg)	19,200,000	
D. Profit without tax (C - B)	10,200,800	
E. Profit (C - B) without depreciation	13,750,000	
F. Payback period (A/E x 13 months)	7.8 months	
G. Rate of return (D/A)	123%	

Notes:
- Cost data are based on 1988 price; Rp. 1,715 = US$1.
- Seedlings for subsequent planting are obtained from initial first planting. Thus, it is included in initial capital cost.
- Payback period (see Tisdell 1972) and rate of return method after Shang (1976).

Source: Interviews with seaweed farmers in Jungut Batu, Nusa Penida, Bali, December 1988.

Table 3. Estimates of annual capital and operating costs for a selected 1-ha seaweed farm in Bali, 1988 (unit: × Rp. 10,000).

Cash outflows	0	1	2	3	4	Year 5	6	7	8	9	10
Capital costs											
Seed	100	-	-	-	-	-	-	-	-	-	-
Nylon plastic (4 mm)	400	-	-	400	-	400	-	400	-	400	-
Nylon plastic (8 mm)	40	-	-	40	-	40	-	40	-	40	-
Bamboo	160	-	-	160	-	160	-	160	-	160	-
Net	1	-	-	1	-	1	-	1	-	1	-
Plastic rafia	30	-	30	30	30	30	30	30	30	30	30
Gum boots	7.5	-	7.5	7.5	7.5	7.5	7.5	7.5	7.5	7.5	7.5
Mask	5	-	5	5	5	5	5	5	5	5	5
Basket	1.5	1.5	3	3	3	3	3	3	3	3	3
Scoop net	0.6	-	0.6	0.6	0.6	0.6	0.6	0.6	0.6	0.6	0.6
Gunny sacks	4	4	8	8	8	8	8	8	8	8	8
Bull hammers	1	-	-	-	-	-	-	-	-	-	-
Iron bar	0.3	-	-	-	-	-	-	-	-	-	-
Knife	0.05	-	-	-	-	-	0.05	-	-	-	-
Axe	0.4	-	-	-	-	-	0.4	-	-	-	-
Wood saw	0.5	-	-	-	-	-	0.5	-	-	-	-
Costs of tying up seeds and setting up bamboo	75	-	-	-	-	-	-	-	-	-	-
Total	826.85	5.5	54.1	655.1	54.1	655.1	55.05	655.1	54.1	655.1	54.1

Continued

Table 3. Continued.

Cash outflows	0	1	2	3	4	Year 5	6	7	8	9	10
Operating cost											
Laborer wage	-	540	540	540	540	540	540	540	540	540	540
License		5	5	5	5	5	5	5	5	5	5
Total	-	545	545	545	545	545	545	545	545	545	545
Cash inflows	0	1,920	1,920	1,920	1,920	1,920	1,920	1,920	1,920	1,920	1,920
Net cash flows (Rp. 1,000)	-8,268.5	13,695	13,209	7,199	13,209	7,199	13,199.5	7,199	13,209	7,199	13,209

Notes: Assumed economic horizon of seaweed farming is 10 years or cycle of 10 years;
Assumed no change in real annual operating cost;
Assumed no change in real price.

Table 4. Internal rate of return (IRR) calculation for a selected seaweed farm (unit: x Rp. 1,000).

Year	NI	Discount rate (155%)	PV	Discount rate (150%)	PV
0	-8,268.5	1.0000	-8,268.5	1.0000	-8,268.5
1	13,695	0.3921	5,369.8	0.4000	5,478
2	13,209	0.1537	1,030.2	0.1600	2,113.4
3	7,199	0.0603	434.2	0.0640	460.7
4	13,209	0.0236	170.3	0.0260	343.4
5	7,199	0.0093	66.8	0.0100	72.0
6	13,199.5	0.0036	48.0	0.0041	54.1
7	7,199	0.0014	10.3	0.0016	11.5
8	13,209	0.0006	7.9	0.0007	9.2
9	7,199	0.0002	1.6	0.0003	2.1
10	13,209	0.00008	1.0	0.0001	1.4
Total			-128.4		277.3

IRR = 150 + 5 (277.3/405.7) = 153.42%

Notes: NI = net income; PV = present value of net income stream.
The initial discount rate is found by trial and error which will make the net present worth of the incremental net benefit stream equal to zero (Gittinger 1982).

seaweed farming is relatively labor-intensive and therefore suited to countries such as Indonesia where labor is relatively abundant. For some farmers, labor expenses (actual outlays) are low since they employ their own family members including children. The opportunity cost of their employment may also be low. In addition, seaweed farming requires few commercial inputs and does not need pharmaceuticals, chemicals or supplementary feed to sustain production (Table 2).

The size of small seaweed farms in Jungut Batu varies between 0.05 and 0.25 ha and the average product harvested per month varies between 200 and 1,500 kg/area holding. Small farmers usually do not hire labor in managing their farms. They rely mainly on family labor to reduce labor outlays. Also, they often obtain seedlings free from neighbors or relatives or gather them from natural stocks. The planting of a 0.25-ha farm requires an initial investment of about Rp. 1,000,000 ($580). Funds to meet the initial capital cost are usually obtained from credit institutions, seaweed collectors or through informal financial sources available in rural areas. Small farmers in Jungut Batu village feel that seaweed culture gives good returns and the initial investment can be paid back in less than one year (pers. comm.).

Conclusion

The 'accounting' rate of return on investment from seaweed farming in Bali for the farm surveyed is 123%/ha/year, which is very high. The IRR of this activity is 153%. This suggests that seaweed farming is a potentially attractive economic investment for coastal rural dwellers in Indonesia. The Indonesian Government should encourage seaweed farming in other coastal areas of Indonesia which are economically and ecologically suited to this type of farming.

However, the above returns must be regarded as above-normal because the farm selected was a model farm with favorable natural conditions. Furthermore, returns have been calculated free of risks. At this stage, we do not have estimates available for risks, but these would no doubt reduce expected returns. A severe cyclone or typhoon in the 10-year period could mean that all capital is lost and might reduce returns to under 100%. Nevertheless, these results support the Padilla and Lampe (1989) contention for the Philippines that the returns to seaweed farming may be high in Southeast Asia.

One may also wonder why the costs of leasing suitable sites is not higher given the attractive level of returns from seaweed farming in Bali. The main reason seems to be that villagers in Bali believe that the seashore areas adjacent to their village belong to their village. Consequently, village chiefs do not determine leasehold allocations on purely economic grounds and by economic competition. This also means that it is difficult or impossible for nonvillagers to obtain rights for seaweed farming in a coastal area adjoining a village. They may be limited to operations on alienated land not being used by villagers.

Some wonder if seaweed farming is likely to be profitable for small farmers. Since small farmers often have surplus labor and are able to gather propugales free, and also sometimes timber and bamboo, they may have some economic advantages compared to large-scale farmers. Furthermore, there appears to be no diseconomies from such scale operations and small growers can often combine seaweed growing with other occupations. The relative non-perishability of seaweed is a particular advantage of seaweed growing compared to many other types of aquaculture.

Acknowledgements

The authors would like to thank two anonymous referees and the editor for fruitful comments and suggestions. Research for this paper was funded in part by the Australian Centre for International Agricultural Research (ACIAR Research Project No. 8823).

References

Abdul Malik, B. and B. Rahardjo. 1988. Seafarming in Indonesia. Workshop proceedings paper on the Appropriate Mariculture and Rural Fisheries, 15-22 October 1988. Philippines.

Anon. 1989. Harga rumput laut meningkat (An increase in the price of seaweed). Ekonomi Indonesia, Jakarta. No. 2249, 6 June 1989: 5-6.

Bradford, D.F. 1975. Constraints on government investment opportunities and the choice of discount rate. Am. Econ. Rev. 65(5): 887-899.

Chapman, V.J. and D.J. Chapman. 1980. Seaweeds and their uses. Chapman and Hall Ltd., New York.

Directorate General of Fisheries. 1988. National planning on fisheries development in the fifth five year development plan (1988/89 - 1993/94), Jakarta.

Gittinger, J.P. 1982. Economic analysis of agricultural projects. The John Hopkins University Press, Second Edition, Baltimore and London.

Mendelsohn, R. 1981. The choice of discount rate for public projects. Am. Econ. Rev. 71(1): 239-241.

Mishan, E.J. 1982. Cost-benefit analysis. Third edition. George Allen and Unwin, London.

Padilla, J.E. and H.C. Lampe. 1989. The economics of seaweed farming in the Philippines. Naga, the ICLARM Quarterly 12(3): 3-5.

Pearce, D.W. and C.A. Nash. 1981. The social appraisal of projects: a text in cost-benefit analysis. Macmillan, London.

Shang, Y.C. 1976. Economic aspects of *Gracilaria* culture in Taiwan. Aquaculture 8: 1-7.

Shang, Y.C. 1981. Aquaculture economics: basic concepts and methods of analysis. Westview Press, Boulder, Colorado.

Tisdell, C. 1972. Microeconomics: the theory of economic allocation. John Wiley and Sons, Sydney.

Manuscript received 17 November 1989; accepted 27 August 1990.

[28]

Aquaculture, capture fisheries and available marine resources: ecological and economic interdependence[1]

Abstract

Anderson (1985) theorizes that development of the aquaculture of a species of fish (one also captured in an open-access fishery) favours the conservation of its wild stocks, if competitive market conditions prevail. However, this theory is subject to significant limitations. While this is less so within his model, it is particularly so in an extended model outlined here. In the extended model, aquaculture development can reduce wild stocks, thereby shifting the supply curve of the capture fishery, or raise the demand for the fish species subject both to aquaculture and capture. Such development can threaten wild stocks and their biodiversity. While aquaculture development might in principle have no impact on the biodiversity of wild stocks, or even raise aquatic biodiversity overall, in the long term the development of aquaculture will probably reduce aquatic diversity both in the wild and overall.

Introduction

Views differ about the likely impact of aquaculture (and of farming or animal husbandry generally) on the survival of species in the wild and about how such activity is likely to affect the stock of available genetic diversity. Some writers see farming (for example, of species threatened in the wild) as a positive force for conservation whereas others regard it as a serious threat to biological conservation. However, the situation is extremely complex. This chapter demonstrates that whether or not farming is a positive force for biological conservation (and adds to or subtracts from the available genetic stock), varies with circumstances, including the scale of aggregated farming activity.

Anderson (1985) argues that if markets are competitive, aquaculture is a positive force for conserving wild stocks of commercially exploited fish. His view is outlined and then shown to require important qualifications in the light of possible supply-side and demand-side interactions.

Anderson's theory that in competitive market conditions aquaculture favours conservation of wild stocks of species

Anderson (1985, p. 1) contends, on the basis of his theory, that market entry of competitive aquaculturists of a fish species subject previously only to capture 'increases natural fish stocks, reduces price and increases total supply. If initially the natural fish stock is at a level below maximum sustainable yield, entry of the aquaculturalist[s] results in an increase in supply from the commercial fishery'. However, this positive result for conservation of wild fish stocks is only true under favourable conditions. The results are not general ones; they rely on the assumption that the supply and demand for captured fish is independent of the supply of aquacultured fish and that aquacultured fish are perfect or close substitutes for captured fish of the same species. Furthermore, even given Anderson's (1985) assumptions, there is one circumstance in which economically viable aquaculture fails to increase natural fish stocks and to save a species, that is both captured and aquacultured, from extinction in the wild.

Let us consider the simplest illustration of Anderson's proposition using a modified form of his Figure 1. In this case, the capture fishery has a single equilibrium and it occurs at E_1 in Figure 1, implying that the stock of the fishery is below the level that yields maximum sustainable yield. The supply curve marked SAS' represents the supply curve of captured fish of a particular species and line DD' represents the market demand curve for these fish. The residual demand curve for aquacultured fish of the same species is marked RGR'. In the absence of aquaculture, the equilibrium at E_1 is stable and the price of the fish is P_3 per unit with supply being X_1^F. Now if aquaculture becomes profitable, total fish supplies can be expected to increase and the surplus from the capture industry may also rise as fish stocks increase with reduced harvesting pressure.

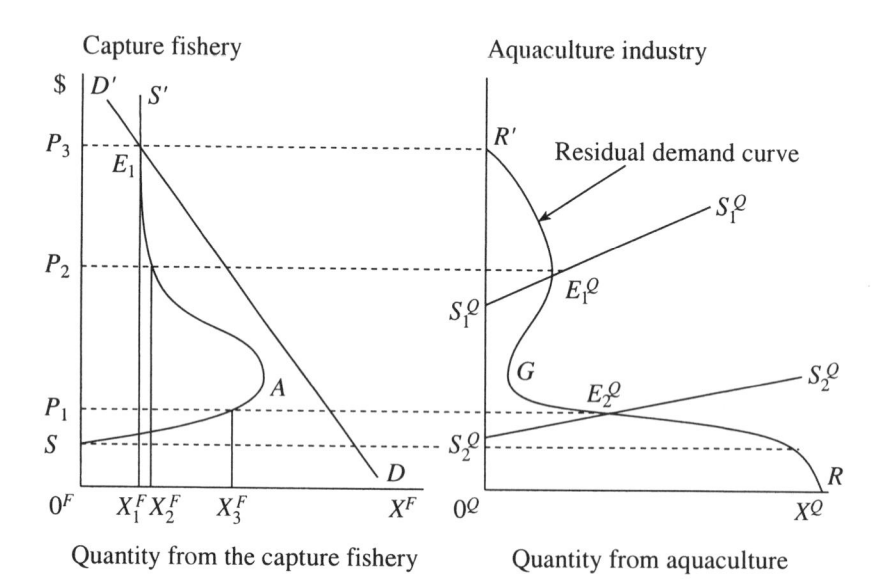

Figure 1 A case in which aquaculture is favourable to biological conservation of wild fish stocks

If, for example, in Figure 1, $S_1^Q S_1^Q$ represents the aquaculture supply curve, the aquaculture industry comes into equilibrium at E_1^Q. The price of the fish species concerned falls from P_3 to P_2 and the capture fisheries supplies rise from X_1^F to X_2^F. Total fish supplies are up. This is also true if the aquaculture supply curve is $S_2^Q S_2^Q$. Now supplies from the capture fishing are below maximum sustainable yield. In fact, if $S_2^Q S_2^Q$ is sufficiently low, supplies from the capture fishers may fall below X_1^F and in the extreme case, exploitation of wild stocks could become completely unprofitable. This, therefore, calls for qualification to Anderson's statement mentioned earlier that if initially the natural resource stock is overexploited, aquaculture results in increased supply from the commercial fishery.

Anderson (1985) also illustrates his theory for a triple equilibrium case for the capture fishery. But in none of the cases that he illustrates does he allow for the possibility that open-access capture fishing could lead to the extinction of wild stocks. In all the cases considered by him, aquaculture increases the size of the wild stock.

However, even under the types of conditions envisaged by Anderson (1985), successful aquaculture may fail to save wild stocks from extinction. While it may save wild stocks from commercial extinction, it need not do so. This can be illustrated by Figure 2.

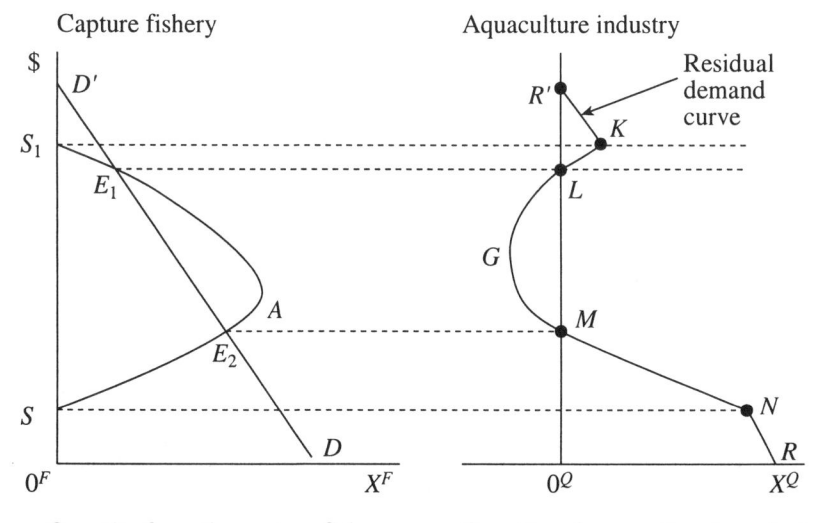

Figure 2   A case in which aquaculture may fail to prevent extinction of wild fish stocks

From Figure 2, it can be observed that if the supply curve from aquaculture cuts the residual demand curve (the demand for the aquacultured product between R' and K, aquaculture fails to prevent extinction of natural stocks given that E_1 is an unstable equilibrium. So even under the type of conditions envisaged by Anderson (1985), aquaculture may fail to have a positive effect in saving wild stocks from extinction. However, it is true that if the supply curve of aquaculture intersects the supply curve of aquaculture in the segment between K and L, its development will be a positive force for conserving wild stocks. This is given the implicit assumption that harvesting of wild stocks will cease at population levels that are so low as to result in elimination of these stocks. At stock levels above this where harvesting continues, survival of the wild population is assumed to occur. In further extension of the argument, this assumption could be varied.

Note that if the aquaculture supply curve does not intersect sections LKR' or RNM of the residual demand curve, the aquaculture industry is not competitive with the capture fishery and cannot survive if the equilibrium price of capture fish falls in the range *LM*.

Observe that, given the assumptions involved in Figure 2, a species survives only in aquaculture if the supply curve for aquaculture intersects the residual demand curve in its segment *R'K*. In this case, all wild stocks disappear. Consequently, further qualification of Anderson's hypothesis is required even accepting his model.

Need to modify Anderson's conclusion to allow for impacts of aquaculture on supply and demand functions for captured fish

However, a potentially serious limitation of Anderson's competitive model is failure to allow for possible impacts of aquaculture on supply and demand functions applicable to the capture fishery. Tisdell (1991, section 6.4) raises this issue in connection with farming generally. While the development of aquaculture need not always affect supply conditions in the capture fishery, in many cases such development shifts the supply curve of the capture fishery to the left. A leftward shift may come about because the aquaculture industry appropriates habitat used by wild stock; competes with wild stocks for food resources, creates health or genetic risks for wild stock and relies on wild stock for seed/fingerlings, broodstock or 'recruits' for aquaculture. While there might be cases in which aquaculture has beneficial effects on the wild stock, e.g. due to nutrient enrichment of the environment as result of aquaculture, these cases might be rare.

Furthermore, aquaculture could raise the overall demand schedule for a fish species (see Asche et al., 2001). This could occur because aquaculture permits greater regularity of market supply of a species subject both to capture and aquaculture and adds to its market promotion. Nevertheless, there is also a small chance that aquaculture might on occasions reduce overall demand for a species, for example if the aquacultured product is not identified and is subject to off flavours (Tisdell, 2001). This problem is akin to the famous lemon versus plum problem (Akerlof, 1970).

Table 1 lists some factors that may cause the supply curve of a capture fishery to move leftwards as a result of aquaculture development, and some factors that may cause the demand curve for a species that is both captured and aquacultured to move to the right.

If the development of aquaculture causes a leftward shift in the supply curve of the capture fishery or an upward shift in market demand for the species both captured and aquacultured, the development of aquaculture may have negative effects on wild stocks and Anderson's conclusions outlined earlier need not hold. For example, when aquaculture development shifts the supply curve for the capture fishery to the left, a lower price brought about by aquaculture supplies may be associated with reduced supply of captured fish. This occurs if the capture fishery is expending so much effort that it is operating at a level resulting in less than maximum yield, that is on the backward-bending portion of its demand curve. This case is illustrated in Figure 3.

In Figure 3, only the position of the capture industry is shown. Curve *SAS* represents the supply curve of this industry and *DD* the demand for its fish before the development of a competing aquaculture industry. Once the aquaculture industry develops, then because of negative supply spillovers, the supply curve of the capture industry shifts leftwards to *S'AS'*. However, assume that the market demand curve for the species

Table 1 Some circumstances in which aquaculture has a negative impact on the supply curve of the capture fishery and positive impacts on demand for the fish species involved

Negative impacts on quantity of wild stocks

- Aquaculture appropriates habitat used by wild stock for breeding, feeding, protection and so on
- Ranching involving collection of broodstock, seed or fingerlings or capture of more mature stock for aquaculture
- Competition for shared food resources, e.g. capture of prey of wild stock to feed aquacultured stock
- Pollution or loss of resources used by wild stock because of spillover or environmental impacts of aquaculture
- Disease and genetic risk to wild stocks from aquacultured stocks

Positive impacts on demand for species as a whole

- Regular and widespread availability of product as a result of aquaculture may stimulate demand for the fish species involved
- Aquaculturalists may add to the market promotion of the fish species as a whole
- The development of aquaculture may enable greater standardization of the fish product. This is usually a plus as far as supermarkets are concerned (Young, 2001) and could have positive effects on demand both for the aquacultured and the captured product

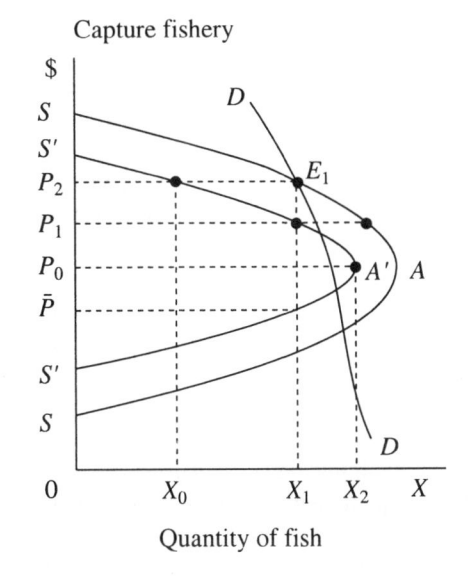

Figure 3   A case in which aquaculture has a negative effect on the capture fishery

involved remains constant. Furthermore, suppose that initially the capture fishery is in equilibrium at E_1.

If aquaculture develops and causes the price of product cultured or captured to be in the range $P_1 < P < P_2$ supplies for the capture fishery fall. This contrasts with Anderson's case in which they rise in such circumstances. Supplies for the capture industry only increase in this case if $\bar{P} < P < P_1$.

Note that even if the capture fishery should attain maximum sustainable yield in the post-aquaculture situation, this yield and the maximum sustainable stock will be lower than in the absence of aquaculture. However, the impact of aquaculture on the yields of the capture fishery and its stock is liable to depend on the scale of aquaculture and the techniques used in aquaculture. Below some threshold of operation, for example, aquaculture might have little or no effect on the capture fishery. If on the other hand, aquaculture is on a considerable scale, it is liable to have negative supply consequences for the parallel capture fishery and can increase the likelihood of elimination of wild stocks. Because of the externalities involved, this may occur irrespective of whether replacement of the wild stock by the cultivated stock is the economically most efficient solution, and irrespective of whether aquaculture results in sustainable production and survival of the cultivated species in the long run.

The demand-side effects on the capture fishery from aquacultural development can be similar to the supply-side effects. An example is given in Figure 4. In this figure, the curve SAE_0S represents the supply curve for the capture industry. For simplicity, this supply curve is assumed to be independent of aquaculture development. DD is assumed to represent the demand for the fish concerned in the absence of aquaculture and D_1D_1 indicates this demand after the development of aquaculture. Initially the industry is in equilibrium at E_0 with fish selling for P_0 per unit and X_0 being supplied by the capture fishery. But imagine that, after aquaculture develops, the price of the fish concerned rises to P_1. This price results in wild stocks being fished to extinction in the case illustrated. In other cases, the price of fish after the development of aquaculture may be higher than P_0, but still intersect the supply curve for the capture fishery. In such cases, supplies from the capture fishery continue but are reduced compared to the pre-aquaculture situation. Once again, this is a consequence not predicted by Anderson's (1985) model.

Thus it is clear that both demand-side and supply-side spillovers from the development of aquaculture can have negative impacts on the biological conservation of wild stocks, even though in some circumstances neutral or positive consequences are possible. While the above modelling assumes, as does Anderson (1985), that captured and cultured fish of the same species are perfect substitutes, this assumption can be relaxed. It is even possible for these products to be complements to some extent and for the type of conservation consequences outlined above to occur.

Concluding comments

While the development of aquaculture can have favourable impacts on the survival of wild fish species and stocks of captured fish, the competitive market model of Anderson (1985) suggests more favourable effects than in fact are likely. Even given Anderson's (1985) model, the development of aquaculture may fail to save a captured fish species from extinction. However, the likelihood of the development of aquaculture having

Capture fishery

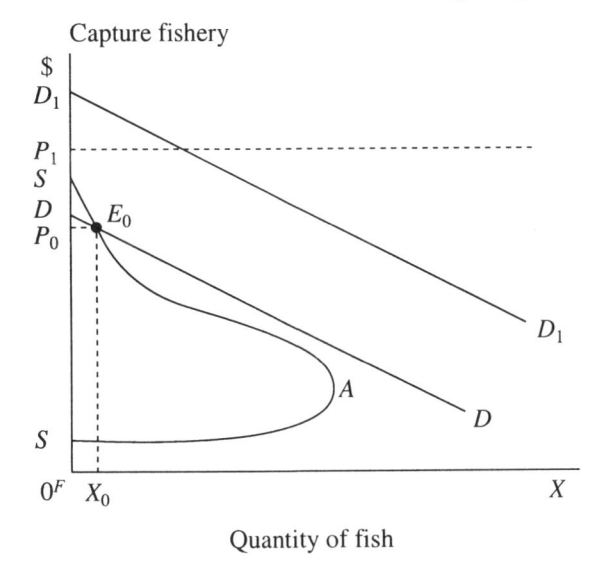

Figure 4 Demand-side effects from the development of aquaculture are liable to put pressure on wild fish stocks and in some cases may result in their elimination. These results are different to those predicted by Anderson's (1985) model

negative consequences for conservation of a species which is also exploited by the capture fishery may increase when the aquaculture industry has negative impacts on the supply of the capture fisheries or raises the demand for the fish species subject to both aquaculture and capture.

Given our experience with the long-term genetic consequences of agriculture, it seems highly likely that as aquaculture develops and expands, this will tend to reduce the wild genetic stock. In addition, although genetic diversity within aquaculture may initially rise, in the very long term, it might be expected to decline after peaking. However, the later development of aquaculture compared to agriculture, especially compared to livestock husbandry, may result in some differences in the evolving extent of animal diversity in aquaculture. The institutional arrangements affecting aquaculture's development today, particularly globalization factors, are quite different to those surrounding the earlier development of livestock husbandry. Hence some differences in patterns of global genetic development in aquaculture and in livestock production might be anticipated.

Note
1. This chapter develops one of the themes contained in a paper presented by the author at the World Aquaculture Conference held in Beijing in April 2002.

References
Akerlof, G. (1970), 'The market for lemons: quality, uncertainty and the market mechanism', *Quarterly Journal of Economics*, **84**, 488–500.
Anderson, J.L. (1985), 'Market interaction between aquaculture and the common-property commercial

fishery', *Marine Resource Economics*, **2**, 1–24.
Asche, F., Bjørndal, T. and Young, J.A. (2001), 'Market interactions for aquaculture products', *Aquaculture Economics and Management*, **5**, 303–18.
Tisdell, C.A. (1991), *The Economics of Environmental Conservation*, Amsterdam: Elsevier Science.
Tisdell, C.A. (2001), 'Externalities, thresholds and marketing of new aquacultural products: theory and examples', *Aquaculture Economics and Management*, **5**, 289–302.
Young, J.A. (2001), 'Communication with cod and others – some perspectives on promotion for expanding markets for fish', *Aquaculture Economics and Management*, **5**.

Name index

Abdul Malik, B. 344
Abeywardena, V. 63, 66
Akerlof, G. 356
Alauddin, M. 45, 46, 47, 48, 49, 51, 57, 58, 98, 218, 249
Alcorn, J.L. 189
Ali, M.S. 246, 250, 252
Anderson, J.L. 17, 353–6, 358, 359
Anderson, J.R. 48, 49, 57
Anderson, L.G. 296
Anderson, N. 233
Anderson, R.C. 277, 278, 284, 290, 303
Andres, L.A. 189, 190
Arrow, K.J. 161
Asche, F. 356
Auld, B.A. 7, 165, 167, 169, 184, 190, 199, 200, 204, 214, 218

Bain, R. 231, 261, 262
Ball, J. 92
Baltaxe, R. 87
Bandara, R. 7
Barker, J.R. 327
Barkley, P.W. 89
Barnett, B.J. 250, 251, 255
Batini, F.E. 91
Baumol, W. 118
Baxter, I.N. 91
Beddington, J.R. 179
Bennett, D. 91
Berglun, D.R. 155
Berryman, A.A. 184
Beverton, R.J.H. 278
Bird, R. 87
Bishop, R. 60
Bowerman, M. 269
Bowers, R.C. 191, 192, 193, 199
Bowes, M.D. 332
Boyce, J.K. 48
Bradford, D.F. 344
Braley, R.D. 317, 322
Branford, J.R. 264
Breckwoldt, R. 87
Brennan, M.J. 155
Brett, R.B. 193
Bromley, D.W. 206
Brougham, R.W. 180
Brown, C.V. 209

Brown, J.L. 294
Brown, M.B. 79
Burdon, J.J. 200
Burgess, R.J. 79
Byrne, J.L. 262

Caltagirone, L.E. 212, 217
Campbell, M.F. 14
Campbell, M.H. 218
Capule, C. 48
Carlson, G.A. 160
Caughley, G. 179
Chan, R.R. 212, 213, 224
Chandler, J.M. 184, 185
Chapman, D.J. 344
Chapman, V.J. 344
Charudattan, R. 190, 207, 209, 214, 225
Chiang, A.C.C. 78, 296
Child, R. 66, 80
Chisaka, H. 155, 161, 169
Chisholm, A.H. 60, 76
Christiansen, F.B. 64, 66
Christy, F.T. Jr 277, 290, 291, 296
Ciriacy-Wantrup, S.V. 60
Clark, C. 64, 66, 73, 134, 262
Coase, R.H. 118
Colt, J. 322
Comitini, S. 278, 280
Conway, G.R. 4, 18, 98, 100, 101, 122, 129, 206, 207, 215
Coomans, P. 66
Coughlan, K.J. 105
Cousens, R. 167
Crawford, C.M. 318, 327
Cremer, K.W. 66
Cromer, R.N. 66
Cross, P. 86
Crowe, A. 281
Cullen, J.M. 198, 200, 201, 221
Cushing, D.H. 279

Dalal, R. 111
Dantwala, M.L. 46
Davidson, R.L. 89
Davis, D.E. 188, 190
Davis, J.S. 219
Davis, R. 7, 238, 239
De Silva, N.T.M.H. 79, 80, 89, 330

361

DeBach, P. 215, 221, 224
Deighton, P.J. 184
Delgardo, C. 139, 148
Deuter, P. 113
Devonshire, P.G. 91
Dew, D.A. 155, 164
Dijkhuisen, A.A. 231
Dillon, J.L. 64
Dixon, W.J. 79
Dodd, A.P. 217
Donald, C.M. 66, 180
Dor, P. 321
Downey, L.A. 64, 66, 67
Downs, A. 272
Dragun, A.K. 3
Duffy, M. 247–8, 252
Dumanski, J.K. 95
Dumsday, R.G. 91–2
Duncan, R.C. 218, 249
Dwyer, G. 113

Eden, T. 66
Edwards, G.W. 218
Ehler, L.E. 225
Ellis P.R. 240
Etherington, D.M. 79
Etzioni, A. 118, 126
Evans 46
Evanson, R.E. 23, 37

Fairbairn, T.I. 294
Fenchel, T.M. 64, 66
Filius, A.M. 90
Finlay, K. 23, 37
Firdausy, C. 15, 331
Firth, F.E. 259
Fitt, W.K. 317, 318, 320, 321, 322, 324
Fleming, P.J. 88
Florence, R.G. 66
Fox, W.W. Jr 14, 278
Freebairn, J.W. 218
Frick, K.E. 184, 185

Garreda, S. 105, 108
Gerald, C.F. 303
Gibson, B.O. 250, 251, 255
Gill, G.J. 57
Gisz, P. 86, 90
Gittinger, J.P. 334, 343, 344
Goeden, R.D. 184
Gomez, A.A. 108, 112
Gopalakrishnan, C. 259
Gregory, E.J. 64, 66
Greig, P.J. 91
Groves, R.H. 198, 200, 201, 221

Gulland, J.A. 296
Gunatilake, G. 294

Hafiz, A. 277, 278, 284, 290, 303
Hambrey, J. 324, 334
Hamid, M.A. 58
Hammond, K. 139, 140
Hampicke, U. 131, 135
Hardjolukito, S. 278, 280
Hardy, G. 27, 30, 39, 168
Harper, J.L. 180
Harris, P. 184, 200, 201, 210, 214
Hartwick, J.M. 332
Harvey, A.M. 66
Haseler, W.H. 200
Hassell, M.P. 180
Hayami, Y. 218, 249
Hazell, P.B.R. 45, 46, 47
Headley, J.C. 203
Heath, S.B. 63, 64, 67, 74
Helliwell, D.R. 85, 87
Henry, G.W. 273
Herdt, R.W. 48, 97–101, 114, 127, 128, 218, 249
Heslinga, G.A. 317–22, 324, 332, 334
Hicks, J.R. 327, 331
Hodge, I. 91
Holling, C.S. 182, 184
Holt, S.J. 278
Hoover, D.C. 272
Hosking, J.R. 184
Howe, C.W. 280
Howes, K.M.W. 86
Hudson, J.P. 25
Huffaker, C.B. 179, 180, 190, 212, 215, 216, 217, 224
Hufschmidt, M.M. 85
Hughes, K.K. 92
Huguenin, J.E. 322

Jackson, P.M. 209
Jain, H.K. 46
James, A.D. 240
James, C. 246, 247
Jarrett, F.G. 249
Johannes, R.E. 279
Johnston, P.J.M. 92
Jones, A.D. 88, 89
Julien, M.H. 212, 213, 224

Karlin, S. 24, 27, 168
Kelso, M.M. 126
Kemp, M.C. 76
Kennedy, J.O.S. 269, 270
Kern, J.D. 212, 213, 224

Kershaw, K.A. 167
Kibria, M.G. 83
Klages, K.H.W. 37
Kohn, R.E. 126
Kristjanson, P. 235, 238
Krutilla, J.V. 60, 332

Laing, J.E. 216, 224
Lampe, H.C. 345, 351
Lampkin, N.H. 130
Lancaster, K. 141
Leibenstein, H. 64
Levin, J. 25, 31, 37
Levins, R. 35, 43
Lindner, R.K. 14, 249
Little, I.M.D. 204, 294
Littlewood, J.E. 27, 30
Lomas, J. 25, 31, 37
Long, J.K. 274
Long, N.V. 76
Lucas, J.S. 318
Luce, R.D. 163, 164
Luck, R.F. 195
Lumley, S. 91
Lyle, J.M. 263
Lynam, J.F. 97–101, 114, 127, 128
Lynch, J.J. 87

MacArthur, R.H. 180
Mackauer, M. 225
Malan, D.E. 180, 185
Mandeville, T.D. 194
Manthriratna, M.A.P.P. 63, 66
Margolis, H. 126
Markowitz, H.M. 49
Marra, M. 160, 252, 254
Marsden, J.S. 200, 213, 219, 222
Marsh, W.E. 236
Marshall, A. 34
Marshall, D.R. 200
May, R.M. 64, 183
McInerny, J.P. 238
McLeod, R. 235, 238
Meade, J.W. 332
Medd, R.W. 155
Meek, A.H. 231, 236
Mehra, S. 45, 46
Mellor, C.J. 4
Mendelsohn, R. 344
Menz, K. 7, 95, 190, 199, 200, 214, 218
Meyers, J.H. 180
Mirrlees, J.A. 294
Mishan, E.J. 204, 273, 332, 344
Mishra, H.R. 120
Moffat, A.S. 246

Moore, R. 86
Moran, D. 123, 133
Morley, F.H.W. 184
Morris, R.S. 231, 233, 236
Muir, W.R. 190
Mukhebi, A.W. 238
Mulcahy, M.J. 91
Mumford, J.D. 215, 224
Munasinghe, H. 296
Munro, J.L. 317, 318, 335
Murshid, K.A.S. 47

Nakamura, K. 180
Nalewaja, J.D. 155
Nash, C.A. 344
Nash, J.F. 270
Nathaniel, W.R.N. 79
Nedham, D.J. 274
Nelson, G.C. 250, 251, 255
Ng, Y. 204
Nichols, E.H. 277
Norton, G.A. 215, 224
Noy-Meir, I. 41, 180
Nulsen, R.A. 91
Nyamudeza, P. 109–111

O'Rourke, D. 284
Odum, E.P. 36, 37
Oelhaf, R.C. 190, 193, 194, 195
Ohgushi, T. 180
Olewiler, N.D. 332
Oram, P.A. 219
Over, H.J. 232
Owen, O.S. 3, 26, 29, 32, 36, 37, 43, 64
Owler, L.W.J. 294

Padel, S. 130
Padilla, J.E. 345, 351
Paoletti, M.G. 253
Parthasarathy, G. 46
Paterson, R. 274
Pearce, D. 97, 123, 124, 133, 344
Pearse, P.H. 64, 73
Perkins, R.C.L. 212
Pernetta, J. 317
Perrin, R.K. 24, 76
Peterman, R.M. 182, 184
Pimentel, D. 246, 250, 252, 253
Plucknett, D.P. 59, 60
Polya, G. 27, 30
Pray, C.E. 57
Pullin, R.S.V. 310, 313
Pushparajah, E. 95

Quandt, R. 118

Quiggin, J. 60
Quimby, P.C. Jr 191, 199, 214

Rabb, R.L. 224
Rahardjo, B. 344
Raiffa, H. 163, 164, 270
Ramakrishnan, P.S. 130
Ramsay, G.C. 241
Randall, A. 60
Rawls, J. 133
Reeves, T.G. 155
Reichelderfer, K.H. 163, 209
Reijntjes, C. 129
Ricker, D.W. 184
Roberts, J.A. 234, 235
Rogers, P.P. 261
Rosegrant, M.W. 24
Rosenzweig, M.L. 180
Rothschild, B.J. 281
Roy 46
Rummery, R.A. 86
Ruttan, V.W. 194
Ryan, J.G. 24, 219

Samuelson, P.A. 76, 83, 194
Santhirasegaram, K. 80
Sar, N.L. 86, 90
Sarma 46
Sathiendrakumar, R. 14, 277, 278, 279, 290, 294, 296
Schaefer, M.B. 13, 14, 278
Schultz, T.W. 129
Schwabe, C.W. 233
Schweizer, E.E. 155
Scott, A. 64, 296
Seckler, D.W. 89
Seipen, G. 87
Shang, Y.C. 327, 329, 332, 335, 343, 345
Shannon, R.E. 79
Shelton, A.M. 253
Shubik, M. 270
Simmonds, F.J. 216, 224
Simon, H. 262
Sinden, J.A. 86, 88, 89, 218
Smith, D.M. 64, 67, 73
Smith, R.W. 64
Smith, V.K. 60
Smyth, A.J. 95, 97, 104, 128
Solis, E.P. 324
Stanistreet, K. 262, 263
Stevens, B. 321
Stinner, R.E. 224
Sturgess, N.H. 262
Sturroch, J.W. 87
Suda, A. 281

Suhardono 234, 235
Sundquist, W.B. 194
Swanson, T.M. 133, 142
Swezey, O.H. 212
Syers, J.K. 95

Templeton, G.E. 188, 214
Thomas, J.F. 91
Tietenberg, T. 97
Tisdell, C.A. 3, 4, 7, 14, 15, 18, 37, 45, 46, 47, 48, 49, 51, 57, 58, 73, 79, 80, 83, 89, 90, 92, 97, 98, 118, 119, 122, 123, 127, 131, 132, 133, 134, 142, 144, 148, 165, 169, 181, 183, 190, 194, 199, 201, 202, 206, 207, 208, 209, 214, 215, 216, 218, 249, 272, 274, 277, 278, 279, 290, 294, 324, 327, 329, 330, 331, 332, 333, 335, 343, 344, 356
Todaro, M.P. 294
Treschev, A.I. 281
Trotman, C.H. 91
Turelli, M. 35

Urquhart, G.M. 231

van den Bosch, R. 224, 225
van Emden, H.F. 179
van Rensburg, C. 86
Vere, D.T. 218
Virgona, J. 273

Walker, H.L. 191, 199
Walker, J. 189, 214
Walsh, C. 90
Wan, H.Y. Jr 76
Watkins, J.W. 269, 270
Watson, T.C. 332, 334
Waugh, G. 296
Weatherspoon, D.M. 155
White, G.G. 180
Whitehead, R.A.S. 64
Wickham, T. 24
Wilkinson, G.N. 23–24, 37
Willey, R.W. 63, 64, 66, 67, 74
Wilson, B.J. 159, 164
Wilson, C. 144
Wilson, F. 215, 224
Winch, D.M. 189
Winder, J.A. 179
Wise, W.S. 201
Wittenberg, R. 327, 329, 335
Wylie, F.R. 92

Xiang, Z. 121

Yamaguchi, M. 332
Young, J.A. 357

Zimmerman, H.G. 180, 185
Zinsstag, J. 232, 233